Novel ZnO-Based Nanostructures: Synthesis, Characterization and Applications

Novel ZnO-Based Nanostructures: Synthesis, Characterization and Applications

Editors

Yamin Leprince-Wang
Guangyin Jing
Basma El Zein

MDPI • Basel • Beijing • Wuhan • Barcelona • Belgrade • Manchester • Tokyo • Cluj • Tianjin

Editors

Yamin Leprince-Wang
Université Gustave Eiffel (UGE)
France

Guangyin Jing
Northwest University
China

Basma El Zein
University of Business and Technology
Saudi Arabia

Editorial Office
MDPI
St. Alban-Anlage 66
4052 Basel, Switzerland

This is a reprint of articles from the Special Issue published online in the open access journal *Crystals* (ISSN 2073-4352) (available at: https://www.mdpi.com/journal/crystals/special_issues/ZnO_nanostructures).

For citation purposes, cite each article independently as indicated on the article page online and as indicated below:

LastName, A.A.; LastName, B.B.; LastName, C.C. Article Title. *Journal Name* **Year**, *Volume Number*, Page Range.

ISBN 978-3-0365-7134-8 (Hbk)
ISBN 978-3-0365-7135-5 (PDF)

Cover image courtesy of Hao LUO

© 2023 by the authors. Articles in this book are Open Access and distributed under the Creative Commons Attribution (CC BY) license, which allows users to download, copy and build upon published articles, as long as the author and publisher are properly credited, which ensures maximum dissemination and a wider impact of our publications.

The book as a whole is distributed by MDPI under the terms and conditions of the Creative Commons license CC BY-NC-ND.

Contents

About the Editors vii

Preface to "Novel ZnO-Based Nanostructures: Synthesis, Characterization and Applications" ix

Yamin Leprince-Wang, Guangyin Jing and Basma El Zein
Novel ZnO-Based Nanostructures: Synthesis, Characterization and Applications
Reprinted from: *Crystals* 2023, 13, 338, doi:10.3390/cryst13020338 1

Ziqian Li, Ningzhe Yan, Yangguang Tian and Hao Luo
Tuning Growth of ZnO Nano-Arrays by the Dewetting of Gel Layer
Reprinted from: *Crystals* 2022, 13, 30, doi:10.3390/cryst13010030 5

Chunfeng Huang, Qi Sun, Zhiling Chen, Dongping Wen, Zongqian Tan, Yaxian Lu, et al.
Controlled Synthesis and Growth Mechanism of Two-Dimensional Zinc Oxide by
Surfactant-Assisted Ion-Layer Epitaxy
Reprinted from: *Crystals* 2022, 13, 5, doi:10.3390/cryst13010005 15

**Zainal Abidin Ali, Iqabiha Shudirman, Rosiyah Yahya, Gopinath Venkatraman,
Abdurahman Hajinur Hirad and Siddique Akber Ansari**
Green Synthesis of ZnO Nanostructures Using *Pyrus pyrifolia*: Antimicrobial, Photocatalytic and
Dielectric Properties
Reprinted from: *Crystals* 2022, 12, 1808, doi:10.3390/cryst12121808 27

Feiling Shen, Ning Cao, Hengyu Li, Zhizheng Wu, Shaorong Xie and Jun Luo
Local Three-Dimensional Characterization of Nonlinear Grain Boundary Length within Bulk
ZnO Using Nanorobot in SEM
Reprinted from: *Crystals* 2022, 12, 1558, doi:10.3390/cryst12111558 43

Ahmed Emara, Amr Yousef, Basma ElZein, Ghassan Jabbour and Ali Elrashidi
Enhanced Broadband Metamaterial Absorber Using Plasmonic Nanorods and Muti-Dielectric
Layers Based on ZnO Substrate in the Frequency Range from 100 GHz to 1000 GHz
Reprinted from: *Crystals* 2022, 12, 1334, doi:10.3390/cryst12101334 49

**Alberto Mendoza-Sánchez, Francisco J. Cano, Mariela Hernández-Rodríguez and
Oscar Cigarroa-Mayorga**
Influence of ZnO Morphology on the Functionalization Efficiency of Nanostructured Arrays
with Hemoglobin for CO_2 Capture
Reprinted from: *Crystals* 2022, 12, 1086, doi:10.3390/cryst12081086 67

Ningzhe Yan, Hao Luo, Yanan Liu, Haiping Yu and Guangyin Jing
Motility Suppression and Trapping Bacteria by ZnO Nanostructures
Reprinted from: *Crystals* 2022, 12, 1027, doi:10.3390/cryst12081027 79

Linda Serairi and Yamin Leprince-Wang
ZnO Nanowire-Based Piezoelectric Nanogenerator Device Performance Tests
Reprinted from: *Crystals* 2022, 12, 1023, doi:10.3390/cryst12081023 91

**Titiradsadakorn Jaithon, Jittiporn Ruangtong, Jiraroj T-Thienprasert and
Nattanan Panjaworayan T-Thienprasert**
Effects of Waste-Derived ZnO Nanoparticles against Growth of Plant Pathogenic Bacteria and
Epidermoid Carcinoma Cells
Reprinted from: *Crystals* 2022, 12, 779, doi:10.3390/cryst12060779 101

Marie Le Pivert, Nathan Martin and Yamin Leprince-Wang
Hydrothermally Grown ZnO Nanostructures for Water Purification via Photocatalysis
Reprinted from: *Crystals* **2022**, *12*, 308, doi:10.3390/cryst12030308 . . . . **115**

K. M. Mohamed, J. John Benitto, J. Judith Vijaya and M. Bououdina
Recent Advances in ZnO-Based Nanostructures for the Photocatalytic Degradation of Hazardous, Non-Biodegradable Medicines
Reprinted from: *Crystals* **2023**, *13*, 329, doi:10.3390/cryst13020329 . . . . **131**

Leslie Patrón-Romero, Priscy Alfredo Luque-Morales, Verónica Loera-Castañeda,
Ismael Lares-Asseff, MaríaÁngeles Leal-Ávila, Jorge Arturo Alvelais-Palacios, et al.
Mitochondrial Dysfunction Induced by Zinc Oxide Nanoparticles
Reprinted from: *Crystals* **2022**, *12*, 1089, doi:10.3390/cryst12081089 . . . . **157**

Suraya Sulaiman, Izman Sudin, Uday M. Basheer Al-Naib and Muhammad Firdaus Omar
Review of the Nanostructuring and Doping Strategies for High-Performance ZnO Thermoelectric Materials
Reprinted from: *Crystals* **2022**, *12*, 1076, doi:10.3390/cryst12081076 . . . . **175**

About the Editors

Yamin Leprince-Wang

Yamin Leprince-Wang is Full Professor at the Gustave Eiffel University (UGE), France, where she is also Head of the Materials Science & Engineering Master's degree course. Her main research interest consists of synthesis and characterization of oxide nanomaterials and their applications in energy and environment fields, such as nanogenerator of electricity, solar cells, chemical sensors, and water & air purification using photocatalysis process. She received B.S. degree from Zhejiang University in China (1985), M.S. and PhD degree from Pierre and Marie Curie University (Paris 6) in France in 1991 and 1995, respectively; then joined UGE (ex. UPEM) in 1995. Chief of the Material Science Department between 2006–2010, and leader of the Laboratoire de Physique des Matériaux Divisés & Interfaces (LPMDI, UMR CNRS) between 2008–2014, she co-authored more than 100 peer reviewed papers, 2 books, and 3 patents.

Guangyin Jing

Guangyin Jing, professor at the School of Physics, head of the Department of Materials Physics, focus on soft matter and biophysics. He got a bachelor's degree in Sun Yat-sen University (2002), and Ph.D in Peking University (2007). Later, he started the postdoctoral fellow in ESPCI-ParisTech working on polymer physics, then another postdoctoral work on fluid dynamics of colloids in Paris-Sud University. Since September 2009, he joined the School of Physics of Northwestern University in Xi'an China, and started to run the Soft Matter Physics Lab (SMPL). His current research interests include active matter, self-propelled particle, fluid dynamics of microswimers and their collective motion, bacterial swimming in biological fluids and the medical applications, which are founded National Natural Science Foundation, Outstanding Youth Foundation of Shaanxi Province Natural Science.

Basma El Zein

Basma El Zein, Director General, Techno-Valley, University of Business and Technology (UBT), Jeddah, Saudi Arabia, she held many positions such as Director General of Governance and Sustainability Center, Dean of Scientific Research at UBT, research scientist at KAUST, Ass.Prof at Dar AL Hekma University and associate researcher at IEMN, France. She has 23 years of Experience in the academic, research, renewable and economic development fields. Prof. El Zein graduated from the University of Lille, France with a Ph.D. in Nanotechnology Engineering with High Distinction for her research Zinc Oxide Nanostructures for Photovoltaic Applications. Her master's degree was from the Lebanese University, Lebanon in the field of Electrical and Electronics Engineering with Distinction. Her research interest is currently in Energy Conversion and Energy Storage. The main objective of Dr. El Zein research is to develop an eco-green Solar Cells with high efficiency and long durability. In addition, she is exploring new materials such as kësterite, perovskite, and protein to be used as a light absorber for Solid State Sensitized Solar Cells. She is also exploring Printed Metal Oxide Batteries. She has published in many international journals and has one granted patent from USA, UK and EPO related to materials for Solar Cells. Dr. El Zein was acknowledged for her remarkable contribution to the field of research, teaching, community service, and industry in Saudi Arabia, and she was granted 2 funded research for her research on Nanomaterials for 3rd generation Solar Cells in 2013 and 2014.

Preface to "Novel ZnO-Based Nanostructures: Synthesis, Characterization and Applications"

Zinc oxide (ZnO) nanomaterials, with practical and low-cost synthesis, provide versatile applications in electric, optical, and mechanoelectrical engineering. More importantly, for environmental and health purposes, ZnO possessing facile tailoring in morphology and mass-producing features offer the possibility and realization for biological and medical applications.

With the great help from Mars (Zhilin) Tan, the three guest editors Yamin Leprince-Wang, Guangyin Jing and Basma El Zein invite the selected researchers to contribute original research articles and reviews. Here, these novel research articles and reviews are included with the focus on the simple introduction of synthesis in novel ways, then the applications on the air and liquid purifications by using the photocatalytic merits of ZnO. Furthermore, antibacterial performance from the ZnO nanostructures were demonstrated by these original contributions in this book. Recently, the antibiotic pollution around us is extremely concerning and is recognized as a profound problem due to drug abuse for antimicrobial issues. Possibilities based on ZnO nanomaterials were introduced in this book as well, which directs the new ways for ZnO towards medical usage. Of course, it is far from the comprehensive and detailed understanding to all of the topics we expect to focus on, rather we show the readers an inspired view to attract their attention.

Yamin Leprince-Wang, Guangyin Jing, and Basma El Zein
Editors

Editorial

Novel ZnO-Based Nanostructures: Synthesis, Characterization and Applications

Yamin Leprince-Wang [1,*], Guangyin Jing [2] and Basma El Zein [3]

1. ESYCOM Lab, CNRS, Université Gustave Eiffel, F-77454 Marne-la-Vallée, France
2. School of Physics, Northwest University, Xi'an 710127, China
3. Department of Electrical Engineering, University of Business and Technology, Jeddah 21432, Saudi Arabia
* Correspondence: yamin.leprince@univ-eiffel.fr

Citation: Leprince-Wang, Y.; Jing, G.; El Zein, B. Novel ZnO-Based Nanostructures: Synthesis, Characterization and Applications. *Crystals* **2023**, *13*, 338. https://doi.org/10.3390/cryst13020338

Received: 14 February 2023
Accepted: 14 February 2023
Published: 16 February 2023

Copyright: © 2023 by the authors. Licensee MDPI, Basel, Switzerland. This article is an open access article distributed under the terms and conditions of the Creative Commons Attribution (CC BY) license (https://creativecommons.org/licenses/by/4.0/).

The Special Issue "Novel ZnO-Based Nanostructures: Synthesis, Characterization and Applications" is a collection of 13 papers, including 3 review papers and 10 original articles dedicated to both experimental research works and numerical simulations on ZnO nanostructures.

On one hand, as a multifunctional material, ZnO possesses unique outstanding properties, such as a wide direct bandgap; high electron mobility; piezoelectric, antibacterial, photocatalytic, and thermoelectric properties; as well as chemical and thermal stability and biocompatibility. On the other hand, ZnO nanomaterials can be obtained using easy and low-cost growth techniques including green synthesis methods, with easy morphological control altering seedlayer composition and/or varying growth conditions. It should be emphasized that ZnO nanostructures can be doped to enhance different physical and chemical properties. These assets make nanostructured ZnO one of the most fascinating nanomaterials, leading to the development of many promising applications around the ZnO nanostructures, as presented in this Special Issue.

Knowing that the properties are closely related to the nanostructures, the study of nanostructures' growth mechanism is important for the desired material properties and the targeted applications. In the case of the widely used classical two-step sol-gel hydrothermal method for ZnO nanoarray growth, the morphology of nanoarrays, such as the number density of the nanorod array and the diameter of nanorods, is often tuned by changing the composition ratio of the seed solution. Li et al. [1] investigated the tuning growth of ZnO nanoarrays by the controlled spreading of gel layer based on the dewetting process. The proposed new mechanism can not only help deepen the understanding of the formation and evolution of the seedlayer, but also provide a new method for the controllable growth of nanomaterials. Continuing with the growth mechanism study, Huang et al. [2] investigated the controlled synthesis and the growth mechanism of two-dimensional (2D) ZnO by surfactant-assisted ion-layer epitaxy (SA-ILE), by controlling the growth parameters, such as the amounts of surfactant, temperature, precursor concentration, and growth time. Their work might guide the development of SA-ILE and pave the way for practical applications of 2D ZnO on photodetectors, sensors, and resistive switching devices.

The green synthesis of ZnO nanoparticles (NPs) has recently gained considerable interests because it is simple, environmentally friendly, and cost-effective. The paper of Jaithon et al. [3] therefore aimed to synthesize ZnO NPs by utilizing bioactive compounds derived from waste materials, mangosteen peels, and water hyacinth crude extracts, and investigated their antibacterial and anticancer activities. This study demonstrated the possibility of using green-synthesized ZnO NPs in the development of antibacterial or anticancer agents. Furthermore, this research raised the prospect of increasing the value of agricultural waste. On the same topic, Ali et al. [4] investigated another green synthesis method for ZnO nanostructures (NS) using Pyrus pyrifolia fruit extract. The obtained ZnO NS demonstrated significant antibacterial activity analyzed by metabolic activity analysis and disc diffusion assay against *Escherichia coli* and Staphylococcus aureus. Furthermore, ZnO NS achieved good photocatalytic activity of organic dye decolorizing. The research findings from this study could offer new insights for developing potential antibacterial and photocatalytic materials.

Regulating the swimming motility of bacteria near surfaces is essential to suppress or avoid bacterial contamination and infection in catheters and medical devices with wall surfaces. In their work, Yan et al. [5] showed that ZnO nanowire arrays could reduce the swimming motility of *Escherichia coli*, thus significantly enhancing the trapping ability for motile bacteria. Additionally, thanks to the wide bandgap nature of ZnO, the UV irradiation rapidly reduces bacterial locomotion due to the production of hydroxyl radicals and singlet oxygen. The authors' findings not only demonstrated the detailed motility of bacterial swimming on nanostructured surfaces, but also provided means for antibacterial applications of ZnO nanomaterials.

ZnO is an excellent photocatalyst able to mineralize a large amount of organic pollutants in water under UV irradiation that can be enlarged to the visible range by doping. With high surface/volume ratio, the ZnO nanostructures showed an enhanced photocatalytic efficiency. Le Pivert et al. [6] investigated the photocatalytic activity of ZnO NSs grown on different substrates, both in classic mode and microfluidic mode. All tests showed the notable photocatalytic efficiency of ZnO NSs with remarkable results obtained from a ZnO-NSs-integrated microfluidic reactor, which exhibited an important enhancement of photocatalytic activity by drastically reducing the photodegradation time. The simultaneous real-time PCR followed by UV-visible spectrometry and high-performance liquid chromatography, coupled with mass spectrometry (HPLC-MS), could reveal both the photodegradation efficiency and the degradation mechanism of the organic dye.

In their study, Mendoza-Sánchez et al. [7] used an accessible thermal oxidation technique to synthesize the nanostructured ZnO arrays. Two different morphologies were achieved, ZnO grass-like and ZnO cane-like nanostructures, allowing them to study the influence of ZnO nanostructure morphology on the functionalization efficiency of nanostructured arrays with hemoglobin for CO_2 capture. After functionalization with hemoglobin, they found that the ZnO grass-like structures exhibited the best efficiency for CO_2 capture, with 98.3% of the initial concentration after 180 min.

A broadband thin-film plasmonic metamaterial absorber (MMA) nanostructure with a multi-dielectric layer operating in the frequency range from 100 GHz to 1000 GHz is introduced and analyzed in the paper presented by Emara et al. [8]. In this work, the ZnO layer was added as a substrate on the top of an Au back reflector to enhance the absorption. The MMA had an average absorption of 84%, with a maximum absorption of 100% and a minimum absorption of approximately 65.9%. The optimized MMA showed good angular stability, as the effect of the incident angle of the electromagnetic wave on the MMA absorption is so small and the absorber is insensitive to polarization for both normal and oblique incidence conditions.

The quick development of wireless sensor networks has required sensor nodes to be self-powered. Driven by this goal, in their work, Serairi and Leprince-Wang [9] demonstrated a ZnO nanowire-array-based piezoelectric nanogenerator (NG) prototype, which can convert mechanical energy into electricity. The NG-device, with an effective area of 0.7 cm^2, was based on high-quality single crystalline ZnO nanowires with an aspect ratio of approximately 15. The NG's performance was tested both in compression mode and in vibration mode, giving the output power density of ~38.47 mW/cm^3 and ~0.9 mW/cm^3, respectively.

Aiming at the problem of a lack of data on the nonlinear morphology to divide uneven grain boundary in bulk ceramics, Shen et al. [10] developed a unique approach of the local three-dimensional characterization of nonlinear grain boundary length within bulk ZnO, using a nanorobot in SEM. SEM can generate an image of the contour shapes of the targeted grain boundaries in the X-Y plane, while the Z-directional relative height differences at different positions can be sequentially probed by the nanorobot. Further, by quantifying the Z-directional relative height differences, it can be verified that irregular characteristics exist in the three-dimensional grain boundary length, which can extend the depth effect on nonlinear bulk conductance. In addition, this method can also obtain nonlinear quantitative topographies to divide grain boundaries into an uneven structure in the analysis of bulk polycrystalline materials.

Today, living organisms are more susceptible to exposure to metallic NPs due to these NPs' constant evolution and applications. ZnO-NPs are one of the most used metallic NPs due to their various interesting properties and large application fields. Therefore, understanding the molecular effects of ZnO-NPs in biological systems is extremely important. Patrón-Romero et al. [11] presented a systematic review that aimed to include the most recent scientific evidence concerning the cytotoxic effect of ZnO-NPs in biological models, with special attention paid to mitochondrial damage. Since mitochondria are among the most complex and relevant organelles for cellular homeostasis, it is indispensable to define the most relevant mechanisms leading to cell dysfunction/death. The contribution and usefulness of ZnO-NPs in medical oncology is of great interest; in particular, their contribution in the therapeutic area is increasingly relevant due to their immense potential in the public health sector. This is because cancer remains among the leading causes of death worldwide. Tumor cells show a different cytotoxic effect compared to healthy cells of the same lineage, and the response varies depending on the exposure time, size, and shape of ZnO-NPs. Furthermore, ZnO-NPs can act as anti-cancer agents against different tumor lines resistant to conventional chemotherapeutic treatments; they provide a talented substitute approach to chemotherapies. This systematic review provides information on correlation and the impact on future research.

ZnO also possesses promising potential in thermoelectric applications, enabling the conversion of waste heat to electrical energies with high thermoelectric performance, due to its high physicochemical stability, tunable properties, and high abundance. ZnO thermoelectric devices can operate at higher temperatures; they have higher conversion efficiency and reliability, and the advantage of costless production. In the review presented by Sulaiman et al. [12], the authors attempted to oversee the approaches to improving ZnO thermoelectric properties, where nanostructuring and doping methods were assessed. The outcomes of the reviewed studies are analyzed and benchmarked in this Special Issue, in order to obtain a preliminary understanding of the involved parameters' influences. According to the extant literature, several strategies have been reported to enhance the thermoelectric properties of ZnO, including nanostructuring synthesis techniques and the doping of foreign particles in ZnO.

Today, antibiotics are pervasive contaminants in aqueous systems that pose an environmental threat to aquatic life and humans. The prolonged and excessive use of antibiotics in our society has led to the presence of excessive amounts of non-biodegradable medicines such as antibiotics, and anti-inflammatory, anti-depressive, and contraceptive drugs in hospital and industrial wastewater, which marks a significant threat to the ecosystem. According to the literature, 33,000 people die directly from drug-resistant bacterial infections in Europe annually, which costs EUR 1.5 billion in health care and productivity losses. Consequently, it is an absolute necessity to develop a sustainable method for effective antibiotic removal from wastewater. In their critical review, Mohamed et al. [13] present and discuss recent advances in the photocatalytic degradation of widely used drugs by ZnO-based nanostructures: namely, (i) antibiotics; (ii) antidepressants; (iii) contraceptives; and (iv) anti-inflammatories. This study endows a comprehensive understanding of the degradation of antibiotics using ZnO-based nanomaterials (bare, metal- and non-metal-doped, as well as nanocomposites) for the effective treatment of wastewater containing antibiotics. In addition, the operational conditions and mechanisms involved during the photocatalytic degradation process are systematically discussed. Finally, particular emphasis is devoted to future challenges and the corresponding outlook with respect to toxic effects following the utilization of ZnO-based nanomaterials.

We hope that this collection of papers will meet the expectations of readers seeking the synthesis methods, enhanced properties, and applications of novel ZnO-based nanostructures; and we hope that this collection can also create inspiration for further research works on the related topics included in this Special Issue.

Author Contributions: Y.L.-W.: writing—original draft, review and editing; G.J. and B.E.Z.: visualization. All authors have read and agreed to the published version of the manuscript.

Acknowledgments: The contributions of all authors are gratefully acknowledged. The authors thank the *Crystals* Editorial Office for their excellent communication, support, and fully professional attitude.

Conflicts of Interest: The authors declare no conflict of interest.

References

1. Li, Z.; Yan, N.; Tian, Y.; Luo, H. Tuning Growth of ZnO Nano-Arrays by the Dewetting of Gel Layer. *Crystals* **2023**, *13*, 30. [CrossRef]
2. Huang, C.; Sun, Q.; Chen, Z.; Wen, D.; Tan, Z.; Lu, Y.; He, Y.; Chen, P. Controlled Synthesis and Growth Mechanism of Two-Dimensional Zinc Oxide by Surfactant-Assisted Ion-Layer Epitaxy. *Crystals* **2023**, *13*, 5. [CrossRef]
3. Jaithon, T.; Ruangtong, J.; T-Thienprasert, J.; T-Thienprasert, N.P. Effects of Waste-Derived ZnO Nanoparticles against Growth of Plant Pathogenic Bacteria and Epidermoid Carcinoma Cells. *Crystals* **2022**, *12*, 779. [CrossRef]
4. Ali, Z.A.; Shudirman, I.; Yahya, R.; Venkatraman, G.; Hirad, A.H.; Ansari, S.A. Green Synthesis of ZnO Nanostructures Using Pyrus pyrifolia: Antimicrobial, Photocatalytic and Dielectric Properties. *Crystals* **2022**, *12*, 1808. [CrossRef]
5. Yan, N.; Luo, H.; Liu, Y.; Yu, H.; Jing, G. Motility Suppression and Trapping Bacteria by ZnO Nanostructures. *Crystals* **2022**, *12*, 1027. [CrossRef]
6. Le Pivert, M.; Martin, N.; Leprince-Wang, Y. Hydrothermally Grown ZnO Nanostructures for Water Purification via Photocatalysis. *Crystals* **2022**, *12*, 308. [CrossRef]
7. Mendoza-Sánchez, A.; Cano, F.J.; Hernández-Rodríguez, M.; Cigarroa-Mayorga, O. Influence of ZnO Morphology on the Functionalization Efficiency of Nanostructured Arrays with Hemoglobin for CO_2 Capture. *Crystals* **2022**, *12*, 1086. [CrossRef]
8. Emara, A.; Yousef, A.; ElZein, B.; Jabbour, G.; Elrashidi, A. Enhanced Broadband Metamaterial Absorber Using Plasmonic Nanorods and Muti-Dielectric Layers Based on ZnO Substrate in the Frequency Range from 100 GHz to 1000 GHz. *Crystals* **2022**, *12*, 1334. [CrossRef]
9. Serairi, L.; Leprince-Wang, Y. ZnO Nanowire-Based Piezoelectric Nanogenerator Device Performance Tests. *Crystals* **2022**, *12*, 1023. [CrossRef]
10. Shen, F.; Cao, N.; Li, H.; Wu, Z.; Xie, S.; Luo, J. Local Three-Dimensional Characterization of Nonlinear Grain Boundary Length within Bulk ZnO Using Nanorobot in SEM. *Crystals* **2022**, *12*, 1558. [CrossRef]
11. Patrón-Romero, L.; Luque-Morales, P.A.; Loera-Castañeda, V.; Lares-Asseff, I.; Leal-Ávila, M.Á.; Alvelais-Palacios, J.A.; Plasencia-López, I.; Almanza-Reyes, H. Mitochondrial Dysfunction Induced by Zinc Oxide Nanoparticles. *Crystals* **2022**, *12*, 1089. [CrossRef]
12. Sulaiman, S.; Sudin, I.; Al-Naib, U.M.B.; Omar, M.F. Review of the Nanostructuring and Doping Strategies for High-Performance ZnO Thermoelectric Materials. *Crystals* **2022**, *12*, 1076. [CrossRef]
13. Mohamed, K.M.; Benitto, J.J.; Vijaya, J.J.; Bououdina, M. Recent Advances in ZnO-Based Nanostructures for the Photocatalytic Degradation of Hazardous, Non-Biodegradable Medicines. *Crystals* **2023**, *13*, 329. [CrossRef]

Disclaimer/Publisher's Note: The statements, opinions and data contained in all publications are solely those of the individual author(s) and contributor(s) and not of MDPI and/or the editor(s). MDPI and/or the editor(s) disclaim responsibility for any injury to people or property resulting from any ideas, methods, instructions or products referred to in the content.

Article

Tuning Growth of ZnO Nano-Arrays by the Dewetting of Gel Layer

Ziqian Li, Ningzhe Yan, Yangguang Tian and Hao Luo *

School of Physics, Northwest University, Xi'an 710069, China
* Correspondence: luo@nwu.edu.cn

Abstract: The classical two-step sol-gel hydrothermal method enables the growth of nanoarrays on various substrates via a seed layer. The morphology of the nanoarrays is often tuned by changing the composition ratio of the seed solution. It is taken for granted that the number density and size of seeds will increase with the proportion of precursors. However, in this work, we found novel two-stage dependencies between the concentration of the precursor (ZnAC) and the geometric parameters (number density and diameter) of ZnO seed particles. The completely opposite dependencies illustrate the existence of two different mechanisms. Especially when the proportion of precursors is low ($\phi_{ZnAC} : \phi_{PVA} < 0.22$), the seed number density and diameter decrease with the increasing precursor concentration. This counterintuitive phenomenon should be caused by the destabilization and dewetting process of the thin film layers during annealing. Based on this new mechanism, we demonstrate the tuning growth of the ZnO seed layer and the nanowire array by annealing time. The number density of the nanorod array can be changed by 10 times, and the diameter of the nanorods can be changed by more than 8 times. The new mechanism we proposed can not only help people deepen their understanding of the formation and evolution of the seed layer but also provide a new way for the controllable growth of nanomaterials.

Keywords: ZnO; nano-structure; dewetting

1. Introduction

Research on ZnO nanomaterials has been booming since the 1990s [1–6]. Due to its excellent characteristics, zinc oxide has always maintained a high degree of attention in the fields of energy, environment, biology, and detection [7–13]. In these decades of research, controllable growth and cost reduction are two aspects to which people have been committed. Among various growth methods, the sol-gel hydrothermal method has great potential for industrial application [14–16]. The sol-gel hydrothermal method is generally divided into two steps. The first step is to grow a seed layer on the target substrate using the sol-gel method, and the second step is to grow the nanomaterials on the seed layer by the hydrothermal method. This seed layer-based growth method is mostly used to prepare array structures. Due to the transition of the seed layer, nanomaterials can be grown on lattice-mismatched or even amorphous materials. At the same time, the hydrothermal method has relatively loose requirements on growth conditions and does not require a high-temperature environment, which is undoubtedly friendly to both substrates and equipment [17–22].

Since the geometric size of ZnO nanomaterials directly affects their performance in various aspects, such as solar power generation, photodegradation, self-cleaning, and antibacterial, people have always been enthusiastic about the controllable growth of nano-materials [23–29]. When using the sol-gel hydrothermal method, the geometric size of the nanomaterials can be tuned in the first and second steps, respectively. In the second step of hydrothermal growth, what can be regulated is usually the size of the monomers that make up the array. In the process of preparing the seed layer in the first step, the size and density of the seed can directly affect the structural parameters of the nanomaterial monomers and arrays [30–33]. Therefore, the geometric parameters of nanomaterials can be regulated from more dimensions through the seed layer. When preparing the seed sol, some polymers will be added to the solution as stabilizers. These macromolecules help form a more uniform film when the solution is coated on the substrate and are completely broken down during the annealing process. Since the temperature at which the precursor is transformed into a seed is different from the decomposition temperature of the polymer stabilizer, the dewetting kinetics during annealing can be exploited to tune the morphology of the seed layer. Müller et al. tuned the shape of gold particles at the nanoscale using the dewetting process of molten gold at high temperatures [34]. Martin et al. achieved highly localized control of pattern formation in two-dimensional nanoparticle assemblies by direct modification of solvent dewetting dynamics [35]. Their works imply that the dewetting process can effectively control the distribution and shape of nanoparticles in two-dimensional space. Huang et al. have reported that uniform single-crystalline nanowire arrays can be grown even on an amorphous seed layer made by low-temperature annealing [36]. This shows that the geometry of the seed layer is far more important than its own crystallinity. In other words, if the polymer stabilizer is not completely vaporized, the nanoparticles can still be used as crystallization nuclei to support the growth of nanomaterials. This implies that the gel layer containing the polymer will undergo a dewetting process from film rupture to complete crystallization during annealing. In this process, the size and distribution of the fragments of the film will change greatly, which means that the morphology of the seed layer will have a large adjustment range.

Therefore, we try to use the rupture and dewetting process of the gel film to achieve a wide range of controllable growth of nano-zinc oxide materials. Here, we used zinc acetate and PVA to prepare different ratios of seed layer solutions, spin-coated on silicon substrates and annealed to prepare ZnO seed layers. ZnO nanoarrays were then grown on the seed layer using a hydrothermal method. We systematically studied the film cracking and dewetting behavior of the gel layer with different precursor concentrations during the annealing process and confirmed the impact of these behaviors on the density of the seed layer/nanowire array by using atomic force microscopy, scanning electron microscopy, and dark field optical microscopy.

2. Materials and Methods

2.1. ZnO Seed Layer Preparation

The chemical materials used are from Alfa Aesar (Ward Hill, MA, USA) and are analytically pure. Silicon wafer substrates (1 cm × 1 cm) were carried out with ultrasonic cleaning in absolute ethanol for 20 min and with plasma cleaning for 5 min to make the surface hydrophilic before use. First, 2.5 g PVA (Mw 89,000–98,000, 99+% hydrolyzed) is dissolved in deionized water to prepare 50 mL PVA solution, with heating and stirring at 82 °C until the solution becomes clear. At the same time, different amounts of zinc acetate dihydrate are dissolved in a proper amount of deionized water to obtain 10 mL of solutions with zinc acetate concentrations of 0.01 g/mL, 0.02 g/mL, 0.04 g/mL, 0.08 g/mL, 0.16 g/mL, 0.26 g/mL, and 0.32 g/mL, respectively. Then, mix and stir 50 mL of PVA solution and 10 mL of zinc acetate solution for 2 h to obtain a sol seed solution. The seed solution is spin-coated on the cleaned silicon substrate at 5000 rpm (1000 rpm/s) for 2 min to obtain a gel layer. The heating plate was preheated to 500 °C to ensure a stable annealing temperature. Then heat the spin-coated sample in the air to form a seed layer.

2.2. ZnO Nanostructures Preparation

Hexamethylene tetramine, zinc nitrate hexahydrate, and ammonia water are used to prepare growth solution for hydrothermal reaction with the concentration of hexamethylene tetramine and zinc nitrate hexahydrate in the growth solution being both 0.1 mol/L. Add 12 mL of ammonia water to clarify the solution after 150 mL of the solution is prepared.

Place the prepared seed coating in the growth solution with growing at 90 °C for 6 h. Then the obtained array sample is washed with deionized water and dried in air.

2.3. Measurement Methods

Atomic force microscope (AFM, Dimension Icon, Bruker, MA, USA) and Scanning electron microscope (SEM, JSM-6700F, JEOL, Tokyo, Japen) were employed to characterize the morphology of the seed layer and nanoarrays, respectively. Dark-field microscopy (LV100, Nikon, Tokyo, Japen) was used to count the seed number density of the seed layer at different annealing times. The geometric parameters of the seeds and nanostructures were obtained using ImageJ from AFM, SEM, and dark-field microscopy images.

3. Results and Discussion

3.1. Morphology of Seed Layer

In order to present the proportion of ZnAC and PVA more intuitively, we express the solution of each seed layer by the ratio of the volume fractions, $\alpha = \phi_{ZnAC} : \phi_{PVA}$. After 40 min annealing at 500 °C, the seed layer samples were characterized by AFM, and the AFM images are shown in Figure 1. When α is relatively small (≤ 0.11), some obvious circular spots appear on the substrate, and the number of these spots increases with the alpha, but the diameter of each spot gradually decreases, as shown in Figure 1a–c. When the concentration of zinc acetate is further increased, those circular spots no longer appear, replaced by a granular film covering the entire substrate. Although the size of these particles gradually increases with the increase of alpha, overall, the particle film is very uniform, as shown in Figure 1d–g. Comparing the circular spots and particles, it can be seen that the size of the spots is larger than that of the particles, but the number of particles far exceeds that of the spots. Moreover, the change in their geometric parameters is relatively regular.

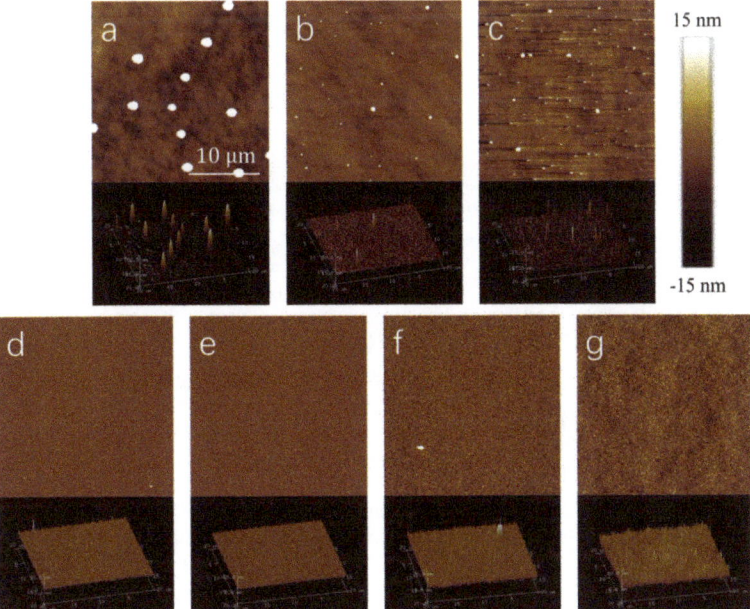

Figure 1. Typical AFM images of annealed seed layer with different ZnAC/PVA volume ratios. (**a**) 0.027:1, (**b**) 0.055:1, (**c**) 0.11:1, (**d**) 0.22:1, (**e**) 0.44:1, (**f**) 0.72:1 and (**g**) 0.87:1. The scanning range of each sample is 20 × 20 µm², and all images use the same scale bar.

As a seed layer, the protrusions on the surface will undoubtedly become the crystallization nuclei for the growth of nanomaterials, so these spots and particles (or be called the seeds) are the focus of our attention. To quantify changes in the morphology of the seed layer, we performed statistics on the number density and diameter of circular spots and particles. In Figure 2a, the change of seed number density with volume ratio can be clearly divided into two regions. Interestingly, in the first region, the number density of seeds increases with the ZnAC/PVA volume ratio, while in the second region, the relationship is negatively correlated. Figure 2b shows the change in seed diameter with α, which is also measured from the AFM image. AFM is very accurate for measuring height information perpendicular to the direction of the substrate, but it is too large for measuring dimensions parallel to the direction of the substrate. However, in our case, what we only need is the relative changes in seed diameter in different samples, so measurements based on AFM images are credible. It can be seen from Figure 2b the variation in seed diameter is also divided into two regions. In the first region, the seed diameter decreased rapidly with the increase in the volume ratio of ZnAC/PVA, while in the second region, the diameter of the seed increased slightly with the increase in the volume ratio. Here "slightly" is the reduction relative to the first region. Focus on the second region, the mean diameter of the seeds increased from 24 nm to 48 nm when the volume ratio of ZnAC/PVA increased from 0.22 to 0.87. It should be noted that both the absolute value of the number density and diameter of the seeds in the first region is much smaller than that in the second region. The opposite dependence and the large difference in number density imply that there are differences in the mechanism of seed formation in these two regions.

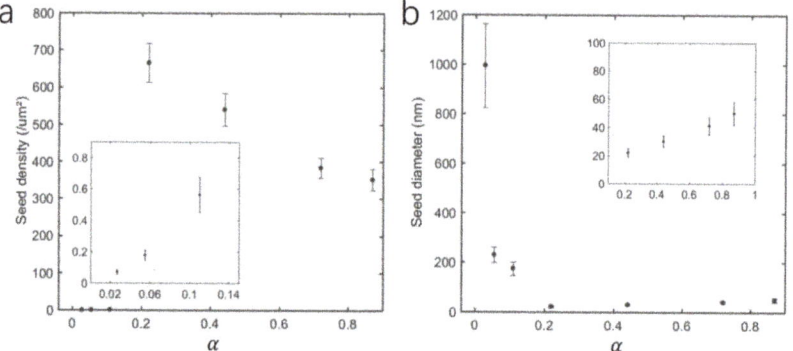

Figure 2. Variation of seed layer morphology parameters with ZnAC/PVA volume ratio α. (a,b) are statistics of the number density and diameter of seeds, respectively.

3.2. Morphology of Nanoarrays

Although the size of seeds has a very wide range of regulation, can the nanostructures grown based on seeds also have a corresponding dependence and regulation range? Each group of samples in Figure 1 was grown by the hydrothermal method under the same conditions for 6 h. SEM was employed to characterize the morphology of the nanoarrays, and typical images are shown in Figure 3. All samples exhibit nanorod/wire array morphology. It can be clearly seen that when α is low, the size of the grown nanorods is very uneven, and the directivity is not good, as shown in Figure 3a,b. When the alpha is increased to 0.11 and above, the uniformity and directivity of the array improve a lot.

To quantitatively investigate the relationship between the nanoarrays and the geometric parameters of the seed layer, we also analyzed the SEM images of the nanoarrays. Here the number density and tip diameter of nanorods were measured and counted. As shown in Figure 4, the number density of nanorods in the array shows a trend of increasing before decreasing with the increase of α, while the diameter of nanorods shows a trend of decreasing before increasing with the increase of α. The variation trend of the number density and diameter of nanorods with alpha is consistent with that of the seed layer, as shown in Figure 2. Moreover, no matter whether it is the seed layer or the nano-array, the turning point, a critical ZnAC/PVA volume ratio α_c, separating the two regions, is between 0.11 and 0.22.

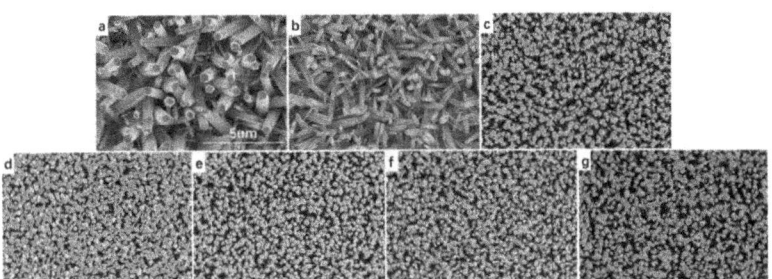

Figure 3. Typical SEM images of ZnO nanoarray based on the seed layer with different ZnAC/PVA volume ratios. (**a**) 0.027:1, (**b**) 0.055:1, (**c**) 0.11:1, (**d**) 0.22:1, (**e**) 0.44:1, (**f**) 0.72:1 and (**g**) 0.87:1. All images use the same scale bar.

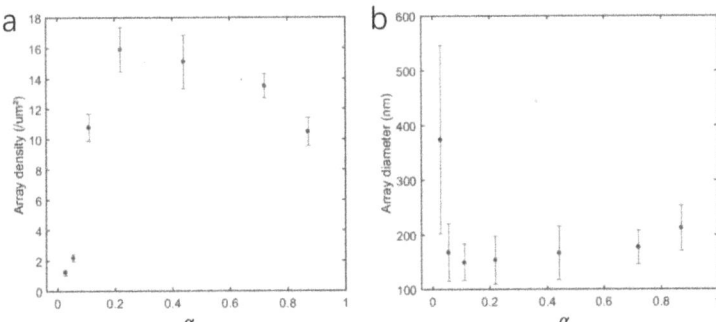

Figure 4. Variation of nanoarrays morphology parameters with ZnAC/PVA volume ratio α. (**a**) and (**b**) are statistics of the number density and diameter of nanorods, respectively.

Here are the following three points to note. (1) In the high α region, the average number density of nanorods is smaller than that of seeds because the adjacent structures will fuse at the early stage of nanomaterial growth. (2) In the low α region, the average number density and diameter of the nanorods are much larger and smaller than the seeds, respectively. This is most likely because the circular spots that appear on the seed layer in the low α region are not a single nucleation site, and due to their large size (diameter can reach 1 micron), it is likely that there are multiple nucleation sites on it. (3) In the low α region, the lower uniformity of ZnO nanoarray is likely to be due to the inhomogeneous distribution of ZnO nanoparticles converted from ZnAC actually used for nucleation. This inhomogeneous distribution is most likely caused by circular spots on the seed layer. The high-density ZnO particle aggregates will fuse into nanorods with larger diameters during the hydrothermal synthesis process.

3.3. Two Mechanisms for Seed Layer

Since the topography of the nanoarrays has a strong dependence on that of the seed layer, and the inhomogeneity of the nanoarrays in the low α region is likely to be caused by round spots on the seed layer, a question is raised: What are the round spots and how do they arise?

PVA accounted for more volume fraction in the ZnAV and PVA blended gel layer. Therefore, we can assume that the properties of the gel film are affected by PVA during annealing, especially in the low alpha region. During the annealing process, the temperature of the substrate is gradually increased, so the PVA will first start to soften (glass transition temperature ~ 80 °C) before the temperature reaches the PVA decomposition temperature (200–250 °C). The softened film becomes unstable and then breaks up due to thermal disturbances. The broken film shrinks under the action of surface tension, which is known as dewetting. This will leave circular spots on the substrate. During the dewetting process, the boundaries of the circular spots shrink gradually, leaving ZnAC molecules or converted ZnO in the shrinking path. However, as the shrinkage rate increases with temperature, the concentration of Zn atoms in the spots is also gradually concentrated, which leads to the uneven density of nucleation sites left on the surface of the substrate, and in turn, affects the uniformity of the subsequently grown nanoarrays. Moreover, when the alpha is higher, the properties of the gel film gradually tend to zinc acetate. The softening degree of the membrane will gradually decrease, and it is not easy to become unstable and broken under the action of thermal disturbance. Therefore, a ZnO seed layer with small particles and uniform distribution can be formed.

3.4. Dewetting Tunned Nanoarray Growth

In order to verify our conjecture and try to use dewetting to regulate the morphology of the seed layer and nanoarray, we designed the following experiments. Two sets of samples below and one set above the critical ZnAC/PVA volume ratio α_c were selected, which were $\alpha = 0.027, 0.11$ and 0.44. Each group of samples was annealed on a hot plate at 500 °C for 10 min, 40 min, and 80 min, respectively. The surface morphology of the annealed samples was observed using a dark-field microscope. Then these samples will undergo hydrothermal reaction in the same environment to grow nano-arrays, and the nano-arrays will be characterized by SEM.

Here we mainly focus on the rupture and dewetting process of the gel film. Due to the large size of the circular spot after the gel film rupture and poor conductivity, AFM and SEM are not applicable. The dark-field microscope is very sensitive to surface fluctuations and is very suitable for observing the breaking process of the gel film. As shown in Figure 5, the bright spots in the images indicate that there is a difference between the height of this position and the base. During the annealing process, a large number of particles appeared in the two samples with low α, while the sample $\alpha = 0.44$ did not change significantly throughout the annealing process. For the two sets of samples with low α, there are changes both in the diameter and number of particles. Since the dark field microscope cannot accurately characterize the size of the particles, here we mainly analyze the number density of the particles.

As shown in Figure 6, as the annealing time increases, the number density of the surface particles of the two groups of low alpha samples first increases and then decreases. This two-stage trend corresponds to the gradual rupture of the film, and more and more "islands" are split, and as the annealing time increases, the residual PVA is gradually decomposed, and the size of the islands is constantly shrinking. This two-stage trend means that the film splits more and more particles during the gradual rupture process at the beginning. Then as the annealing time increases, the PVA is gradually decomposed, and the size of the particles continues to shrink. The particles will gradually disappear from small to large until the PVA is completely decomposed and only ZnO particles remain. Another interesting phenomenon is that the number density of particles on the $\alpha = 0.11$ sample is larger than that of $\alpha = 0.027$, and at 80 min, when the number density of $\alpha = 0.027$ has returned to a small value, the number density of $\alpha = 0.11$ is still increasing. This phenomenon is consistent with the results of AFM. The reasons for this phenomenon are multifaceted and complex. It may be because the increase in the volume fraction of zinc acetate changes the viscosity, surface tension of the solution, the stiffness of the film after softening, and so on.

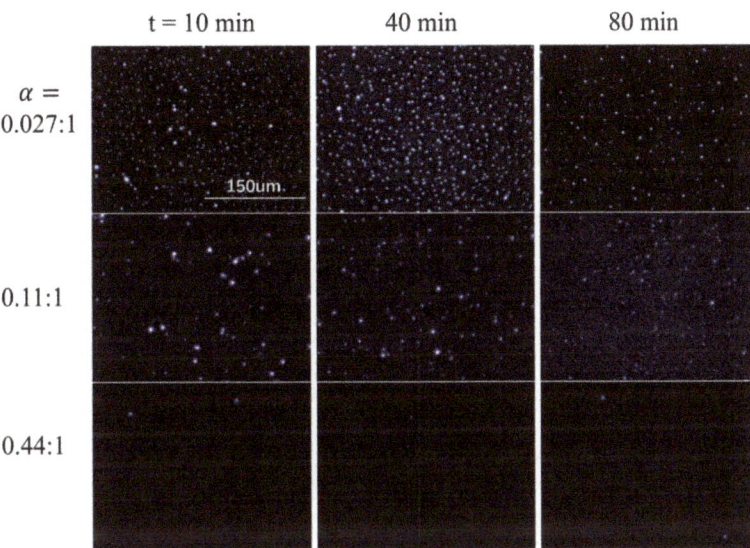

Figure 5. Dark field microscope images of gel layers with different alpha values after different annealing times.

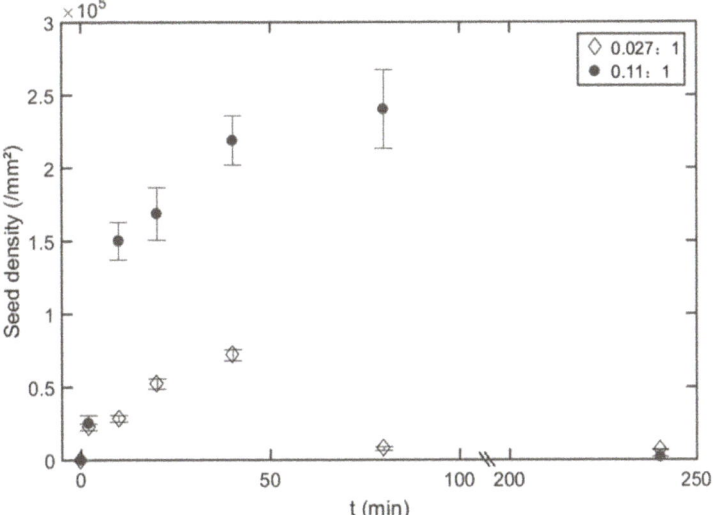

Figure 6. Variation of number density of particles with annealing time.

Since the process of dewetting after gel film rupture can be controlled by the annealing time, the nanoarrays grown based on the seed layer must also be controlled. The ZnO nanoarrays shown in Figure 7 were obtained by hydrothermal growth of the seed layer in Figure 6 under the same conditions. The large difference in morphology of samples $\alpha = 0.027$ and the stability of sample $\alpha = 0.44$ are in line with expectations obtained from their seed layer. However, for sample 0.11, the difference between the seed layer annealed for 40 min and 80 min is not obvious on the ZnO array. This means that when the number density of seeds is greater than a certain value, the nanoarray is no longer sensitive to the number density of seeds. The same conclusion can be obtained more intuitively from statistical data, as shown in Figure 8. For a high α sample ($\alpha = 0.44$), array uniformity is better, but tunability is lost. The sample $\alpha = 0.11$ is more balanced, with good uniformity and some tunability. If the effect of A and annealing time is considered comprehensively, the number density of the nanorod array can be changed by 10 times (2~20/μm^2), and the diameter of the nanorods can be changed by more than 8 times (50~400 nm).

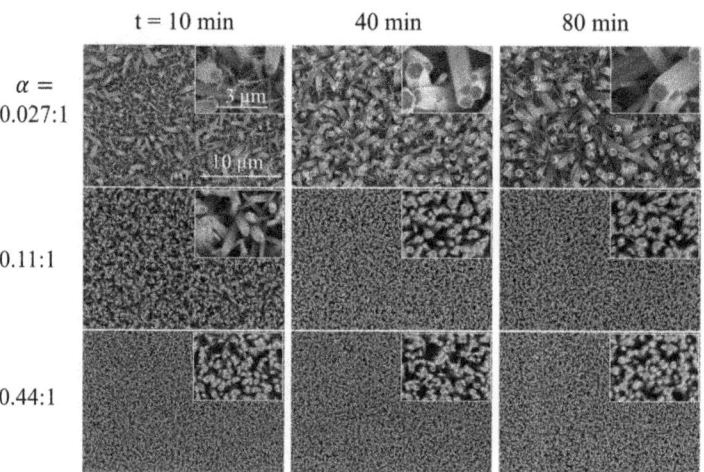

Figure 7. SEM images of nanoarrays obtained by different zinc acetate/PVA volume ratios and annealing times.

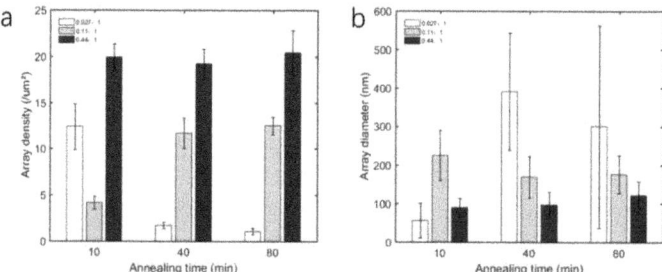

Figure 8. Variation of nanoarrays morphology parameters with ZnAC/PVA volume ratio α and annealing time. (**a**,**b**) are statistics of the number density and diameter of nanorods, respectively.

It is true that the dewetting process of seeds can regulate the morphology of nanoarrays, but in reverse thinking, when we want the product to have high stability, we should try to avoid the occurrence of the de-wetting process, that is, choose a high alpha value. However, if high alpha cannot be selected; for example, the precursor is very expensive, and we have to pay more patience to let the dewetting process be completed.

4. Conclusions

In summary, based on the classical sol-gel hydrothermal method, we systematically investigated the effect of the volume ratio α between the precursor ZnAC and the stabilizer PVA in the seed layer solution on the morphology of seeds and nanoarrays. We found that there are two mechanisms in the annealing process of the seed layer. When alpha is less than or equal to 0.22, the fragmentation and dewetting process of the gel film is crucial, and the morphology of the seed layer and nanoarray is very sensitive to alpha; when alpha is greater than or equal to 0.44, the gel film will not be broken, and the morphology of the seed layer and nano-array will only change slightly with alpha. In addition, by adjusting the annealing time, we controlled the dewetting process of the seed layer and then successfully regulated the morphology of the nanoarrays. The number density of the nanorod array can be changed from $2/\mu m^2$ to $20/\mu m^2$, and the diameter of the nanorods can be changed from 50 nm to 400 nm. Our work demonstrates that the film rupture and dewetting process is an important mechanism in the formation of the seed layer by the sol-gel method. This not only helps people to complete the understanding of the mechanism of the sol-gel hydrothermal method but also provides an important reference for the controllable growth of nanomaterials.

Author Contributions: Conceptualization, Z.L., N.Y. and H.L.; methodology, Z.L., N.Y. and H.L.; software, Z.L. and H.L.; investigation, N.Y., Y.T. and H.L.; resources, Z.L.; data curation, Z.L.; writing—review and editing, N.Y., Y.T. and H.L.; supervision, N.Y., Y.T. and H.L.; project administration, N.Y., Y.T. and H.L.; funding acquisition, H.L.; All authors have read and agreed to the published version of the manuscript.

Funding: This research received funding from National Natural Science Foundation of China (61902316).

Institutional Review Board Statement: Not applicable.

Informed Consent Statement: Not applicable.

Data Availability Statement: Not applicable.

Acknowledgments: The authors gratefully acknowledge the help from P. Li.

Conflicts of Interest: The authors declare no conflict of interest.

References

1. Zhang, H.; Ma, X.; Xu, J.; Niu, J.; Yang, D. Arrays of ZnO nanowires fabricated by a simple chemical solution route. *Nanotechnology* **2003**, *14*, 423. [CrossRef]
2. Greyson, E.C.; Babayan, Y.; Odom, T.W. Directed growth of ordered arrays of small-diameter ZnO nanowires. *Adv. Mater.* **2004**, *16*, 1348–1352. [CrossRef]
3. Fan, H.J.; Lee, W.; Scholz, R.; Dadgar, A.; Krost, A.; Nielsch, K.; Zacharias, M. Arrays of vertically aligned and hexagonally arranged ZnO nanowires: A new template-directed approach. *Nanotechnology* **2005**, *16*, 913. [CrossRef]
4. Weintraub, B.; Zhou, Z.; Li, Y.; Deng, Y. Solution synthesis of one-dimensional ZnO nanomaterials and their applications. *Nanoscale* **2010**, *2*, 1573–1587. [CrossRef]
5. Zhu, P.; Weng, Z.; Li, X.; Liu, X.; Wu, S.; Yeung, K.; Wang, X.; Cui, Z.; Yang, X.; Chu, P.K. Biomedical applications of functionalized ZnO nanomaterials: From biosensors to bioimaging. *Adv. Mater. Interfaces* **2016**, *3*, 1500494. [CrossRef]
6. Le Pivert, M.; Poupart, R.; Capochichi-Gnambodoe, M.; Martin, N.; Leprince-Wang, Y. Direct growth of ZnO nanowires on civil engineering materials: Smart materials for supported photodegradation. *Microsyst. Nanoeng.* **2019**, *5*, 57. [CrossRef]
7. Le Pivert, M.; Piebourg, A.; Bastide, S.; Duc, M.; Leprince-Wang, Y. Direct One-Step Seedless Hydrothermal Growth of ZnO Nanostructures on Zinc: Primary Study for Photocatalytic Roof Development for Rainwater Purification. *Catalysts* **2022**, *12*, 1231. [CrossRef]
8. Wang, J.; Chen, R.; Xiang, L.; Komarneni, S. Synthesis, properties and applications of ZnO nanomaterials with oxygen vacancies: A review. *Ceram. Int.* **2018**, *44*, 7357–7377. [CrossRef]
9. Rong, P.; Ren, S.; Yu, Q. Fabrications and applications of ZnO nanomaterials in flexible functional devices-a review. *Crit. Rev. Anal. Chem.* **2019**, *49*, 336–349. [CrossRef]
10. Chaudhary, S.; Umar, A.; Bhasin, K.; Baskoutas, S. Chemical sensing applications of ZnO nanomaterials. *Materials* **2018**, *11*, 287. [CrossRef]
11. Djurišić, A.B.; Chen, X.; Leung, Y.H.; Ng, A.M.C. ZnO nanostructures: Growth, properties and applications. *J. Mater. Chem.* **2012**, *22*, 6526–6535. [CrossRef]
12. Theerthagiri, J.; Salla, S.; Senthil, R.; Nithyadharseni, P.; Madankumar, A.; Arunachalam, P.; Maiyalagan, T.; Kim, H.-S. A review on ZnO nanostructured materials: Energy, environmental and biological applications. *Nanotechnology* **2019**, *30*, 392001. [CrossRef]
13. Kumar, R.; Umar, A.; Kumar, G.; Nalwa, H.S. Antimicrobial properties of ZnO nanomaterials: A review. *Ceram. Int.* **2017**, *43*, 3940–3961. [CrossRef]
14. Khan, M.F.; Ansari, A.H.; Hameedullah, M.; Ahmad, E.; Husain, F.M.; Zia, Q.; Baig, U.; Zaheer, M.R.; Alam, M.M.; Khan, A.M. Sol-gel synthesis of thorn-like ZnO nanoparticles endorsing mechanical stirring effect and their antimicrobial activities: Potential role as nano-antibiotics. *Sci. Rep.* **2016**, *6*, 1–12. [CrossRef]
15. Al Abdullah, K.; Awad, S.; Zaraket, J.; Salame, C. Synthesis of ZnO nanopowders by using sol-gel and studying their structural and electrical properties at different temperature. *Energy Procedia* **2017**, *119*, 565–570. [CrossRef]

16. Torres, F.D.C.G.; López, J.L.C.; Rodríguez, A.S.L.; Gallardo, P.S.; Morales, E.R.; Hernández, G.P.; Guillen, J.C.D.; Flores, L.L.D. Sol–gel/hydrothermal synthesis of well-aligned ZnO nanorods. *Boletín Soc. Española Cerámica Y Vidr.* 2022; in press.
17. Zhang, J.; Que, W. Preparation and characterization of sol–gel Al-doped ZnO thin films and ZnO nanowire arrays grown on Al-doped ZnO seed layer by hydrothermal method. *Sol. Energy Mater. Sol. Cells* **2010**, *94*, 2181–2186. [CrossRef]
18. Tian, J.-H.; Hu, J.; Li, S.-S.; Zhang, F.; Liu, J.; Shi, J.; Li, X.; Tian, Z.-Q.; Chen, Y. Improved seedless hydrothermal synthesis of dense and ultralong ZnO nanowires. *Nanotechnology* **2011**, *22*, 245601. [CrossRef]
19. Bai, S.-N.; Wu, S.-C. Synthesis of ZnO nanowires by the hydrothermal method, using sol–gel prepared ZnO seed films. *J. Mater. Sci. Mater. Electron.* **2011**, *22*, 339–344. [CrossRef]
20. Chevalier-César, C.; Capochichi-Gnambodoe, M.; Leprince-Wang, Y. Growth mechanism studies of ZnO nanowire arrays via hydrothermal method. *Appl. Phys. A* **2014**, *115*, 953–960. [CrossRef]
21. Zhang, Z.; Lv, Y.; Yan, J.; Hui, D.; Yun, J.; Zhai, C.; Zhao, W. Uniform ZnO nanowire arrays: Hydrothermal synthesis, formation mechanism and field emission performance. *J. Alloy. Compd.* **2015**, *650*, 374–380. [CrossRef]
22. Alshehri, N.A.; Lewis, A.R.; Pleydell-Pearce, C.; Maffeis, T.G. Investigation of the growth parameters of hydrothermal ZnO nanowires for scale up applications. *J. Saudi Chem. Soc.* **2018**, *22*, 538–545. [CrossRef]
23. Tokumoto, M.S.; Pulcinelli, S.H.; Santilli, C.V.; Briois, V. Catalysis and temperature dependence on the formation of ZnO nanoparticles and of zinc acetate derivatives prepared by the sol−gel route. *J. Phys. Chem. B* **2003**, *107*, 568–574. [CrossRef]
24. Tak, Y.; Yong, K. Controlled growth of well-aligned ZnO nanorod array using a novel solution method. *J. Phys. Chem. B* **2005**, *109*, 19263–19269. [CrossRef] [PubMed]
25. Hossain, M.; Zhang, Z.; Takahashi, T. Novel micro-ring structured ZnO photoelectrode for dye-sensitized solar cell. *Nano-Micro Lett.* **2010**, *2*, 53–55. [CrossRef]
26. Evans, J.E.; Jungjohann, K.L.; Browning, N.D.; Arslan, I. Controlled growth of nanoparticles from solution with in situ liquid transmission electron microscopy. *Nano Lett.* **2011**, *11*, 2809–2813. [CrossRef] [PubMed]
27. Blandin, P.; Maximova, K.A.; Gongalsky, M.B.; Sanchez-Royo, J.F.; Chirvony, V.S.; Sentis, M.; Timoshenko, V.Y.; Kabashin, A.V. Femtosecond laser fragmentation from water-dispersed microcolloids: Toward fast controllable growth of ultrapure Si-based nanomaterials for biological applications. *J. Mater. Chem. B* **2013**, *1*, 2489–2495. [CrossRef] [PubMed]
28. Harish, V.; Ansari, M.M.; Tewari, D.; Gaur, M.; Yadav, A.B.; García-Betancourt, M.-L.; Abdel-Haleem, F.M.; Bechelany, M.; Barhoum, A. Nanoparticle and Nanostructure Synthesis and Controlled Growth Methods. *Nanomaterials* **2022**, *12*, 3226. [CrossRef]
29. He, Z.; Zhang, Z.; Bi, S. Nanoparticles for organic electronics applications. *Mater. Res. Express* **2020**, *7*, 012004. [CrossRef]
30. Baruah, S.; Dutta, J. Effect of seeded substrates on hydrothermally grown ZnO nanorods. *J. Sol-Gel Sci. Technol.* **2009**, *50*, 456–464. [CrossRef]
31. Chen, C.-L.; Rosi, N.L. Preparation of unique 1-D nanoparticle superstructures and tailoring their structural features. *J. Am. Chem. Soc.* **2010**, *132*, 6902–6903. [CrossRef]
32. Prieto, P.; Nistor, V.; Nouneh, K.; Oyama, M.; Abd-Lefdil, M.; Díaz, R. XPS study of silver, nickel and bimetallic silver–nickel nanoparticles prepared by seed-mediated growth. *Appl. Surf. Sci.* **2012**, *258*, 8807–8813. [CrossRef]
33. Lee, W.C.; Fang, Y.; Kler, R.; Canciani, G.E.; Draper, T.C.; Al-Abdullah, Z.T.; Alfadul, S.M.; Perry, C.C.; He, H.; Chen, Q. Marangoni ring-templated vertically aligned ZnO nanotube arrays with enhanced photocatalytic hydrogen production. *Mater. Chem. Phys.* **2015**, *149*, 12–16. [CrossRef]
34. Müller, C.M.; Mornaghini, F.C.F.; Spolenak, R. Ordered arrays of faceted gold nanoparticles obtained by dewetting and nanosphere lithography. *Nanotechnology* **2008**, *19*, 485306. [CrossRef] [PubMed]
35. Martin, C.P.; Blunt, M.O.; Pauliac-Vaujour, E.; Stannard, A.; Moriarty, P.; Vancea, I.; Thiele, U. Controlling pattern formation in nanoparticle assemblies via directed solvent dewetting. *Phys. Rev. Lett.* **2007**, *99*, 116103. [CrossRef]
36. Huang, J.-S.; Lin, C.-F. Influences of ZnO sol-gel thin film characteristics on ZnO nanowire arrays prepared at low temperature using all solution-based processing. *J. Appl. Phys.* **2008**, *103*, 014304. [CrossRef]

Disclaimer/Publisher's Note: The statements, opinions and data contained in all publications are solely those of the individual author(s) and contributor(s) and not of MDPI and/or the editor(s). MDPI and/or the editor(s) disclaim responsibility for any injury to people or property resulting from any ideas, methods, instructions or products referred to in the content.

Article

Controlled Synthesis and Growth Mechanism of Two-Dimensional Zinc Oxide by Surfactant-Assisted Ion-Layer Epitaxy

Chunfeng Huang, Qi Sun, Zhiling Chen, Dongping Wen, Zongqian Tan, Yaxian Lu, Yuelan He and Ping Chen *

Center on Nanoenergy Research, Guangxi Key Laboratory for Relativistic Astrophysics, School of Physical Science and Technology, Guangxi University, Nanning 530004, China
* Correspondence: chenping@gxu.edu.cn

Citation: Huang, C.; Sun, Q.; Chen, Z.; Wen, D.; Tan, Z.; Lu, Y.; He, Y.; Chen, P. Controlled Synthesis and Growth Mechanism of Two-Dimensional Zinc Oxide by Surfactant-Assisted Ion-Layer Epitaxy. *Crystals* **2023**, *13*, 5. https://doi.org/10.3390/cryst13010005

Academic Editors: Yamin Leprince-Wang, Guangyin Jing and Basma El Zein

Received: 30 November 2022
Revised: 14 December 2022
Accepted: 16 December 2022
Published: 20 December 2022

Copyright: © 2022 by the authors. Licensee MDPI, Basel, Switzerland. This article is an open access article distributed under the terms and conditions of the Creative Commons Attribution (CC BY) license (https://creativecommons.org/licenses/by/4.0/).

Abstract: Two-dimensional (2D) zinc oxide (ZnO) has attracted much attention for its potential applications in electronics, optoelectronics, ultraviolet photodetectors, and resistive sensors. However, little attention has been focused on the growth mechanism, which is highly desired for practical applications. In this paper, the growth mechanism of 2D ZnO by surfactant-assisted ion-layer epitaxy (SA-ILE) is explored by controlling the amounts of surfactant, temperature, precursor concentration, and growth time. It is found that the location and the number of nucleation sites at the initial stages are restricted by the surfactant, which absorbs Zn^{2+} ions via electrostatic attraction at the water-air interface. Then, the growth of 2D ZnO is administered by the temperature, precursors, and growth time. In other words, the temperature is connected with the diffusion of solute ions and the number of nucleation sites. The concentration of precursors determines the solute ions in solution, which plays a dominant role in the growth rate of 2D ZnO, while growth time affects the nucleation, growth, and dissolution processes of ZnO. However, if the above criteria are exceeded, the nucleation sites significantly increase, resulting in multiple 2D ZnO with tiny size and multilayers. By optimizing the above parameters, 2D ZnO nanosheets with a size as large as 20 μm are achieved with 10×10^{-5} of the ratio of sodium oleyl sulfate to Zn^{2+}, 70 °C, 50 mM of precursor concentration, and 50 min of growth time. 2D ZnO sheets, are confirmed by scanning electron microscope (SEM), energy-dispersive X-ray spectrometer (EDS), X-ray photoelectron spectroscopy (XPS), and Raman spectrum. Our work might guide the development of SA-ILE and pave the platform for practical applications of 2D ZnO on photodetectors, sensors, and resistive switching devices.

Keywords: 2D ZnO; surfactant-assisted ion-layer epitaxy; growth mechanism; controlled synthesis

1. Introduction

Zinc oxide (ZnO) has received considerable attention for its excellent properties of wide-bandgap (3.3 eV) [1], high exciton binding energy (60 meV) [2], non-centrosymmetric crystal structure [3], and high electron mobility [4,5], presenting wide applications in photodetectors [6–9], gas sensors [10–14], photoelectric catalytic devices [15–20], piezoelectric devices [21–25], resistive sensors [26–28], and short-channel high-performance transistors [29]. However, compared with zero-dimensional (0D) and one-dimensional (1D) ZnO, the development of two-dimensional (2D) ZnO has been deserted, though the high surface area and tunable bandgap promote the exploitation of 2D ZnO in new fields [30–36].

2D ZnO possesses a great deal of excellent performance beyond bulk, 1D, and 0D ZnO. For example, decreasing the thickness of 2D ZnO increases the bandgap due to the quantum confinement effect, yielding promising potential in deep-ultraviolet photodetectors and photocatalysis [31,35]. The mechanical stability is also better than 1D ZnO, overcoming the low durability of piezoelectric nanogenerators based on 1D ZnO [37]. Notably, additional non-centrosymmetric structure appears when the thickness of 2D ZnO is reduced to a few atomic layers [38]. That is, piezoelectric performance is largely improved with a higher output voltage and current. Despite the enormous potential applications of 2D ZnO, development is slow due to the immature growth of 2D ZnO.

Various methods have been tried to grow 2D ZnO. Magnetron sputtering and atomic layer deposition were first developed to grow 2D ZnO, but the excessive power produced lots of defects and a small size of monocrystalline ZnO [39–44]. Subsequently, laser deposition was exploited, and the thickness of 2D ZnO was lowered to 5 nm, but with poor crystal quality and morphology [45,46]. Surfactant-assisted ion-layer epitaxy (SA-ILE) was proposed and rapidly disseminated to tackle the challenges of tiny size and irregular shape. 2D ZnO nanosheets with a regular shape and thickness of 2 nm were first realized at the water-air interface, revealing a novel approach to obtain 2D materials from non-van der Waals solids [47]. The method was subsequently extended to the water-air interface to develop 2D ZnO with high Zn vacancy concentrations and strong ferromagnetism [48]. Later, different kinds of anionic surfactants were used to modulate the growth of 2D ZnO, including sodium oleyl sulfate (SOS) [28] and sodium dodecyl sulfate (SDS) [35]. Thus, SA-ILE is considered a simple and feasible method to grow 2D ZnO with a lower layer number and higher crystal quality. However, the growth mechanism of 2D ZnO by SA-ILE has not been reported in detail, which is highly desired for practical applications.

Herein, the nucleation and growth of 2D ZnO by SA-ILE is investigated by exploring the factors of amount of surfactants, temperature, precursor concentration, and growth time. Actually, the nucleation and location of 2D ZnO are determined by the amount and distribution of surfactants, which absorb Zn^{2+} ions with electrostatic interaction between Zn^{2+} and air. Then, the growth of 2D ZnO is influenced by Zn^{2+}/OH^- at the water-air interface. The temperature shows an important effect on its growth due to the temperature dependence of the chemical reaction rate and the nucleation potential. Moreover, the precursor concentration decides the quantity of zinc ion at the water-air interface. It is noteworthy that the thickness of 2D ZnO will increase greatly and the size will decrease rapidly as the precursor concentration increases. The growth time affects the amounts of Zn^{2+}/OH^- attracted to the water-air interface, hence, the size and thickness of 2D ZnO. However, excessive growth time leads to the dissolution of 2D ZnO, resulting in the generation of irregular, thick, and small ZnO. Finally, by optimizing the mentioned parameters, 2D ZnO nanosheets with regular triangle morphology and high surface cleanliness were synthesized at the water-air interface, which were verified by scanning electron microscope (SEM), as well as Raman and X-ray photoelectron spectroscopy (XPS) spectra. This work lays a foundation for understanding the mechanism of 2D ZnO by SA-ILE and provides a road for improvement and optimization of 2D ZnO in the application of deep-ultraviolet photodetectors, 2D spintronics, and piezoelectric devices.

2. Materials and Methods

2.1. Preparation of 2D ZnO Nanosheets

For a typical synthesis process, precursor with the ratio 1:1 of $Zn(NO_3)_2 \cdot 6H_2O$ and hexamethylenetetramine (HMT) was dissolved in deionized water (50 mM, 17 mL). Then, surfactants of sodium oleyl sulfate, dissolved in chloroform (0.1 vol%, 30 µL), were added at the water-air interface formed after 20 min. Subsequently, the solution was heated at 70 °C for 50 min, and 2D ZnO would appear and grow in the water-air interface after completion of the reaction. SiO_2/Si substrate was used to scoop the nanosheets from the water–air interface for subsequent characterization. Due to the small volume of the surfactant and the large water-air interface, the surfactant could be separated discretely.

2.2. Materials Characterization

The morphology of the synthesized nanosheets was investigated by scanning electron microscope (SEM, Carl Zeiss, Jena, Germany, Sigma 500) and optical microscope (OM, OLYMPUS, BX43F). The chemical bonding state was investigated by X-ray Photoelectron Spectroscopy (XPS, Thermo Fisher Scientific, Waltham, MA, USA, ESCALAB 250Xi). The Raman spectra were recorded with Raman spectroscopy with a 532 nm laser (Raman,

HORIBA Jobin Yvon, Paris, France, LABRAM HR EVO). The characterizations for all of the materials were performed at room temperature.

3. Results and Discussion

For a typical synthesis process of 2D ZnO by SA-ILE, precursor was dissolved in deionized water. Then, surfactant was added at the water-air interface after 20 min. Subsequently, the solution was heated at a certain temperature for a long growth time, and 2D ZnO grew in the water-air interface. In this synthesis process, the precursor, surfactant, growth temperature, and time play important roles for the 2D ZnO. Thus, we will discussion the factors of precursor, surfactant, growth temperature, and time of the growth of 2D ZnO as following.

3.1. Effect of the Ratio of Sodium Oleyl Sulfate to Zn^{2+} on 2D ZnO Growth

Surfactant is first chosen to enable the growth of 2D ZnO because it provides the dominant roles for nucleation and growth. Sodium oleyl sulfate is adopted as a surfactant for its excellent dispersion performance at the water-air interface. The amounts of surfactants are increased gradually. When the ratio of sodium oleyl sulfate to Zn^{2+} is lower than 6×10^{-5}, no obvious ZnO is present. Once increased, the ratio of sodium oleyl sulfate to Zn^{2+} to 6×10^{-5}, scattered 2D ZnO with a size around 2.1 μm first appears at the position of the surfactant location with triangle morphology (Figure 1a), indicating the nucleation and growth of ZnO. Then, the 2D ZnO grows, increasing the ratio to 10×10^{-5} (Figure 1b,c). The size of 2D ZnO is around 8.4 μm, indicating the growth rate is 4 times the ratio of 10×10^{-5}. However, the size reduces a great deal while increasing the ratio of sodium oleyl sulfate to Zn^{2+} after 10×10^{-5}, accompanied by a rough surface and increased thickness of ZnO (Figure 1d–f). As the increment of the ratio of sodium oleyl sulfate to Zn^{2+}, the group of oleyl sulfate becomes crumpled at the water-air interface due to the strong intermolecular repulsion [49]. In addition, random nucleation and stacking appear on the surface of the 2D ZnO, resulting in rough surfaces and boundaries. The rough surface was analyzed by energy-dispersive X-ray spectroscopic (EDS). The EDS mapping images show the uniform distribution of the Cl, N, S, and O elements (Figure 1g), while the prominent points at the surface of the ZnO are identified as Zn^{2+} and Na^+ (Insets in Figure 1d–f), which are the nucleation sites for the next layer. The larger amount of surfactants, the greater the number of nucleation sites. As a result, the number of ZnO increases, but the size reduces and the thickness increases. Therefore, the surfactant determines the location of nucleation and the number of nucleation sites.

3.2. Effect of Temperature on 2D ZnO Growth

Temperature is an important factor for the growth of 2D ZnO because the reaction ratio and growth are closely related to temperature. When increasing the temperature from 64 °C to 70 °C, the size of the 2D ZnO enlarged from 4 μm to 10 μm (Figure 2a–d). Then, the rough surface with prominent points appeared while increasing the temperature further (Figure 2e,f). The increased size with increasing temperature is derived from the fast response time to form the nucleation of 2D ZnO, as in Equations (1)–(6).

$$Zn(NO_3)_2 \cdot 6H_2O + H_2O \rightarrow Zn^{2+} + 2NO_3^- + 7H_2O \qquad (1)$$

$$(CH_2)_6N_4 + 6H_2O \rightarrow 6HCHO + 4NH_3 \qquad (2)$$

$$NH_3 + H_2O \rightarrow NH_4^+ + OH^- \qquad (3)$$

$$Zn^{2+} + 2OH^- \rightarrow Zn(OH)_2 \qquad (4)$$

$$Zn(OH)_2 + 2OH^- \rightarrow Zn(OH)_4^{2-} \qquad (5)$$

$$Zn(OH)_4^{2-} \rightarrow ZnO + H_2O + 2OH^- \qquad (6)$$

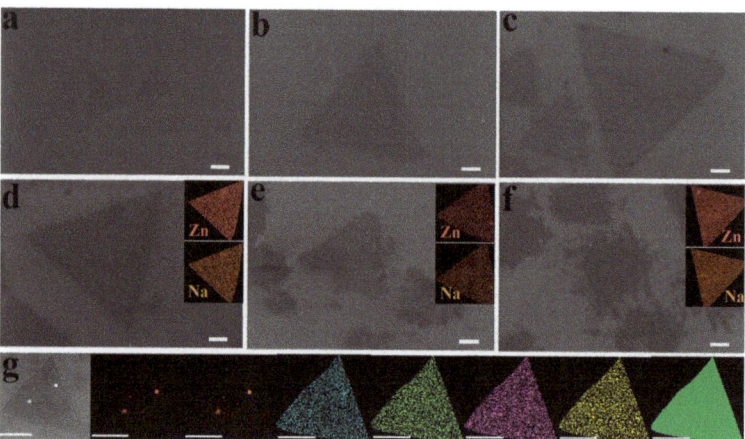

Figure 1. Controlling the growth of 2D ZnO by the amount of surfactant. SEM images of 2D ZnO synthesized with the ratio of sodium oleyl sulfate to Zn^{2+} is around (**a**) 6×10^{-5}, (**b**) 8×10^{-5}, (**c**) 10×10^{-5}, (**d**–**f**) 11×10^{-5}, 13×10^{-5}, and 14×10^{-5}. Insets are the corresponding EDS mapping images of Zn and Na elements. Scale bar is 1 μm. (**g**) SEM image and EDS mapping images for Zn, Na, C, Cl, N, S, and O elements, respectively. Scale bar is 2.5 μm.

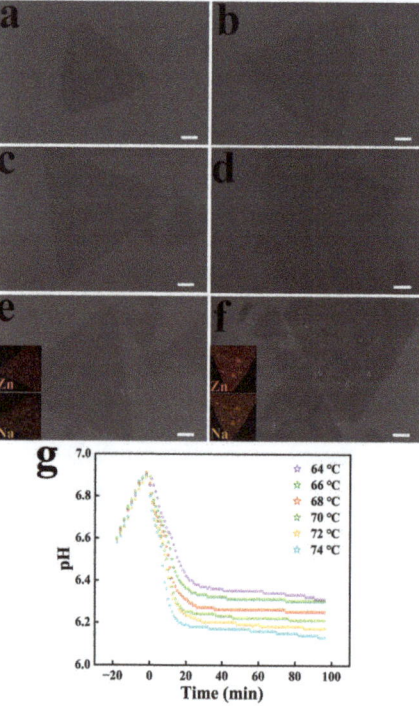

Figure 2. Controlling the growth of 2D ZnO by temperature. SEM images of 2D ZnO synthesized with the temperature at (**a**) 64 °C, (**b**) 66 °C, (**c**) 68 °C, (**d**) 70 °C, (**e**,**f**) 72 °C and 74 °C. Insets are the corresponding EDS mapping images of Zn and Na elements. Scale bar is 1 μm. (**g**) Dependence of pH values on times with different temperatures.

When increasing the temperature, the chemical reactions of Equations (1)–(5) are accelerated because the promoted diffusive transfer of the Zn^{2+}/OH^- to the water-air interface for the higher temperature accelerates the molecular motion. As a result, $Zn(OH)_4^{2-}$ is quickly produced, prompting the nucleation growth of 2D ZnO for the decomposition of $Zn(OH)_4^{2-}$, as in Equation (6). The accelerated chemical reactions can be confirmed by pH with increasing temperature (Figure 2g). Because the fast conformation of $Zn(OH)_4^{2-}$, leads to a more rapid consumption of OH^- in the solution, pH will be reduced with increasing temperature. Figure 2g shows that the pH values are reduced and rapidly decrease with the increment of the temperature, indicating the accelerated chemical reactions with increasing temperature. Thus, the size of 2D ZnO gradually increased with the temperature from 64 °C to 70 °C. Furthermore, the sharp triangle angle changed with decreasing pH, and a truncated triangle emerged. The reason for this might be the imbalance between the zinc and the hydroxide at the water-air interface for the decrease of the concentration of OH^-. Meanwhile, the rough surface with prominent points is connected with the increasing nucleation of the 2D ZnO, again with increasing temperature. On one hand, classical nucleation theory predicts that as the temperature increases, the nucleation potential barrier decreases [50], causing an increase in the number of nuclei of 2D ZnO. On the other hand, the high temperature also promotes the diffusive transfer of Zn^{2+}/OH^- and nucleation on the 2D ZnO surface. The nucleation on the 2D ZnO surface is presented by EDS mapping images (Insets in Figure 2e,f), and the nucleation on 2D ZnO results in a rough surface. Therefore, temperature administrates the diffusive transfer of Zn^{2+}/OH^-, nucleation, and the growth of the 2D ZnO.

3.3. Effect of Precursor Concentration on 2D ZnO Growth

According to chemical reaction kinetics, the rate of the radical reaction is proportional to the concentration of each reactant. Therefore, precursor concentration should play an important role in the growth of 2D ZnO. Thus, we investigated the effect of precursor concentration on the growth of 2D ZnO to further clarify the growth mechanism of 2D ZnO. The shape of 2D ZnO was triangular at the precursor concentration of 40 mM, but the 2D ZnO nanosheets are sparsely distributed (Figure 3a). This phenomenon is attributed to the lower precursor concentration, which causes a lower zinc concentration. Furthermore, fewer zinc ions are attracted to the water-air interface by the oleyl sulfate anions, which is not conducive to the nucleation and growth of 2D ZnO. When the precursor concentration was increased from 40 mM to 50 mM, the size of the 2D ZnO gradually increased (Figure 3b,c). The increment of precursor concentration induces an increase of Zn^{2+}/OH^- in the solution, promoting the diffusive transfer of Zn^{2+}/OH^- from the solution to the water-air interface. Thus, the growth rate of 2D ZnO is accelerated and the size of the 2D ZnO increases. Subsequently, when the precursor concentration is incremented from 50 mM to 60 mM, the size of the 2D ZnO was seen to increase further, accompanied by the surface stacking of the 2D ZnO (Figure 3d,e), i.e., originated from nucleation on the 2D ZnO surface with higher precursor concentration and Zn^{2+}/OH^- in the solution. Finally, the size is reduced and the layering of 2D ZnO is increased with increasing precursor of more than 65 mM (Figure 3f). The excessive precursor induces the rapid spread of Zn^{2+}/OH^- from the solution to the water-air interface and a massive amount of nucleation of the 2D ZnO, leading to a small size. The larger number of Zn^{2+}/OH^- also prompts the nucleation on the first layer of 2D ZnO, resulting in a thick and irregular shape. In a word, the concentration of the precursor controls the amount of Zn^{2+}/OH^- at the water-air interface, as well as the size and thickness of the 2D ZnO.

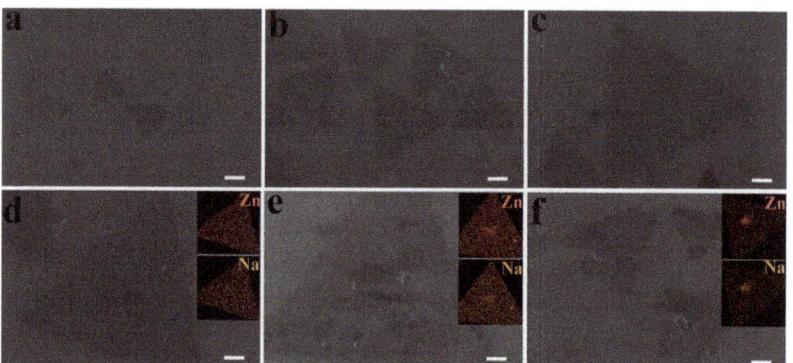

Figure 3. Controlling the growth of 2D ZnO by precursor concentration. SEM images of 2D ZnO synthesized with precursor concentration of (**a**) 40 mM, (**b**) 45 mM, (**c**) 50 mM, (**d**–**f**) 55 mM, 60 mM and 65 mM. Insets are the corresponding EDS mapping images of Zn and Na elements. Scale bar is 1 μm.

3.4. Effect of Growth Time on 2D ZnO Growth

In order to clarify the growth mechanism of 2D ZnO, it is particularly important to investigate the effect of different growth times on 2D ZnO. The size of the 2D ZnO increased continuously when the growth time was incremented from 40 min to 50 min (Figure 4a–c). As the reaction time increased, the number of Zn^{2+} attracted to the water-air interface by the anionic surfactant increased, leading to the increment of $Zn(OH)_4^{2-}$. Then, the $Zn(OH)_4^{2-}$ gradually decomposed, and ZnO formed and grew at the water-air interface, causing the enlarged size of the 2D ZnO. However, when increasing the growth time from 55 min, a rough surface with prominent points appeared, and the dissolving phenomenon emerged (Figure 4d–f). The rough surface could also be confirmed as clusters of Zn^{2+} and Na^+ by the EDS mapping images (Insets in Figure 4d–f), which can be seen as the nucleation of the next layer of 2D ZnO. The dissolving phenomenon of the 2D ZnO might be connected to the following reasons. Due to the amphoteric character and large surface area of 2D ZnO, the 2D ZnO cannot maintain stability in alkaline solutions for a long time and start to dissolve with increasing time [51]. Another possible scenario is poor crystalline quality, which accelerates the dissolving process [52]. In a word, a suitable growth time promotes the gathering of $Zn(OH)_4^{2-}$ and the growth of 2D ZnO. However, a long growth time will induce dissolution of 2D ZnO.

3.5. Characterization

Thus, the growth process of 2D ZnO, modulated by the surfactant, temperature, precursor, and growth time, can be concluded as follows (Table 1 and Figure 5a–e). Firstly, the surfactant dissolves to form a group of oleyl sulfate, which cannot be dissolved in the water and floats on the solution's surface. The ionized surfactants are negatively charged, which absorb Zn^{2+} at the water-air interface. Then, Zn^{2+} ions react to yield $Zn(OH)_4^{2-}$ and 2D ZnO. Thus, the location of the surfactant determines the sites, number of nucleation sites, and growth of the 2D ZnO. Secondly, temperature is connected with the conformation rate of $Zn(OH)_4^{2-}$. The lower the temperature, the slower the diffusion rate of the Zn^{2+}/OH^- and the formation of $Zn(OH)_4^{2-}$, and vice versa. Moreover, the higher temperature also causes a low potential barrier of nucleation, resulting in a large number of nucleation sites of 2D ZnO. However, the excessive number of nucleation sites will lead to a small size of 2D ZnO. Thirdly, the concentration of precursor controls the amount of Zn^{2+}/OH^- in the solution, which dominates the rates of diffusion and the transfer of Zn^{2+}/OH^- to the water-air interface by chemical gradient, as well as the rates of the nucleation and growth of 2D ZnO. Additionally, an excessive concentration of precursor induces rapid diffusion and transfer of Zn^{2+}/OH^-, which generates a great deal of new nucleation sites and a small size of 2D ZnO. Fourthly, the growth time influences the amount of Zn^{2+}/OH^- at the water-air interface. A suitable growth time encourages continuous growth and a large size of 2D ZnO. However, growth time that is too long will result in the dissolution of 2D ZnO.

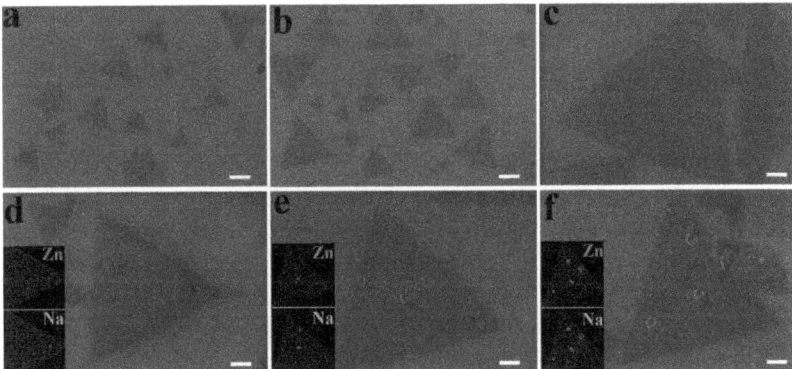

Figure 4. Controlling the growth of 2D ZnO by growth time. Controlling the growth of 2D ZnO with a time of (**a**) 40 min, (**b**) 45 min, (**c**) 50min, (**d–f**) 55 min, 60 min and 65 min. Insets are the corresponding EDS mapping images of Zn and Na elements. Scale bar is 1 μm.

Table 1. Effect of the above parameters on 2D ZnO growth.

Parameters	Value	Size (μm)	Morphology
The ratio of sodium oleyl sulfate to Zn^{2+}	6×10^{-5}	2.1	triangle
	8×10^{-5}	6.5	triangle
	10×10^{-5}	8.4	triangle
	11×10^{-5}	7.2	triangle + rough
	13×10^{-5}	4.6	triangle + rough
	14×10^{-5}	3.7	triangle + rough
Temperature (°C)	64	5.2	triangle
	66	7.5	triangle
	68	7.6	triangle
	70	8.5	triangle
	72	7.8	triangle + rough
	74	7.6	truncated triangle + rough
precursor concentration (mM)	40	1.8	triangle
	45	3.5	triangle
	50	8.6	triangle
	55	10.1	triangle + rough
	60	13	triangle + rough
	65	3.8	triangle + rough

Table 1. Cont.

Parameters	Value	Size (µm)	Morphology
Growth time (min)	40	1.5	triangle
	45	2.1	triangle
	50	8.9	triangle
	55	7.7	triangle + rough
	60	7.3	triangle + rough
	65	7.2	triangle + rough

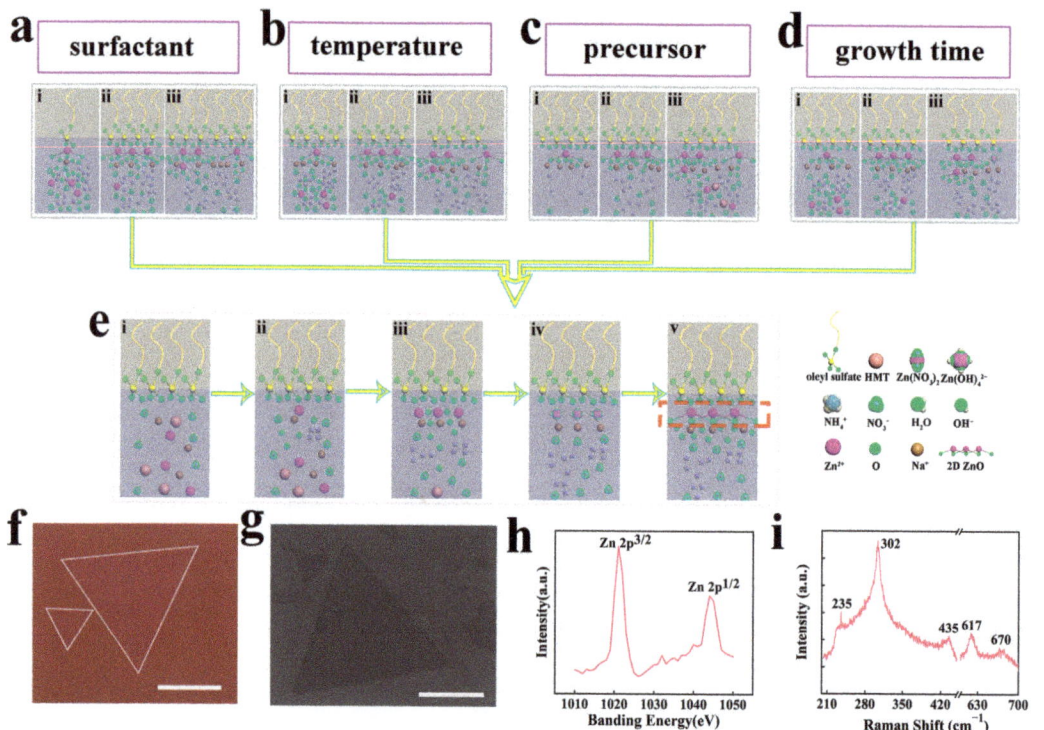

Figure 5. Mechanism analysis of 2D ZnO synthesis by SA-ILE. Schematic diagram of the synthesis of 2D ZnO while controlling (**a**) the ratio of sodium oleyl sulfate to Zn^{2+} (i): lower than 10×10^{-5}, (ii): 10×10^{-5}, (iii): higher than 10×10^{-5}, (**b**) temperature (i): lower than 70 °C, (ii): 70 °C, (iii): higher than 70 °C. (**c**) precursor (i): lower than 50 mM, (ii): 50 mM, (iii): higher than 60 mM and (**d**) growth time (i): lower than 50 min, (ii): 50 min, (iii): higher than 50 min. (**e**) Diagram illustrates the nucleation and growth of 2D ZnO. The red dotted box indicated the formation of 2D ZnO at the water-air interface. (**f**) Optical image of 2D ZnO. Scale bar is 5 µm. (**g**) SEM image of 2D ZnO. Scale bar is 5 µm. (**h**) XPS spectrum of 2D ZnO. (**i**) Raman spectrum of 2D ZnO.

By carefully controlling the factors of surfactant, temperature, precursor, and growth time, the 2D ZnO nanosheets with a smooth surface and a regular triangle shape are achieved. As can be seen from the optical and SEM images, the triangle shape and high cleanliness of the 2D ZnO are present (Figure 5f,g). The crystal structure can be characterized by XPS and Raman spectra. Figure 5h shows 2 peaks centered around 1021 eV and 1044 eV, corresponding to the binding energies for Zn 2p3/2 and Zn 2p1/2, respectively, revealing that the Zn element is in a +2 oxidation state [53]. The Raman spectrum gives 5 peaks, centered around 235 cm^{-1}, 302 cm^{-1}, 435 cm^{-1}, 617 cm^{-1}, and 670 cm^{-1}, respectively. The peaks centered at 235 cm^{-1} may correspond to undesired adulteration [54]. The peaks at 302 cm^{-1}, 617 cm^{-1}, and 670 cm^{-1} originate from symmetric stretching vibration, E1 (LO) modes, and the TA + LO intrinsic mode of ZnO [55,56]. The peak centered at 435 cm^{-1} corresponds to the E2 (high) of ZnO, corresponding to the energy band characteristic peak of wurtzite ZnO and confirming the formation of wurtzite ZnO (Figure 5i) [57,58].

4. Conclusions

In summary, we have achieved the large size, triangle shape, and high cleanliness of 2D ZnO by precisely controlling the concentration of the surfactant, precursor, temperature, and growth time. The surfactant administrates the location of the nucleation and the number of nucleation sites. With the ratio of sodium oleyl sulfate to Zn^{2+} increasing from 6×10^{-5} to 10×10^{-5}, the size of the 2D ZnO grows from 2.1 µm to 8.4 µm. However, the size reduces a great deal while increasing the ratio, accompanied by rough surface and increased thickness of the ZnO. The temperature controls the diffusive transfer of Zn^{2+}/OH$^-$, nucleation, and the growth of the 2D ZnO. The size of the 2D ZnO increases, and the surface is smooth, with the temperature increased from 64 °C to 70 °C. However, truncated triangles appeared as the temperature increases higher than 70 °C. The concentration of the precursor dominates the amount of Zn^{2+}/OH$^-$ at the water-air interface, as well as the size and thickness of the 2D ZnO. The size of the 2D ZnO grows from 1.8 µm to 13 µm as the precursor concentration increases from 40 mM to 60 mM. Yet, the rough surface emerges once it exceeds 50 mM. The suitable growth time promotes the gathering of Zn(OH)$_4^{2-}$ and the growth of the 2D ZnO. The size of the 2D ZnO increases with the growth time from 40 min to 50 min. However, long growth time will induce dissolution of 2D ZnO. Finally, the best parameters for the growth of 2D ZnO is 10×10^{-5} of the ratio of sodium oleyl sulfate to Zn^{2+}, 70 °C, 50 mM of precursor concentration, and 50 min of growth time. Our work clarifies the mechanism of the SA-ILE growth of 2D ZnO, promotes the further development of SA-ILE, and paves the way for applications of 2D ZnO in ultraviolet photodetectors, 2D spintronics, and piezoelectric devices.

Author Contributions: P.C. and C.H. conceived the ideas. C.H. performed the synthesis. C.H., Q.S., Z.C., D.W., Z.T., Y.L. and Y.H. conducted the characterization. C.H. and P.C. analyzed the results. C.H. and P.C. write the manuscript. All authors have read and agreed to the published version of the manuscript.

Funding: The work is supported by the National Natural Science Foundation of China (Nos. 11704081) and Guangxi Natural Science Foundation (Nos. 2020GXNSFAA297182, 2020GXNSFAA297041, 2017GXNSFBA198229).

Data Availability Statement: Not applicable.

Conflicts of Interest: The authors declare no conflict of interest.

References

1. Izaki, M.; Chizaki, R.; Saito, T.; Murata, K.; Sasano, J.; Shinagawa, T. Hybrid ZnO/Phthalocyanine Photovoltaic Device with Highly Resistive ZnO Intermediate Layer. *ACS Appl. Mater. Interfaces* **2013**, *5*, 9386–9395. [CrossRef] [PubMed]
2. Cole, J.J.; Wang, X.; Knuesel, R.J.; Jacobs, H.O. Patterned Growth and Transfer of ZnO Micro and Nanocrystals with Size and Location Control. *Adv. Mater.* **2008**, *20*, 1474–1478. [CrossRef]
3. Lee, P.-C.; Hsiao, Y.-L.; Dutta, J.; Wang, R.-C.; Tseng, S.-W.; Liu, C.-P. Development of porous ZnO thin films for enhancing piezoelectric nanogenerators and force sensors. *Nano Energy* **2020**, *82*, 105702. [CrossRef]
4. Hsu, Y.F.; Xi, Y.Y.; Tam, K.H.; Djurišić, A.B.; Luo, J.; Ling, C.C.; Cheung, C.K.; Ng, A.M.C.; Chan, W.K.; Deng, X.; et al. Undoped p-Type ZnO Nanorods Synthesized by a Hydrothermal Method. *Adv. Funct. Mater.* **2008**, *18*, 1020–1030. [CrossRef]
5. Kang, Z.; Gu, Y.; Yan, X.; Bai, Z.; Liu, Y.; Liu, S.; Zhang, X.; Zhang, Z.; Zhang, X.; Zhang, Y. Enhanced photoelectrochemical property of ZnO nanorods array synthesized on reduced graphene oxide for self-powered biosensing application. *Biosens. Bioelectron.* **2015**, *64*, 499–504. [CrossRef] [PubMed]
6. Ji, Y.; Wu, L.; Liu, Y.; Yang, Y. Chemo-phototronic effect induced electricity for enhanced self-powered photodetector system based on ZnO nanowires. *Nano Energy* **2021**, *89*, 106449. [CrossRef]

7. Wang, P.; Wang, Y.; Ye, L.; Wu, M.; Xie, R.; Wang, X.; Chen, X.; Fan, Z.; Wang, J.; Hu, W. Ferroelectric Localized Field-Enhanced ZnO Nanosheet Ultraviolet Photodetector with High Sensitivity and Low Dark Current. *Small* **2018**, *14*, 1800492. [CrossRef]
8. Meng, J.; Li, Q.; Huang, J.; Pan, C.; Li, Z. Self-powered photodetector for ultralow power density UV sensing. *Nano Today* **2022**, *43*, 101399. [CrossRef]
9. Peng, Y.; Lu, J.; Wang, X.; Ma, W.; Que, M.; Chen, Q.; Li, F.; Liu, X.; Gao, W.; Pan, C. Self-powered high-performance flexible GaN/ZnO heterostructure UV photodetectors with piezo-phototronic effect enhanced photoresponse. *Nano Energy* **2022**, *94*, 106945. [CrossRef]
10. Hoffmann, M.W.G.; Mayrhofer, L.; Casals, O.; Caccamo, L.; Hernandez-Ramirez, F.; Lilienkamp, G.; Daum, W.; Moseler, M.; Waag, A.; Shen, H.; et al. A highly selective and self-powered gas sensor via organic surface functionalization of p-Si/n-ZnO diodes. *Adv. Mater.* **2014**, *26*, 8017–8022. [CrossRef]
11. Shan, H.; Liu, C.; Liu, L.; Wang, L.; Zhang, X.; Chi, X.; Bo, X.; Wang, K. Excellent ethanol sensor based on multiwalled carbon nanotube-doped ZnO. *Chin. Sci. Bull.* **2014**, *59*, 374–378. [CrossRef]
12. Yuan, H.; Aljneibi, S.A.A.A.; Yuan, J.; Wang, Y.; Liu, H.; Fang, J.; Tang, C.; Yan, X.; Cai, H.; Gu, Y.; et al. ZnO Nanosheets Abundant in Oxygen Vacancies Derived from Metal-Organic Frameworks for ppb-Level Gas Sensing. *Adv. Mater.* **2019**, *31*, e1807161. [CrossRef] [PubMed]
13. Fan, C.; Shi, J.; Zhang, Y.; Quan, W.; Chen, X.; Yang, J.; Zeng, M.; Zhou, Z.; Su, Y.; Wei, H.; et al. Fast and recoverable NO_2 detection achieved by assembling ZnO on $Ti_3C_2T_x$ MXene nanosheets under UV illumination at room temperature. *Nanoscale* **2022**, *14*, 3441–3451. [CrossRef] [PubMed]
14. Li, D.L.; Lu, J.F.; Zhang, X.J.; Peng, X.L.; Li, J.; Yang, Y.T.; Hong, B.; Wang, X.Q.; Jin, D.F.; Jin, H.X. Reversible switching from P- to N-Type NO_2 sensing in ZnO Rods/rGO by changing the NO_2 concentration, temperature, and doping ratio. *J. Phys. Chem. C* **2022**, *126*, 14470–14478. Available online: https://pubs.acs.org/doi/full/10.1021/acs.jpcc.2c03616 (accessed on 29 November 2022). [CrossRef]
15. Park, S.J.; Das, G.S.; Schuett, F.; Adelung, R.; Mishra, Y.K.; Tripathi, K.M.; Kim, T. Visible-light photocatalysis by car-bon-nano-onion-functionalized ZnO tetrapods: Degradation of 2,4-dinitrophenol and a plant-model-based ecological assess-ment. *NPG Asia Mater.* **2019**, *11*. [CrossRef]
16. Wang, F.; Zhao, D.; Xu, Z.; Zheng, Z.; Zhang, L.; Shen, D. Monochromatic visible light-driven photocatalysis realized on 2D ZnO shell arrays. *J. Mater. Chem. A* **2013**, *1*, 9132–9137. [CrossRef]
17. Song, S.; Song, H.; Li, L.; Wang, S.; Chu, W.; Peng, K.; Meng, X.; Wang, Q.; Deng, B.; Liu, Q.; et al. A selective Au-ZnO/TiO_2 hybrid photocatalyst for oxidative coupling of methane to ethane with dioxygen. *Nat. Catal.* **2022**, *5*, 1032–1042. [CrossRef]
18. Peng, F.; Lin, J.; Li, H.; Liu, Z.; Su, Q.; Wu, Z.; Xiao, Y.; Yu, H.; Zhang, M.; Wu, C.; et al. Design of piezoelectric ZnO based catalysts for ammonia production from N_2 and H_2O under ultrasound sonication. *Nano Energy* **2022**, *95*, 107020. [CrossRef]
19. Liu, W.; Liu, H.; Liu, Y.; Dong, Z.; Luo, L. Surface Plane Effect of ZnO on the Catalytic Performance of Au/ZnO for the CO Oxidation Reaction. *J. Phys. Chem. C* **2022**, *126*, 14155–14162. [CrossRef]
20. Gleißner, R.; Noei, H.; Chung, S.; Semione, G.D.L.; Beck, E.E.; Dippel, A.-C.; Gutowski, O.; Gizer, G.; Vonk, V.; Stierle, A. Copper Nanoparticles with High Index Facets on Basal and Vicinal ZnO Surfaces. *J. Phys. Chem. C* **2021**, *125*, 23561–23569. [CrossRef]
21. Kim, H.; Yun, S.; Kim, K.; Kim, W.; Ryu, J.; Nam, H.G.; Han, S.M.; Jeon, S.; Hong, S. Breaking the elastic limit of piezoelectric ceramics using nanostructures: A case study using ZnO. *Nano Energy* **2020**, *78*, 105259. [CrossRef]
22. Yin, B.; Qiu, Y.; Zhang, H.; Lei, J.; Chang, Y.; Ji, J.; Luo, Y.; Zhao, Y.; Hu, L. Piezoelectric performance enhancement of ZnO flexible nanogenerator by a NiO-ZnO p-n junction formation. *Nano Energy* **2015**, *14*, 95–101. [CrossRef]
23. Park, J.; Ghosh, R.; Song, M.S.; Hwang, Y.; Tchoe, Y.; Saroj, R.K.; Ali, A.; Guha, P.; Kim, B.; Kim, S.-W.; et al. Individually addressable and flexible pressure sensor matrixes with ZnO nanotube arrays on graphene. *NPG Asia Mater.* **2022**, *14*, 40. [CrossRef]
24. Yu, Q.; Ge, R.; Wen, J.; Du, T.; Zhai, J.; Liu, S.; Wang, L.; Qin, Y. Highly sensitive strain sensors based on piezotronic tunneling junction. *Nat. Commun.* **2022**, *13*, 778. [CrossRef]
25. An, C.; Qi, H.; Wang, L.; Fu, X.; Wang, A.; Wang, Z.L.; Liu, J. Piezotronic and piezo-phototronic effects of atomically-thin ZnO nanosheets. *Nano Energy* **2020**, *82*, 105653. [CrossRef]
26. Yalishev, V.; Yuldashev, S.U.; Kim, Y.S.; Park, B.H. The role of zinc vacancies in bipolar resistance switching of Ag/ZnO/Pt memory structures. *Nanotechnology* **2012**, *23*, 375201. [CrossRef]
27. Wang, X.; Qian, H.; Guan, L.; Wang, W.; Xing, B.; Yan, X.; Zhang, S.; Sha, J.; Wang, Y. Influence of metal electrode on the performance of ZnO based resistance switching memories. *J. Appl. Phys.* **2017**, *122*, 154301. [CrossRef]
28. Yin, X.; Wang, Y.; Chang, T.; Zhang, P.; Li, J.; Xue, P.; Long, Y.; Shohet, J.L.; Voyles, P.M.; Ma, Z.; et al. Memristive Behavior Enabled by Amorphous–Crystalline 2D Oxide Heterostructure. *Adv. Mater.* **2020**, *32*, 2000801. [CrossRef]
29. Wang, L.; Liu, S.; Gao, G.; Pang, Y.; Yin, X.; Feng, X.; Zhu, L.; Bai, Y.; Chen, L.; Xiao, T.; et al. Ultrathin Piezotronic Transistors with 2 nm Channel Lengths. *ACS Nano* **2018**, *12*, 4903–4908. [CrossRef]
30. Jang, E.S.; Won, J.-H.; Hwang, S.-J.; Choy, J.-H. Fine tuning of the face orientation of ZnO crystals to optimize their photo-catalytic activity. *Adv. Mater.* **2006**, *18*, 3309–3312. [CrossRef]
31. Gupta, S.P.; Pawbake, A.S.; Sathe, B.R.; Late, D.J.; Walke, P.S. Superior humidity sensor and photodetector of mesoporous ZnO nanosheets at room temperature. *Sens. Actuators B Chem.* **2019**, *293*, 83–92. [CrossRef]

32. Wang, L.; Liu, S.; Zhang, Z.; Feng, X.; Zhu, L.; Guo, H.; Ding, W.; Chen, L.; Qin, Y.; Wang, Z.L. 2D piezotronics in atomically thin zinc oxide sheets: Interfacing gating and channel width gating. *Nano Energy* **2019**, *60*, 724–733. [CrossRef]
33. Gao, J.; Meng, H.; Hu, Q.; Chang, J.; Feng, L. 2D ZIF-derived ZnO nanosheets—An example for improving semiconductor metal oxide detector performance in gas chromatography through material design strategy. *Sens. Actuators B Chem.* **2019**, *307*, 127580. [CrossRef]
34. Ren, K.; Yu, J.; Tang, W. Two-dimensional ZnO/BSe van der waals heterostructure used as a promising photocatalyst for water splitting: A DFT study. *J. Alloys Compd.* **2019**, *812*, 152049. [CrossRef]
35. Yu, H.; Liao, Q.; Kang, Z.; Wang, Z.; Liu, B.; Zhang, X.; Du, J.; Ou, Y.; Hong, M.; Xiao, J.; et al. Atomic-thin ZnO sheet for visible-blind ultraviolet photodetection. *Small* **2020**, *16*, 2005520. [CrossRef]
36. Kumar, M.; Bhatt, V.; Kim, J.; Abhyankar, A.C.; Chung, H.-J.; Singh, K.; Bin Cho, Y.; Yun, Y.J.; Lim, K.S.; Yun, J.-H. Holey engineered 2D ZnO-nanosheets architecture for supersensitive ppm level H_2 gas detection at room temperature. *Sens. Actuators B Chem.* **2021**, *326*, 128839. [CrossRef]
37. Rafique, S.; Kasi, A.K.; Aminullah; Kasi, J.K.; Bokhari, M.; Shakoor, Z. Fabrication of Br doped ZnO nanosheets piezoelectric nanogenerator for pressure and position sensing applications. *Curr. Appl. Phys.* **2020**, *21*, 72–79. [CrossRef]
38. Mahmood, N.; Khan, H.; Tran, K.; Kuppe, P.; Zavabeti, A.; Atkin, P.; Ghasemian, M.B.; Yang, J.; Xu, C.; Tawfik, S.A.; et al. Maximum piezoelectricity in a few unit-cell thick planar ZnO—A liquid metal-based synthesis approach. *Mater. Today* **2021**, *44*, 69–77. [CrossRef]
39. Samavati, A.; Nur, H.; Ismail, A.F.; Othaman, Z. Radio frequency magnetron sputtered ZnO/SiO_2/glass thin film: Role of ZnO thickness on structural and optical properties. *J. Alloys Compd.* **2016**, *671*, 170–176. [CrossRef]
40. Akhtaruzzaman; Hossain, M.I.; Islam, M.A.; Shahiduzzaman; Muhammad, G.; Hasan, A.K.M.; Tsang, Y.H.; Sopian, K. Nanophotonic-structured front contact for high-performance perovskite solar cells. *Sci. China Mater.* **2022**, *65*, 1727–1740. [CrossRef]
41. Walter, T.N.; Lee, S.; Zhang, X.; Chubarov, M.; Redwing, J.M.; Jackson, T.N.; Mohney, S.E. Atomic layer deposition of ZnO on MoS_2 and WSe_2. *Appl. Surf. Sci.* **2019**, *480*, 43–51. [CrossRef]
42. Wang, Y.; Kang, K.-M.; Kim, M.; Park, H.-H. Film thickness effect in c-axis oxygen vacancy-passivated ZnO prepared via atomic layer deposition by using H_2O_2. *Appl. Surf. Sci.* **2020**, *529*, 147095. [CrossRef]
43. Lu, J.; Wang, W.; Liang, J.; Lan, J.; Lin, L.; Zhou, F.; Chen, K.; Zhang, G.; Shen, M.; Li, Y. Contact Resistance Reduction of Low Temperature Atomic Layer Deposition ZnO Thin Film Transistor Using Ar Plasma Surface Treatment. *IEEE Electron Device Lett.* **2022**, *43*, 890–893. [CrossRef]
44. Parmar, D.H.; Pina, J.M.; Zhu, T.; Vafaie, M.; Atan, O.; Biondi, M.; Najjariyan, A.M.; Hoogland, S.; Sargent, E.H. Controlled crystal plane orientations in the ZnO transport layer enable high-responsivity, low-dark-current infrared photodetectors. *Adv. Mater.* **2022**, *34*, 2200321. [CrossRef] [PubMed]
45. Kraemer, A.; Engel, S.; Sangiorgi, N.; Sanson, A.; Bartolome, J.F.; Graef, S.; Mueller, F.A. ZnO thin films on single carbon fibres fabricated by Pulsed Laser Deposition (PLD). *Appl. Surf. Sci.* **2017**, *399*, 282–287. [CrossRef]
46. Li, B.; Ding, L.; Gui, P.; Liu, N.; Yue, Y.; Chen, Z.; Song, Z.; Wen, J.; Lei, H.; Zhu, Z.; et al. Pulsed Laser Deposition Assisted van der Waals Epitaxial Large Area Quasi-2D ZnO Single-Crystal Plates on Fluorophlogopite Mica. *Adv. Mater. Interfaces* **2019**, *6*, 1901156. [CrossRef]
47. Wang, F.; Seo, J.-H.; Luo, G.; Starr, M.B.; Li, Z.; Geng, D.; Yin, X.; Wang, S.; Fraser, D.G.; Morgan, D.; et al. Na-nometre-thick single-crystalline nanosheets grown at the water-air interface. *Nat. Commun.* **2016**, *7*, 10444. [CrossRef] [PubMed]
48. Yin, X.; Wang, Y.; Jacobs, R.; Shi, Y.; Szlufarska, I.; Morgan, D.; Wang, X. Massive Vacancy Concentration Yields Strong Room-Temperature Ferromagnetism in Two-Dimensional ZnO. *Nano Lett.* **2019**, *19*, 7085–7092. [CrossRef]
49. Yin, X.; Shi, Y.; Wei, Y.; Joo, Y.; Gopalan, P.; Szlufarska, I.; Wang, X. Unit Cell Level Thickness Control of Single-Crystalline Zinc Oxide Nanosheets Enabled by Electrical Double-Layer Confinement. *Langmuir* **2017**, *33*, 7708–7714. [CrossRef]
50. Kim, S.; Iida, K.; Kuromiya, Y.; Seto, T.; Higashi, H.; Otani, Y. Effect of Nucleation Temperature on Detecting Molecular Ions and Charged Nanoparticles with a Diethylene Glycol-Based Particle Size Magnifier. *Aerosol Sci. Technol.* **2014**, *49*, 35–44. [CrossRef]
51. Mao, J.; Li, J.-J.; Ling, T.; Liu, H.; Yang, J.; Du, X.-W. Facile synthesis of zinc hydroxide carbonate flowers on zinc oxide nanorods with attractive luminescent and optochemical performance. *Nanotechnology* **2011**, *22*, 245607. [CrossRef] [PubMed]
52. Li, P.; Liu, H.; Lu, B.; Wei, Y. Formation Mechanism of 1D ZnO Nanowhiskers in Aqueous Solution. *J. Phys. Chem. C* **2010**, *114*, 21132–21137. [CrossRef]
53. Fang, M.; Liu, Z.W. Controllable size and photoluminescence of ZnO nanorod arrays on Si substrate prepared by micro-wave-assisted hydrothermal method. *Ceram. Int.* **2017**, *43*, 6955–6962. [CrossRef]
54. Luo, C.-Q.; Ho, L.-P.; Ling, F.C.-C. The origin of additional modes in Raman spectra of ZnO:Sb films. *Physica B* **2020**, *593*, 412256. [CrossRef]
55. Das, P.K.; Biswal, R.; Choudhary, R.J.; Sathe, V.; Ganesan, V.; Khan, S.A.; Mishra, N.C.; Mallick, P. Effect of 120 MeV Au^{9+} ion irradiation on the structure and surface morphology of ZnO/NiO heterojunction. *Surf. Interface Anal.* **2018**, *50*, 954–961. [CrossRef]
56. Sharma, G.N.; Dutta, S.; Singh, S.K.; Chatterjee, R. Growth and optical properties of nano-textured (110) Pb($Zr_{0.52}Ti_{0.48}$)O_3/(001) ZnO hetero-structure on oxidized silicon substrate. *J. Mater. Sci.-Mater. Electron.* **2017**, *28*, 5058–5062. [CrossRef]

57. Husain, S.; Alkhtaby, L.A.; Giorgetti, E.; Zoppi, A.; Miranda, M.M. Effect of Mn doping on structural and optical properties of sol gel derived ZnO nanoparticles. *J. Lumin.* **2013**, *145*, 132–137. [CrossRef]
58. Zamiri, R.; Kaushal, A.; Rebelo, A.; Ferreira, J. Er doped ZnO nanoplates: Synthesis, optical and dielectric properties. *Ceram. Int.* **2014**, *40*, 1635–1639. [CrossRef]

Disclaimer/Publisher's Note: The statements, opinions and data contained in all publications are solely those of the individual author(s) and contributor(s) and not of MDPI and/or the editor(s). MDPI and/or the editor(s) disclaim responsibility for any injury to people or property resulting from any ideas, methods, instructions or products referred to in the content.

Article

Green Synthesis of ZnO Nanostructures Using *Pyrus pyrifolia*: Antimicrobial, Photocatalytic and Dielectric Properties

Zainal Abidin Ali [1,2,*], Iqabiha Shudirman [3], Rosiyah Yahya [3], Gopinath Venkatraman [4,5], Abdurahman Hajinur Hirad [6] and Siddique Akber Ansari [7]

1. Pusat Asasi Sains, Universiti Malaya, Kuala Lumpur 50603, Malaysia
2. Centre for Ionics Universiti Malaya (C.I.U.M), Department of Physics, Faculty of Science, Universiti Malaya, Kuala Lumpur 50603, Malaysia
3. Department of Chemistry, Universiti Malaya, Kuala Lumpur 50603, Malaysia
4. Universiti Malaya Centre for Proteomics Research, Universiti Malaya, Kuala Lumpur 50603, Malaysia
5. Department of Biochemistry, Saveetha Dental College, Saveetha Institute of Medical & Technical Sciences, Saveetha University, Chennai 600 077, India
6. Department of Botany and Microbiology, College of Science, King Saud University, P.O. Box 2455, Riyadh 11451, Saudi Arabia
7. Department of Pharmaceutical Chemistry, College of Pharmacy, King Saud University, P.O. Box 2454, Riyadh 11451, Saudi Arabia
* Correspondence: zaba_87@um.edu.my

Abstract: In this study, zinc oxide nanostructures (ZnO NS) were synthesized using *Pyrus pyrifolia* fruit extract. Biophysical characterization results confirmed that the synthesized materials are crystalline wurtzite ZnO structures. Field emission scanning electron microscopy (FESEM) revealed that the ZnO NS are cubical, and the sizes range 20–80 nm. Transmission electron microscopy (TEM) and XRD results revealed a crystal lattice spacing of 0.23 nm and (101) the crystalline plane on ZnO NS. UV-Visible spectrophotometer results showed an absorbance peak at 373 nm. The ZnO NS demonstrated significant antibacterial activity analyzed by metabolic activity analysis and disc diffusion assay against *Escherichia coli* and *Staphylococcus aureus*. FESEM analysis confirmed the bacterial membrane disruption and the release of cytoplasmic contents was studied by electron microscopy analysis. Further, ZnO NS achieved good photocatalytic activity of decolorizing 88% of methylene blue (MB) in 60 min. The dielectric constant and loss of ZnO were found to be 3.19 and 2.80 at 1 kHz, respectively. The research findings from this study could offer new insights for developing potential antibacterial and photocatalytic materials.

Keywords: antibacterial; zinc oxide nanoparticles; photocatalytic; methylene blue; *Pyrus pyrifolia*; dielectric

1. Introduction

Nanotechnology involves the application of materials in the range of 1–100 nm. The applications span from coatings [1], gas sensing devices [2], solar cells [3], batteries [4], environmental catalysts and antimicrobials [5,6]. Metal and metal oxide nanomaterials have improved the efficiency and performance of such devices. Consequently, researchers have given ZnO much attention due to its biocompatibility, low cost, high photocatalytic efficiency and antimicrobial potential. Indeed, it is recognized as a multifunctional material, as it has played a significant role in various fields such as biomedical (e.g., antimicrobials, anticancer, tissue engineering) [7–9], cosmetics industries [10] and photocatalysts [11]. ZnO nanostructures are II–VI semiconductors with a wide bandgap energy of 3.3 eV and a high excitation binding energy of 60 meV. Thus, the materials are suitable for large electrical fields, high temperatures and high-power functionalities such as photovoltaic cells and chemical sensors. ZnO nanocrystals mostly show a wurtzite structure with lattice parameters of $a = 0.325$ nm and $c = 0.520$ nm [12]. The scarcity of clean water has been a serious issue debated globally. Water contamination is severely inflicted everywhere.

This is due to the irresponsible and uncontrolled development of industries that release chemicals into streams [13]. For example, textile industries have used large amounts of dyes and water in their textile coloring process. The release of dye wastewater from this process substantially pollutes the aqueous environment, which was predicted to be 15–20% of the total industrial pollution [14]. The introduction of carcinogenic dye pollutants to the environment cause lethal side effects to human and aquatic organisms [15]. Therefore, finding ways to remove wastewater dyes before discharging them into the environment is necessary.

In recent years, ZnO has been intensively researched as a potential material to treat dye effluents. In this process, ZnO absorbs UV light with a wavelength equal to or less than 385 nm to generate radicals that, in turn, degrade the dyes. Recent studies demonstrated that the photodegradation of methylene blue (MB) was up to 98.3% [16]. Factors such as morphology [17], shape, size and concentrations of the ZnO can directly affect the efficiency of the photocatalytic activities [18]. For instance, Barnes et al. reported that lowering the concentration of ZnO from 1 to 0.1 g/L reduced the photodegradation performance from 18 to 7% [19]. Recognizing the simplicity and wide potential of ZnO as a photocatalytic agent, numerous approaches are being used to synthesize ZnO nanostructures. However, green synthesis is more advantageous than the conventional approach and hazardous chemical-free procedures [20,21] because it is eco-friendly. This technique is similar to chemical reduction, except that the extracts of natural products replace the reducing and stabilizing agent. Furthermore, studies have shown that the nature of biological elements and the concentrations of extracts could influence the size, shape, and optical properties of nanostructures [22–25].

ZnO is also gaining interest due to its electronic polarizability. This allows ZnO to be researched as a dielectric material. Dielectric materials are mainly applied in developing flexible electronic devices [26]. In addition, the dielectric properties of the developed material strongly depend on the synthesis conditions [27]. Lanje et al. reported that the ZnO obtained via the precipitation method has a dielectric constant of 14.52 at 100 kHz [28]. On the other hand, ZnO synthesized using starch as a stabilizing agent reported a dielectric constant in the range of 4–5 at the same frequency [27]. At the same time, numerous research studies have investigated the phytosynthesis of ZnO, but very few have discussed its dielectric or electrical properties. In the present work, we develop a facile method for the synthesis of ZnO nanostructures by using a fruit extract of *Pyrus pyrifolia*. To the best of our knowledge, this is the first study of ZnO NS synthesis using *P. pyrifolia* fruits. To study the interaction between ZnO nanostructures and the bacteria and the decolorization ability of MB, we studied antimicrobial assays and the photocatalytic reaction. Notably, the dielectric properties of ZnO NS have also been reported in this study.

2. Materials and Methods

2.1. Chemicals and Reagents

All chemicals used in this study were of analytical grade and used without further purification. The zinc nitrate ($Zn(NO_3)_2 \cdot 6H_2O$), Ethidium bromide (EB) and MB were purchased from Sigma Aldrich (Burlington, MA, USA). Acridine orange (AO) was obtained from VWR AMRESCO Life science, Radnor, PA, USA. Distilled water (DW) was used in all experiments.

2.2. Preparation of Extracts

Fresh fruits of *Pyrus pyrifolia* were obtained from a supermarket in Kuala Lumpur, Malaysia. The fruits were washed twice with DW to remove any dust and impurities, then 100 g of the fruit was cut into small pieces and ground with 100 mL DW. The resulting saturated extract was filtered through the Buchner funnel by vacuum filtration and further filtration procedures were carried out (e.g., using gravity filtration) to ensure a clear extract was attained.

2.3. Synthesis of ZnO Nanostructures

For the synthesis of ZnO nanostructures, 0.1 M Zn(NO$_3$)$_2$.6H$_2$O was prepared by dissolving 7.43 g of zinc nitrate in 250 mL DW and sonicated for 30 min to achieve complete dissolution. Aqueous fruit extract in the amount of 50 mL was introduced into the above solution, magnetically stirred for 15 min and left at room temperature for 24 h. Afterwards, the reaction solution was heated at 80 °C till the volume was reduced to 3/4 of its original volume and the color changed to a deep yellow paste [29]. Next, the paste was collected in a clean crucible and calcined at 450 °C for 60 min using a small benchtop muffle furnace (Barnstead/Thermolyne furnace 1400, Thermo Scientific, Waltham, MA, USA). The calcined materials were washed with DW and ethanol to remove the impurities. Finally, the resultant materials were dried for 12 h in a vacuum oven at 70 °C to obtain the white zinc oxide powder.

2.4. Characterization of the ZnO

X-ray diffraction (XRD) analysis was performed using Ultima IV (Rigaku, Tokyo, Japan) at a scan speed of 2° min^{-1} and a wavelength of 1.5406 Å in the 2θ range of 20–90 degrees. The crystallite size of the resultant ZnO NS was calculated using the Scherrer Equation (1).

$$D = \frac{\kappa \lambda}{\beta \cos\theta} \tag{1}$$

where D is the average crystallite size (in Å), κ is the shape factor, λ is the X-ray wavelength of X-ray (1.5406 Å) Cu-Kα radiation, β is the full width at half maximum (FWHM) of the diffraction peak and θ is the Bragg angle [30].

The UV-Vis absorbance spectra were obtained using a UV-1700 Spectrometer (Shimadzu, Kyoto, Japan) with measurements in the wavelength range of 300–500 nm. Fourier transform infrared (FT-IR) spectroscopy analysis was carried out to detect the possible functional groups involved in the synthesis of ZnO nanostructures. A PerkinElmer, Waltham, MA, USA, Frontier FT-IR Spectrophotometer in the attenuated total reflectance (ATR) mode in the range of 4000–500 cm^{-1} was used. Raman spectra were recorded in the backscattering geometry using a 632 nm HeNe laser with a LabRAM HR Evolution spectrometer, Kyoto, Japan. The morphological properties were characterized using FESEM (JEOL JSM 6701-F, Peabody, MA, USA) equipped with EDX analysis. EDX tests were carried out to identify the element and obtain the weight/atomic ratio of each element in the synthesized nanomaterials. Transmission electron microscopy (TEM) analysis was carried out on a (JEOL 2010, Peabody, MA, USA) instrument operated at an accelerating voltage of 200 kV. The Brunauer-Emmett-Teller (BET) nitrogen adsorption-desorption (Nova 2000E) was used to calculate the specific surface areas using desorption data. The sample was prepared using 200 mg of ZnO NS, which was subsequently degassed at 200 °C for 2 h to remove the moisture. Nitrogen gas was introduced as an adsorbent into the sample cell and the pressure changes due to the adsorption process were monitored via pressure transducers. When the saturation pressure was achieved, the sample was removed from the nitrogen atmosphere and heated to release the adsorbed nitrogen, which was then quantified.

2.5. Dielectric Studies

The dielectric studies of ZnO NS were carried out using a chemical impedance analyzer (Model: Hioki Im3590, Nagano, Japan). The as-synthesized ZnO nanostructure was compressed into pellets 13 mm in diameter and 0.539 mm in thickness by applying a force of 10 tons with a hydraulic press. The dielectric constant and loss were measured in the frequency range of 1 Hz–100 kHz.

2.6. Photocatalytic Decolorization of Methylene Blue

We placed 2.5 mg/L MB dye (50 mL) and 10 mg catalyst ZnO NS in a glass beaker and the suspension was magnetically stirred for 30 min in the dark to reach equilibrated adsorption between the ZnO NS and MB. Then the suspension was irradiated under UV light sources (Philips, λ = 365 nm, 6 W) with continuous stirring. Another beaker containing the same concentration of MB dye without ZnO was prepared as a control and underwent the same treatment as the one with ZnO. The samples (5 mL) were taken out at regular intervals of 0, 15, 30, 45 and 60 min, centrifuged for 5 min at 4000 rpm and the absorbance was measured using UV-Vis spectroscopy. The decolorization efficiencies of the dyes were estimated from the following equation [31]:

$$\text{Decolorization}(\%) = \frac{A_o - A_t}{A_o} \times 100 \quad (2)$$

where A_o represents the absorbance of dye before illumination and A_t denotes the absorbance of the dye after a specific irradiation time.

2.7. Antimicrobial Assay of ZnO NS

2.7.1. Disc Diffusion Assay

The antibacterial activity of the ZnO NS was studied using disc diffusion assay. Briefly, Gram-positive bacteria B. subtilis (ATCC 23857) and Gram-negative bacteria E. coli (ATCC 25922) were used to perform the antimicrobial assays. The freshly grown bacterial single colonies were spread in the Mueller-Hinton agar plate. The different concentrations (100, 200, 300 and 400 µg/mL) of ZnO NS solution (20 µL) were impregnated on the paper disc (6 mm diameter) and labelled as 1, 2, 3 and 4. The disc in the center was loaded with 20 µL DW and served as a control. The plates were then incubated for 24 h at 37 °C. The bacterial inhibition zones observed around the discs were measured and tabulated.

2.7.2. Resazurin Assay Based Minimum Inhibitory Concentration (MIC) Determination

The MICs of the ZnO NS were determined by a standard broth microdilution assay in a 96 well plate. Briefly, the concentration of overnight grown bacterial pathogens was adjusted to 1×10^5 CFU/mL. The ZnO NS (500 µg/mL) were serially diluted (2-fold) in 175 µL of MHB with 10 µL of selected bacterial inoculum. The wells containing only MHB and the MHB containing bacteria were negative and positive controls, respectively. Then, the plates were incubated for 24 h at 37 °C. Next, the MIC was determined by visually observing the turbidity. The lowest concentration of ZnO NS treated bacteria wells without turbidity were considered as the MIC [32]. Meanwhile, another set of experimental wells were flooded with 10 µL of resazurin and incubated for 60 min. The sample wells turned from blue to pink, indicating bacterial viability, and the minimal dosage of ZnO NS treated wells remained blue, indicating MIC [33].

2.7.3. Analysis of Bacterial Morphological Changes

The morphological changes of bacteria caused by the ZnO NS were examined by scanning electron microscopy. Briefly, P. aeruginosa and B. subtilis were treated with ZnO NS and incubated at 37 °C for 2 h. Then the cells were washed with PBS, followed by glutaraldehyde fixation for 6 h. The samples were then washed with H_2O and dehydrated with increasing concentrations of ethanol and acetone. Lastly, the samples were processed for gold coating and viewed under a field emission scanning electron microscope (FESEM).

3. Results and Discussion

3.1. ZnO NS Synthesis and UV-Vis Spectrum

The phytochemical constituents in the P. pyrifolia fruit extract chelate with the metallic zinc ions (Zn^{2+}) and form yellow-colored zinc coordinated complex sediment in the reaction medium [34]. Then, the obtained Zn complex was decomposed by calcining it at 450 °C for 60 min to get the nanostructured zinc oxide materials. The schematic diagram of P. pyrifolia mediated ZnO NS synthesis is represented in (Figure 1a).

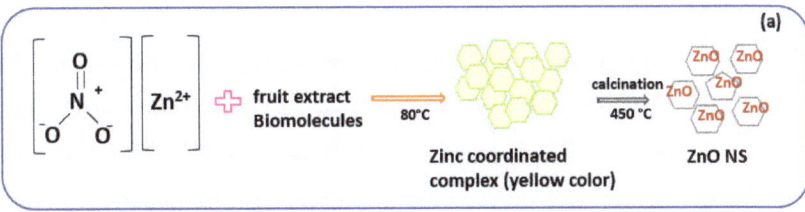

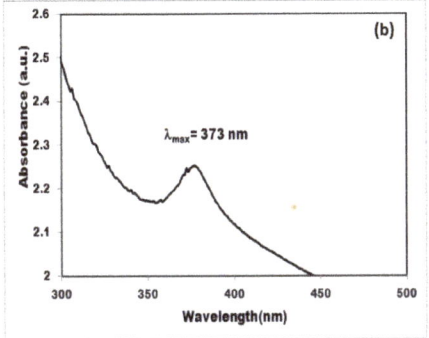

Figure 1. Schematic representation of zinc oxide nanostructures synthesis using *Pyrus pyrifolia* fruit extract (**a**). UV-Vis spectrum of the synthesized zinc oxide nanostructure solution (**b**).

Generally, UV-Vis spectral analysis is used to confirm the formation of metal and metal oxide nanoparticles. The synthesized aqueous ZnO NS solution showed that the synthesized nanoparticles exhibit an excitation wavelength of 373 nm (Figure 1b), which is almost similar to the range of results in the literature [11,22]. The band gap energy (E) of the synthesized ZnO nanostructure was estimated by applying the peak at 373 nm, using Equation (2).

$$E = \frac{hc}{\lambda_{max}} \quad (3)$$

where h is the Planck's constant and c is the speed of light in vacuum. E of the resultant ZnO NS was calculated to be 3.32 eV. This result is in accordance with the previous report on the synthesis of ZnO nanoparticles [35].

3.2. X-ray Diffraction Analysis of ZnO NS

The X-Ray diffraction peaks of the synthesized ZnO NS are shown in Figure 2. The X-ray diffraction patterns of the resultant ZnO NS showed different diffraction peaks at the 2θ values of 31.87°, 34.49°, 36.36°, 47.49°, 56.75°, 62.94°, 66.29°, 68.06°, 69.29°, 75.58° and 76.85°, which can be indexed to the (1 0 0), (0 0 2), (1 0 1), (1 0 2), (1 1 0), (1 0 3), (2 0 0), (1 1 2), (2 0 1), (0 0 4) and (2 0 2) planes, respectively. These observed diffraction peaks are highly matched with the hexagonal phase of the wurtzite ZnO structure (JCPDS card number: 36–1451). The sharp diffraction peaks demonstrate the high crystalline nature of the formation of ZnO particles. No obvious presence of unknown peaks implies a high purity of the synthesized ZnO NS. Our results are in accordance with an earlier report [36]. The crystallite size of the resultant ZnO NS was calculated using the diffraction peak of 36.36° and found to be 17 nm.

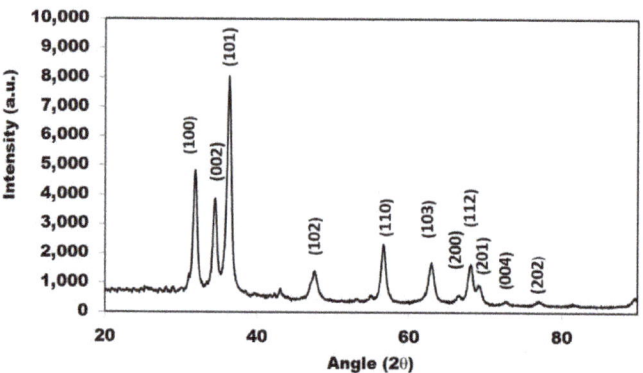

Figure 2. XRD pattern of the synthesized zinc oxide nanostructures.

3.3. FESEM and EDX Pattern of ZnO NS

The FESEM images showed that the diameters of the samples of ZnO nanostructures are in the range of 20–80 nm (Figure 3a,b). The ZnO is cubical, and the shape is almost uniform throughout the sample. The formation of clusters by the nanoparticles is mostly likely attributed to agglomeration, which is common in nano-sized materials. Furthermore, the elemental analysis spectrum results revealed the presence of zinc and oxygen in the densely populated nanoparticles region (Figure 3c). The elemental weight percentage is shown in the inset of (Figure 3c).

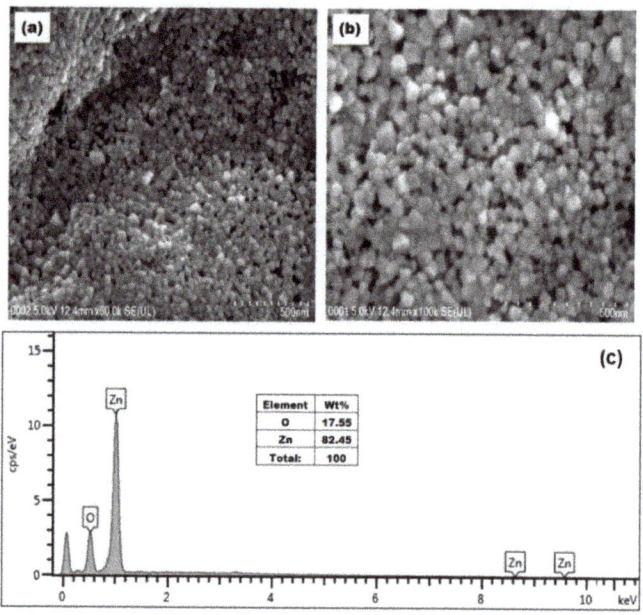

Figure 3. Scanning electron microscopic images of zinc oxide nanostructures at different magnifications (a,b). EDX pattern of zinc oxide nanostructures (c).

3.4. TEM and BET Measurement of Resultant Nanomaterials

To study the structure and morphology of the synthesized ZnO NS, the TEM analysis was employed, as shown in (Figure 4a), and thus the mean size of the particles can be analyzed. It is apparent from the TEM image that the shape of the ZnO particles is modified from a spindle structure to a nearly spherical form due to the involvement of the extract in the crystallization process. This further asserts the findings of Bayrami et al. [37]. The results showed the lattice fringes spacing was measured to be 0.23 nm, which can be attributed to the space between two planes (101) of the wurtzite ZnO NS [38]. Moreover, this result corroborates the crystal lattice spacing determined from the above XRD peak 2θ value of 36.36° with a corresponding lattice plane of (101).

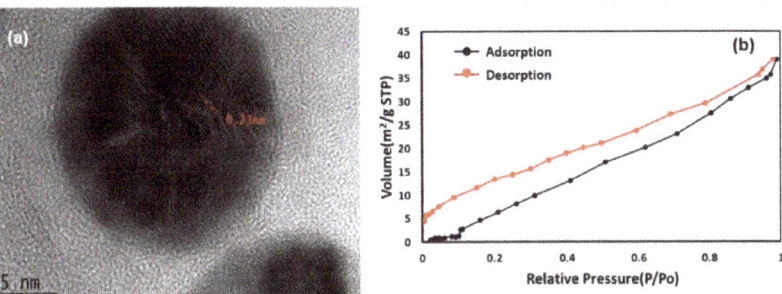

Figure 4. Transmission electron microscopy images of zinc oxide nanostructures (**a**), BET adsorption-desorption isotherms of resultant zinc oxide nanostructures (**b**).

To examine the surface properties of resultant ZnO NS, the Brunauer-Emmett-Teller (BET) nitrogen adsorption-desorption isotherms were measured, as shown in (Figure 4b). In both cases, it is evident that the volume of the sample increases monotonously with increasing relative pressure. Particularly, at low pressure (between 0 and 0.1 P/P$_0$), the adsorption can be largely attributed to its microporous filling. A nearly similar trend is noticed for desorption. However, with the enhancement in pressure, a gentle hysteresis loop ensues, which is a habitual hallmark of mesoporous materials [39]. The BET measurements estimate the surface area of ZnO to be 21.4 m^2/g. Furthermore, relating the ZnO NS surface area with their photocatalytic performances, two trends of results have been reported, as the photocatalytic activity of ZnO nanoparticles increased with both an increase in specific surface area [40] and a decrease in specific surface area [41]. Moreover, it has been reported that, although the ZnO NS have a larger surface area and it has no significant effect on the pore diameter, it may result in a larger active site [42]. The obtained data curve denotes typical type II isotherms, which relate to the nonporous characteristic of solids [37].

3.5. FTIR and Raman Measurement of ZnO NS

FTIR spectra of the *P. pyrifolia* fruit extract and synthesized ZnO nanostructures before and after calcination were showed in (Figure 5a, A,B,C). A vibration band of fruit extract showed peaks at 3382, 1638, 1488, 1381 and 845 cm^{-1}. The peak at 3382 cm^{-1} corresponds to the O-H stretching functional group [43]. The peaks at 1638 and 1488 cm^{-1} can be attributed to the carbonyl (C=O) functional group and bending vibration of the sp^2 C=C aromatic ring [44], respectively. The obtained peaks at 1381 and 845 cm^{-1} are probably due to the alkene group of C-H stretching and C-N amine [45], respectively. Phenolic contents in the extract of pear may also be involved in the formation of ZnO NS. Momeni et al. suggested that the peaks that range from 3500 to 3100, 1720, 1605, 1395 and 1100 cm^{-1} can be linked to the free OH in the extract, thus forming hydrogen bonds, a carbonyl group (C=O), a stretching C=C aromatic ring and C-OH and C-H stretching vibrations. Correspondingly, these indicate the presence of phenolic structures in the plant extract [46]. The FTIR spectrum of synthesized zinc complex before calcination showed peaks at 3412, 1635 and 1385 cm^{-1}. A comparison of FTIR spectra revealed a slight shift in the peaks of the extract and the zinc complex. This is anticipated due to the adsorption of the plant extract (C-H stretching vibration, O-H and carbonyl groups) onto the zinc surface, which may be involved in the synthesis of the nanoparticles. Further, the FTIR spectrum of the

ZnO NS revealed no significant peaks due to the decomposition of bioactive functional groups during calcination [5].

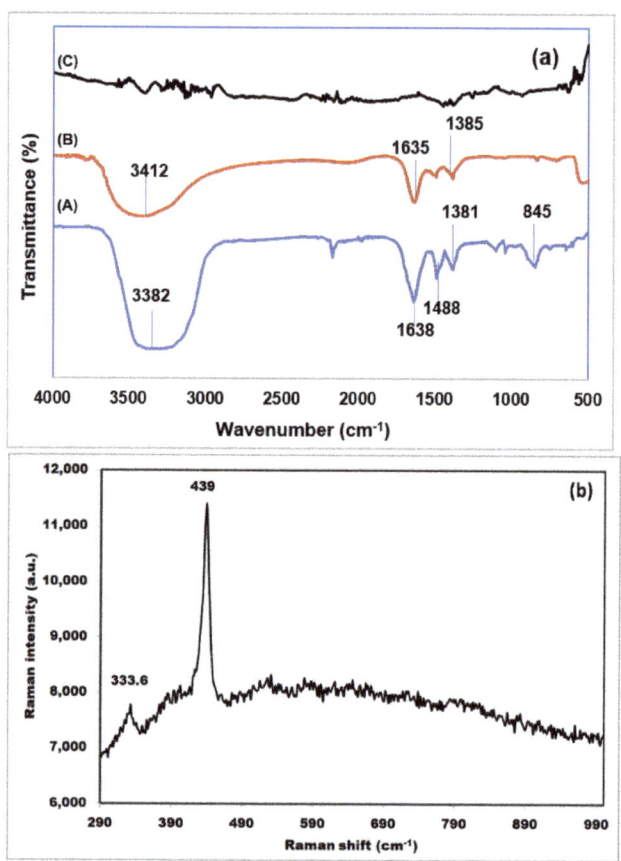

Figure 5. (a) FTIR spectrum of *P. pyrifolia* fruit extract (A), ZnO NS before calcination (B), calcined ZnO NS (C) and Raman spectrum of ZnO nanostructures (b).

The Raman spectrum of the ZnO nanostructures from 290 cm^{-1} to 990 cm^{-1} is shown in Figure 5b. The obtained spectrum has a Raman peak comparable to ZnO nanocrystals presented in an earlier report [47] with a slight shift (in the range of 1–2 cm^{-1}) caused by different crystal sizes [48]. Theoretically, the wurtzite crystal structure of ZnO belongs to the C6v4, possessing 2 formula units in each primitive cell with all the atoms lodging the C3V sites [49]. The major, sharp peak labelled as E2 at 439 cm^{-1} is recognized as Raman active optical phonon mode, which is the characteristic of the wurtzite hexagonal phase ZnO. Raman modes at 333.6 and 439 cm^{-1} are denoted as 2E2 and E2 modes, respectively [47,48].

3.6. ZnO NS Photocatalytic Decolorization of Methylene Blue

The photocatalytic decolorization efficiency of ZnO NS on MB was observed using different amounts of ZnO catalyst (1 mg, 5 mg, 10 mg and 20 mg) in 50 mL of 2.5 mg/L MB solution under UV light for 60 min. The decolorization was analyzed by measuring the absorbance peak of the MB. As shown in Figure 6a, a lower absorbance peak indicates that more MB is decolorized and vice-versa. These results showed that by using 10 mg of ZnO, the absorbance peak was the lowest compared to the other tested ZnO concentrations, inferring the highest amount of MB decolorized at this particular concentration. This may be attributed to the increase of the catalyst concentration, which subsequently increases the production of ROS and accelerates the number of active sites on the ZnO NS for the reaction.

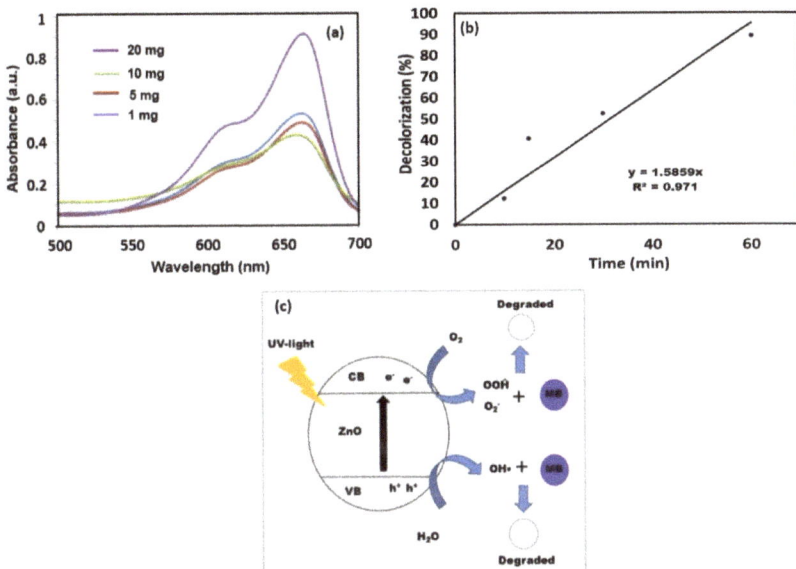

Figure 6. The absorption spectrum of photodegradation efficiency of MB (a), degradation percentage of MB (b) and schematic illustration of photodegradation of methylene blue by ZnO NS (c).

As seen in Figure 6b, the decolorization exhibits a linear behavior indicating that degradation is directly proportional to the amount of ZnO. However, it was also observed that when the catalyst loading was further increased to 20 mg, the absorbance peak of the MB was the highest, corresponding to the lowest decolorization of MB. The decrease in the photocatalytic decolorization efficiency is probably due to the agglomeration of the catalyst particles. As a result, the specific surface area decreased and subsequently decreased the number of active sites [50]. Moreover, a high quantity of ZnO NS would lower the opacity, turbidity of the suspension and light scattering of the catalyst particles. The more significant amount of nanoparticle suspension may have increased UV shading to hinder photocatalytic activity [19,51]. This would decrease the path of irradiation through the sample [52]. Therefore, in our case, the most effective decolorization for MB was recorded with 10 mg of ZnO NS catalyst.

Figure 6c represents the photocatalytic mechanism for MB in the presence of ZnO. Adsorption and adhesion of the MB dye molecules on the surface of ZnO result in the degradation of the MB. It has been extensively discussed that the photocatalytic decolorization of MB by semiconductors, such as ZnO, can occur due to hydroxyl radicals ($^{\bullet}$OH) [53,54]. The $^{\bullet}$OH can be formed either from (i) the highly hydroxylated ZnO surface or (ii) by direct oxidation of dye pollutants under UV irradiation. Moreover, there is also a possibility that the $^{\bullet}$OH co-occurs by both methods. The photo decolorization process starts when ZnO absorbs UV light of energy equal to or higher than its bandgap (3.37 eV). This promotes the formation of free electrons (e^-) and holes (h^+) in the conduction and valence bands, respectively. These electrons can either recombine with the holes (and scatter the captivated energy as heat), or the electron-hole pairs can contribute to redox reactions. In the case of participating redox reactions, the electron-hole pairs can generate $^{\bullet}$OH either from the reaction of h^+ with water or with OH^- anions [50]. On the other hand, the response of dissolved O_2 and e^- will produce superoxide ($\bullet O_2^-$) and may also proceed to make OOH.

All of these active oxygen species ($\bullet O_2^-$) and free hydroxyl radicals ($\bullet OH$, $\bullet OOH$) could also be involved in the photodegradation of the MB [16].

3.7. Dielectric Studies

The dielectric constant is a measure of the capability of a material to stock electrical energy in an electric field. It is a ratio of the material permittivity to the free space permittivity. Permittivity (ε) is a measure of the ability of a material to be polarized by an electric field. An efficient dielectric material supports polarization with minimal dissipation or loss of energy. The dissipation of energy in the form of heat, as the movement of charges in an alternating electromagnetic field occurs, as polarization switches direction. This is known as the dielectric loss (D) or $tan\ \delta$ (loss tangent). It is proportional to the amount of energy stored and dissipated due to the presence of an applied electric field. The dielectric constant and loss were assessed in the frequency range from 1 Hz to 100 kHz in the present study. The dielectric constant, also known as relative permittivity, is determined using the equation:

$$\varepsilon_r = \frac{C \times d}{\varepsilon_0 A}$$

where C is the capacitance of the sample, d and A are the thickness and the area of the sample pellet, respectively, and ε_0 is the dielectric permittivity of vacuum (8.854 × 10 F/m). The ε_r and $tan\ \delta$ for the synthesized ZnO were found to be 3.19 and 2.80 at 1 kHz, respectively. A recent report of the synthesized ZnO NPs by the co-precipitation method demonstrated the value of the dielectric constant and loss to be approximately 12 and 0.01, respectively [55], while ZnO NPs synthesized using sol-gel observed the dielectric constant and loss to be 40 and 50, respectively [56]. Apart from the preparation conditions that can influence the value of the dielectric properties [27], it is noteworthy to mention that the compression force used in preparing the pellet must also be considered because different compression forces will result in different void spaces between the particles [57], and void space affects electrical measurements. Therefore, a direct comparison of the values is difficult, as many aspects and factors come into play. Figure 7a,b represents the deviation of the dielectric constant (ε) and dielectric loss ($tan\ \delta$) with respect to frequency at room temperature (300 K). The values of ε_r and D were found to decline with increasing frequency. The decrement rate was observed to be quicker at a lower frequency and slower at a higher frequency. The decrease of the dielectric constant at high frequencies is typical because any species contributing to polarization will have their space charges reduced under the applied field at higher frequencies [58]. Polarization could arise from electronic dislodgment, ionic displacement, dipole orientation and space charge displacement [56].

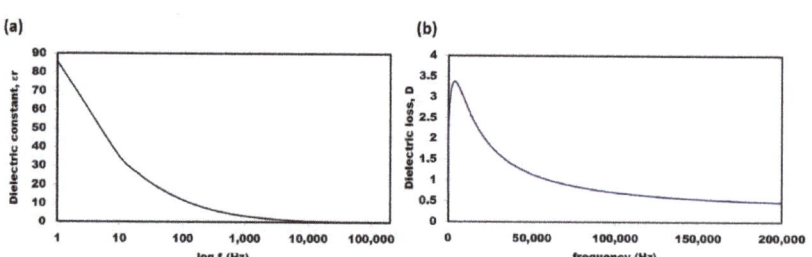

Figure 7. Electrical properties of ZnO NS; (a) dielectric constant and (b) dielectric loss.

3.8. Antibacterial Activity of ZnO NS

3.8.1. MIC Determination and Metabolic Activity

Although industrial effluents pollute the environment, the emergence of pathogenic bacterial drug resistance epitomizes the high risk to public health. Nanomaterials are considered alternate antimicrobials due to their unique physiochemical properties. To determine the MIC level of ZnO NS against *P. aeruginosa* and *B. subtilis*, a broth microdilution assay was performed. The turbidity observation results showed that the ZnO NS significantly inhibited the growth of *P. aeruginosa* and *B. subtilis* with MIC values of 125 and 250 µg/mL (Figure 8, Row B and D), respectively. A similar pattern of greater and lesser antibacterial activity was observed against *P. aeruginosa* and *B. subtilis*, which may be due to the differential cell wall structure of Gram-positive and Gram-negative bacteria [59]. ZnO NPs demonstrated a prominent antibacterial effect; their combinations are used in food additives due to their non-toxic nature to humans within FDA approved concentrations [60]. The turbid white color appearance of increasing ZnO NS concentrations slightly interfered with the determination of MIC and whether bacterial growth caused the turbidity. Furthermore, the samples that were incubated with resazurin after 60 min showed that the metabolically active bacterial cells appeared to change colors, from blue (resazurin) to pink (resorufin), as shown in Figure 8 (Row A and C). The lowest dosage of ZnO NS (125 and 250 µg/mL) exposed *P. aeruginosa* and *B. subtilis*, remaining unchanged in its blue color, were determined as MIC. Moreover, these results are in line with the MIC of turbidity observation analysis.

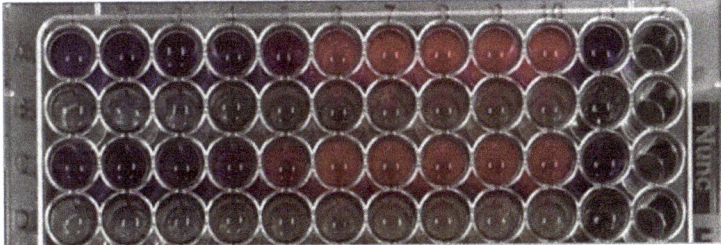

Figure 8. Antibacterial activity resazurin assay (**A,C**) and turbidity observation (**B,D**) of ZnO NS against *P. aeruginosa* and *B. subtilis*, respectively.

3.8.2. Disc Diffusion Assay

The antibacterial activity of ZnO NS was studied against *B. subtilis* and *P. aeruginosa* by disc diffusion assay. The assay results showed clear growth inhibition on the plates against both the tested bacteria (Figure 9a,d). From the results, the DW loaded control disc in the center, and for lesser concentrations of ZnO NS impregnated discs (1. 100 µg/mL and 2. 200 µg/mL), did not show any inhibition against both the tested bacteria. In contrast, a clear zone of inhibition was observed at increasing concentrations of ZnO NS discs (3. 300 µg/mL and 4. 400 µg/mL). A higher inhibitory zone was observed at 8 mm for both bacteria at a 400 µg/mL dosage of ZnO NS. These results showed an increased bacterial inhibitory effect as ZnONS dosage was increased, which was correlated with an earlier report [61].

To study the ZnO NS effect on bacterial cells, the ZnO NS treated cells were imaged and compared with the control bacteria. Figure 9b,e shows the bacteria without any treatment, which demonstrated a rod-like shape with a smooth cell membrane surface. After 2 h ZnO NS treatment, our observations show that both bacteria underwent structural changes, including membrane damage, pits and holes on the cell membranes (Figure 9c,f). Furthermore, the leakage of cytoplasmic content was observed, which led to bacterial death. The bactericidal effect of ZnO is ascribed to multiple reasons, such as cell wall and cell membrane damage and the release of zinc ions and their ability to produce ROS, which causes oxidative stress to the bacteria [62]. It was reported that the release of Zn^{2+} ions accelerates the ROS generation in the bacterial surface and may involve oxidizing glutathione and induce lipid peroxidation, which subsequently causes bacterial lysis [63].

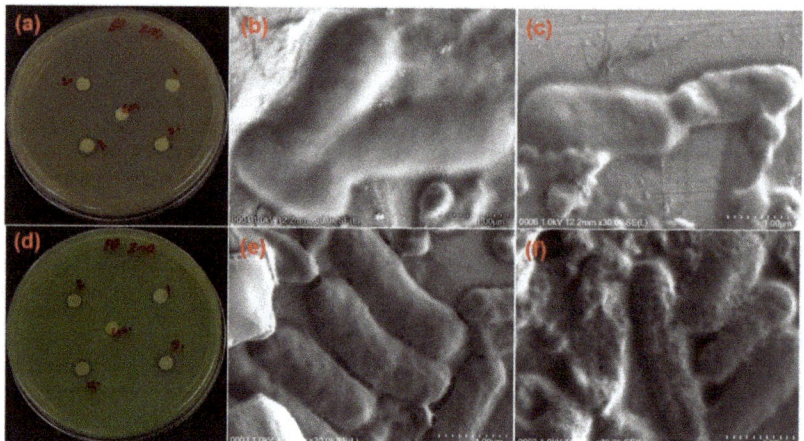

Figure 9. Antibacterial disc diffusion assay of ZnO NS against *B. subtilis* and *P. aeruginosa* (**a**,**d**). Discs 1, 2, 3, 4 and control were loaded with 20 μL of ZnO NS (100, 200, 300, 400 μg/mL) and DW. FESEM image of bacterial control (**b**,**e**) and after ZnO NS treatment (**c**,**f**).

4. Conclusions

ZnO nanostructures were successfully synthesized using *Pyrus pyrifolia* fruit extract as a reducing agent by the green synthesis route. The structural, morphological and optical properties of the ZnO nanostructures were analyzed by FESEM, UV-Vis, FTIR and Raman. The XRD pattern result confirmed the wurtzite structure of ZnO nanostructures. FESEM analysis revealed the average size of ZnO NS in the range of 20–80 nm. Flavonoids/limonoids/carotenoids, proteins and other functional groups in the fruit extract are likely responsible for forming ZnO nanostructures. Further, the ZnO NS demonstrated significant antibacterial activity against *B. subtilis*, and *P. aeruginosa*, which was confirmed by metabolic assay and morphological analysis. MB dye was effectively decolorized under UV light by controlling the concentration and catalyst loading of the MB. The synthesized ZnO NS exhibits a typical pattern of dielectric constant and loss of ZnO with respect to the frequency. The results of this study may provide new insights into the utilization of green-synthesized ZnO NS for developing novel antimicrobial combinations to treat bacterial infections, and for environmental photocatalysts to remove pollutant antibiotics.

Author Contributions: Conceptualization, investigation and writing original draft preparation, Z.A.A.; validation, R.Y.; formal analysis, I.S.; data curation, A.H.H.; writing, review and editing, G.V.; project administration, S.A.A. All authors have read and agreed to the published version of the manuscript.

Funding: The research was supported by RU Geran Fakulti Program (GPF040A-2020) from the University of Malaya, Malaysia. The authors also thank the Deanship of Scientific Research at King Saud University for funding this work through research group No (RG-1441-362).

Institutional Review Board Statement: Not applicable.

Informed Consent Statement: Not applicable.

Data Availability Statement: Not applicable.

Acknowledgments: The authors extend their appreciation to the Deanship of Scientific Research at King Saud University for funding this work through research group No RG-1441-362.

Conflicts of Interest: The authors declare no conflict of interest.

References

1. Asmatulu, R.; Claus, R.; Mecham, J.; Corcoran, S. Nanotechnology-associated coatings for aircrafts. *Mater. Sci.* **2007**, *43*, 415–422. [CrossRef]
2. Kalantar-zadeh, K.; Fry, B. *Nanotechnology-Enabled Sensors*; Springer Science & Business Media: Berlin/Heidelberg, Germany, 2007.
3. Im, J.-H.; Jang, I.-H.; Pellet, N.; Grätzel, M.; Park, N.-G. Growth of $CH_3NH_3PbI_3$ cuboids with controlled size for high-efficiency perovskite solar cells. *Nat. Nanotechnol.* **2014**, *9*, 927–932. [CrossRef] [PubMed]
4. Lee, S.W.; Yabuuchi, N.; Gallant, B.M.; Chen, S.; Kim, B.-S.; Hammond, P.T.; Shao-Horn, Y. High-power lithium batteries from functionalized carbon-nanotube electrodes. *Nat. Nanotechnol.* **2010**, *5*, 531. [CrossRef] [PubMed]
5. Sakthi Mohan, P.; Sonsuddin, F.; Mainal, A.B.; Yahya, R.; Venkatraman, G.; Vadivelu, J.; Al-Farraj, D.A.; Al-Mohaimeed, A.M.; Alarjani, K.M. Facile In-Situ Fabrication of a Ternary $ZnO/TiO_2/Ag$ Nanocomposite for Enhanced Bactericidal and Biocompatibility Properties. *Antibiotics* **2021**, *10*, 86. [CrossRef] [PubMed]
6. Gopinath, V.; MubarakAli, D.; Vadivelu, J.; Manjunath Kamath, S.; Syed, A.; Elgorban, A.M. Synthesis of biocompatible chitosan decorated silver nanoparticles biocomposites for enhanced antimicrobial and anticancer property. *Process Biochem.* **2020**, *99*, 348–356. [CrossRef]
7. Mani, V.M.; Nivetha, S.; Sabarathinam, S.; Barath, S.; Das, M.P.A.; Basha, S.; Elfasakhany, A.; Pugazhendhi, A. Multifunctionalities of mycosynthesized zinc oxide nanoparticles (ZnONPs) from Cladosporium tenuissimum FCBGr: Antimicrobial additives for paints coating, functionalized fabrics and biomedical properties. *Prog. Org. Coat.* **2022**, *163*, 106650. [CrossRef]
8. Gopinath, V.; Kamath, S.M.; Priyadarshini, S.; Chik, Z.; Alarfaj, A.A.; Hirad, A.H. Multifunctional applications of natural polysaccharide starch and cellulose: An update on recent advances. *Biomed. Pharmacother.* **2022**, *146*, 112492. [CrossRef]
9. Prasad, K.S.; Prasad, S.K.; Veerapur, R.; Lamraoui, G.; Prasad, A.; Prasad, M.N.N.; Singh, S.K.; Marraiki, N.; Syed, A.; Shivamallu, C. Antitumor Potential of Green Synthesized ZnONPs Using Root Extract of Withania somnifera against Human Breast Cancer Cell Line. *Separations* **2021**, *8*, 8. [CrossRef]
10. Lu, P.-J.; Huang, S.-C.; Chen, Y.-P.; Chiueh, L.-C.; Shih, D.Y.-C. Analysis of titanium dioxide and zinc oxide nanoparticles in cosmetics. *J. Food Drug Anal.* **2015**, *23*, 587–594. [CrossRef]
11. Suresh, D.; Nethravathi, P.C.; Udayabhanu; Rajanaika, H.; Nagabhushana, H.; Sharma, S.C. Green synthesis of multifunctional zinc oxide (ZnO) nanoparticles using Cassia fistula plant extract and their photodegradative, antioxidant and antibacterial activities. *Mater. Sci. Semicond. Process.* **2015**, *31*, 446–454. [CrossRef]
12. Özgür, Ü.; Alivov, Y.I.; Liu, C.; Teke, A.; Reshchikov, M.; Doğan, S.; Avrutin, V.; Cho, S.-J.; Morkoc, H. A comprehensive review of ZnO materials and devices. *J. Appl. Phys.* **2005**, *98*, 11. [CrossRef]
13. Fanun, M. *The Role of Colloidal Systems in Environmental Protection*; Elsevier: Amsterdam, The Netherlands, 2014.
14. Foroutan, R.; Peighambardoust, S.J.; Boffito, D.C.; Ramavandi, B. Sono-Photocatalytic Activity of Cloisite 30B/ZnO/Ag_2O Nanocomposite for the Simultaneous Degradation of Crystal Violet and Methylene Blue Dyes in Aqueous Media. *Nanomaterials* **2022**, *12*, 3103. [CrossRef]
15. Wei, X.; Chen, D.; Wang, L.; Ma, Y.; Yang, W. Carboxylate-functionalized hollow polymer particles modified polyurethane foam for facile and selective removal of cationic dye. *Appl. Surf. Sci.* **2022**, *579*, 152153. [CrossRef]
16. Siripireddy, B.; Mandal, B.K. Facile green synthesis of zinc oxide nanoparticles by Eucalyptus globulus and their photocatalytic and antioxidant activity. *Adv. Powder Technol.* **2017**, *28*, 785–797. [CrossRef]
17. Marquez, J.R.; Rodríguez, C.M.B.; Herrera, C.M.; Rosas, E.R.; Angel, O.Z.; Pozos, O.T. Effect of surface morphology of ZnO electrodeposited on photocatalytic oxidation of methylene blue dye part I: Analytical study. *Int. J. Electrochem. Sci.* **2011**, *6*, 4059–4069.
18. Mclaren, A.; Valdes-Solis, T.; Li, G.; Tsang, S.C. Shape and size effects of ZnO nanocrystals on photocatalytic activity. *J. Am. Chem. Soc.* **2009**, *131*, 12540–12541. [CrossRef]
19. Barnes, R.J.; Molina, R.; Xu, J.; Dobson, P.J.; Thompson, I.P. Comparison of TiO_2 and ZnO nanoparticles for photocatalytic degradation of methylene blue and the correlated inactivation of gram-positive and gram-negative bacteria. *J. Nanopart. Res.* **2013**, *15*, 1432. [CrossRef]
20. Ramesh, M.; Anbuvannan, M.; Viruthagiri, G. Green synthesis of ZnO nanoparticles using Solanum nigrum leaf extract and their antibacterial activity. *Spectrochim. Acta Part A Mol. Biomol. Spectrosc.* **2015**, *136*, 864–870. [CrossRef]
21. Suresh, D.; Shobharani, R.; Nethravathi, P.; Kumar, M.P.; Nagabhushana, H.; Sharma, S. Artocarpus gomezianus aided green synthesis of ZnO nanoparticles: Luminescence, photocatalytic and antioxidant properties. *Spectrochim. Acta Part A Mol. Biomol. Spectrosc.* **2015**, *141*, 128–134. [CrossRef]
22. Sangeetha, G.; Rajeshwari, S.; Venckatesh, R. Green synthesis of zinc oxide nanoparticles by aloe barbadensis miller leaf extract: Structure and optical properties. *Mater. Res. Bull.* **2011**, *46*, 2560–2566. [CrossRef]
23. Philip, D. Green synthesis of gold and silver nanoparticles using Hibiscus rosa sinensis. *Phys. E Low-Dimens. Syst. Nanostruct.* **2010**, *42*, 1417–1424. [CrossRef]
24. Kora, A.J.; Sashidhar, R.; Arunachalam, J. Gum kondagogu (Cochlospermum gossypium): A template for the green synthesis and stabilization of silver nanoparticles with antibacterial application. *Carbohydr. Polym.* **2010**, *82*, 670–679. [CrossRef]
25. Kumar, V.; Yadav, S.K. Plant-mediated synthesis of silver and gold nanoparticles and their applications. *J. Chem. Technol. Biotechnol.* **2009**, *84*, 151–157. [CrossRef]

26. Chittibabu, S.K.; Chintagumpala, K.; Chandrasekhar, A. Porous dielectric materials based wearable capacitance pressure sensors for vital signs monitoring: A review. *Mater. Sci. Semicond. Process.* **2022**, *151*, 106976. [CrossRef]
27. Zamiri, R.; Singh, B.; Belsley, M.S.; Ferreira, J. Structural and dielectric properties of Al-doped ZnO nanostructures. *Ceram. Int.* **2014**, *40*, 6031–6036. [CrossRef]
28. Lanje, A.S.; Sharma, S.J.; Ningthoujam, R.S.; Ahn, J.-S.; Pode, R.B. Low temperature dielectric studies of zinc oxide (ZnO) nanoparticles prepared by precipitation method. *Adv. Powder Technol.* **2013**, *24*, 331–335. [CrossRef]
29. Jayachandran, A.; Aswathy, T.R.; Nair, A.S. Green synthesis and characterization of zinc oxide nanoparticles using Cayratia pedata leaf extract. *Biochem. Biophys. Rep.* **2021**, *26*, 100995. [CrossRef]
30. Cullity, B.D. *Elements of X-ray Diffraction*; Addison-Wesley Publishing: Boston, MA, USA, 1956.
31. Karnan, T.; Selvakumar, S.A.S. Biosynthesis of ZnO nanoparticles using rambutan (*Nephelium lappaceum* L.) peel extract and their photocatalytic activity on methyl orange dye. *J. Mol. Struct.* **2016**, *1125*, 358–365. [CrossRef]
32. Khan, M.S.; Ranjani, S.; Hemalatha, S. Synthesis and characterization of Kappaphycus alvarezii derived silver nanoparticles and determination of antibacterial activity. *Mater. Chem. Phys.* **2022**, *282*, 125985. [CrossRef]
33. Chakansin, C.; Yostaworakul, J.; Warin, C.; Kulthong, K.; Boonrungsiman, S. Resazurin rapid screening for antibacterial activities of organic and inorganic nanoparticles: Potential, limitations and precautions. *Anal. Biochem.* **2022**, *637*, 114449. [CrossRef]
34. Selvanathan, V.; Aminuzzaman, M.; Tan, L.X.; Win, Y.F.; Cheah, E.S.G.; Heng, M.H.; Tey, L.-H.; Arullappan, S.; Algethami, N.; Alharthi, S.S.; et al. Synthesis, characterization, and preliminary in vitro antibacterial evaluation of ZnO nanoparticles derived from soursop (*Annona muricata* L.) leaf extract as a green reducing agent. *J. Mater. Res. Technol.* **2022**, *20*, 2931–2941. [CrossRef]
35. Sadiq, H.; Sher, F.; Sehar, S.; Lima, E.C.; Zhang, S.; Iqbal, H.M.N.; Zafar, F.; Nuhanović, M. Green synthesis of ZnO nanoparticles from Syzygium Cumini leaves extract with robust photocatalysis applications. *J. Mol. Liq.* **2021**, *335*, 116567. [CrossRef]
36. Vijayakumar, S.; Vaseeharan, B.; Malaikozhundan, B.; Shobiya, M. Laurus nobilis leaf extract mediated green synthesis of ZnO nanoparticles: Characterization and biomedical applications. *Biomed. Pharmacother.* **2016**, *84*, 1213–1222. [CrossRef]
37. Bayrami, A.; Alioghli, S.; Rahim Pouran, S.; Habibi-Yangjeh, A.; Khataee, A.; Ramesh, S. A facile ultrasonic-aided biosynthesis of ZnO nanoparticles using *Vaccinium arctostaphylos* L. leaf extract and its antidiabetic, antibacterial, and oxidative activity evaluation. *Ultrason. Sonochem.* **2019**, *55*, 57–66. [CrossRef]
38. Zhou, H.; Wu, X.-Y.; Zhao, Q.; Liu, K.-K.; Dong, L.; Shan, C.-X. One-step synthesis of multi-colored ZnO nanoparticles for white light-emitting diodes. *J. Lumin.* **2022**, *252*, 119425. [CrossRef]
39. Li, W.; Yang, H.; Jiang, X.; Liu, Q. Highly selective CO_2 adsorption of ZnO based N-doped reduced graphene oxide porous nanomaterial. *Appl. Surf. Sci.* **2016**, *360*, 143–147. [CrossRef]
40. Prerna; Agarwal, H.; Goyal, D. Photocatalytic degradation of textile dyes using phycosynthesised ZnO nanoparticles. *Inorg. Chem. Commun.* **2022**, *142*, 109676. [CrossRef]
41. Kusiak-Nejman, E.; Wojnarowicz, J.; Morawski, A.W.; Narkiewicz, U.; Sobczak, K.; Gierlotka, S.; Lojkowski, W. Size-dependent effects of ZnO nanoparticles on the photocatalytic degradation of phenol in a water solution. *Appl. Surf. Sci.* **2021**, *541*, 148416. [CrossRef]
42. Kuang, D.; Liu, L.; Mead, J.L.; Deng, L.; Luo, H.; Wang, S. Facile synthesis and excellent microwave absorption performance of ultra-small ZnO-doped onion-like carbon nanoparticles. *Mater. Res. Bull.* **2023**, *157*, 112007. [CrossRef]
43. Gnanasangeetha, D.; Thambavani, D.S. Biogenic production of zinc oxide nanoparticles using Acalypha indica. *J. Chem. Biol. Phys. Sci.* **2013**, *4*, 238.
44. Nasrollahzadeh, M.; Sajadi, S.M.; Rostami-Vartooni, A.; Hussin, S.M. Green synthesis of CuO nanoparticles using aqueous extract of *Thymus vulgaris* L. leaves and their catalytic performance for N-arylation of indoles and amines. *J. Colloid Interface Sci.* **2016**, *466*, 113–119. [CrossRef] [PubMed]
45. Vankudoth, S.; Dharavath, S.; Veera, S.; Maduru, N.; Chada, R.; Chirumamilla, P.; Gopu, C.; Taduri, S. Green synthesis, characterization, photoluminescence and biological studies of silver nanoparticles from the leaf extract of Muntingia calabura. *Biochem. Biophys. Res. Commun.* **2022**, *630*, 143–150. [CrossRef] [PubMed]
46. Momeni, S.S.; Nasrollahzadeh, M.; Rustaiyan, A. Green synthesis of the Cu/ZnO nanoparticles mediated by *Euphorbia prolifera* leaf extract and investigation of their catalytic activity. *J. Colloid Interface Sci.* **2016**, *472*, 173–179. [CrossRef] [PubMed]
47. Ojha, A.K.; Srivastava, M.; Kumar, S.; Hassanein, R.; Singh, J.; Singh, M.K.; Materny, A. Influence of crystal size on the electron–phonon coupling in ZnO nanocrystals investigated by Raman spectroscopy. *Vib. Spectrosc.* **2014**, *72*, 90–96. [CrossRef]
48. Procek, M.; Pustelny, T.; Stolarczyk, A. Influence of External Gaseous Environments on the Electrical Properties of ZnO Nanostructures Obtained by a Hydrothermal Method. *Nanomaterials* **2016**, *6*, 227. [CrossRef]
49. Arguello, C.; Rousseau, D.; Porto, S.; Cheesman, L.; Scott, J. Rayleigh scattering of linearly polarized light from optically active quartz. *Appl. Opt.* **1968**, *7*, 1913–1915. [CrossRef]
50. Mohamed, R.; Mkhalid, I.; Baeissa, E.; Al-Rayyani, M. Photocatalytic degradation of methylene blue by Fe/ZnO/SiO_2 nanoparticles under visiblelight. *J. Nanotechnol.* **2012**, *2012*, 329082. [CrossRef]
51. Maness, P.-C.; Smolinski, S.; Blake, D.M.; Huang, Z.; Wolfrum, E.J.; Jacoby, W.A. Bactericidal activity of photocatalytic TiO_2 reaction: Toward an understanding of its killing mechanism. *Appl. Environ. Microbiol.* **1999**, *65*, 4094–4098. [CrossRef]
52. Hayat, K.; Gondal, M.A.; Khaled, M.M.; Ahmed, S. Kinetic study of laser-induced photocatalytic degradation of dye (alizarin yellow) from wastewater using nanostructured ZnO. *J. Environ. Sci. Health Part A* **2010**, *45*, 1413–1420. [CrossRef]

53. Irani, M.; Mohammadi, T.; Mohebbi, S. Photocatalytic Degradation of Methylene Blue with ZnO Nanoparticles; a Joint Experimental and Theoretical Study. *J. Mex. Chem. Soc.* **2016**, *60*, 218–225. [CrossRef]
54. Nosaka, Y.; Nosaka, A. Understanding Hydroxyl Radical (˙OH) Generation Processes in Photocatalysis. *ACS Energy Lett.* **2016**, *1*, 356–359. [CrossRef]
55. Ashokkumar, M.; Muthukumaran, S. Electrical, dielectric, photoluminescence and magnetic properties of ZnO nanoparticles co-doped with Co and Cu. *J. Magn. Magn. Mater.* **2015**, *374*, 61–66. [CrossRef]
56. Mote, V.; Purushotham, Y.; Dole, B. Structural, morphological, physical and dielectric properties of Mn doped ZnO nanocrystals synthesized by sol–gel method. *Mater. Des.* **2016**, *96*, 99–105. [CrossRef]
57. Omar, K.; Ooi, M.J.; Hassin, M. Investigation on dielectric constant of zinc oxide. *Mod. Appl. Sci.* **2009**, *3*, 110. [CrossRef]
58. Gul, I.; Abbasi, A.; Amin, F.; Anis-ur-Rehman, M.; Maqsood, A. Structural, magnetic and electrical properties of $Co_{1-x}Zn_xFe_2O_4$ synthesized by co-precipitation method. *J. Magn. Magn. Mater.* **2007**, *311*, 494–499. [CrossRef]
59. Sasi, S.; Fathima Fasna, P.H.; Bindu Sharmila, T.K.; Julie Chandra, C.S.; Antony, J.V.; Raman, V.; Nair, A.B.; Ramanathan, H.N. Green synthesis of ZnO nanoparticles with enhanced photocatalytic and antibacterial activity. *J. Alloys Compd.* **2022**, *924*, 166431. [CrossRef]
60. Liu, J.; Huang, J.; Hu, Z.; Li, G.; Hu, L.; Chen, X.; Hu, Y. Chitosan-based films with antioxidant of bamboo leaves and ZnO nanoparticles for application in active food packaging. *Int. J. Biol. Macromol.* **2021**, *189*, 363–369. [CrossRef]
61. Şahin, B.; Aydin, R.; Soylu, S.; Türkmen, M.; Kara, M.; Akkaya, A.; Çetin, H.; Ayyıldız, E. The effect of thymus syriacus plant extract on the main physical and antibacterial activities of ZnO nanoparticles synthesized by SILAR method. *Inorg. Chem. Commun.* **2022**, *135*, 109088. [CrossRef]
62. Naserian, F.; Mesgar, A.S. Development of antibacterial and superabsorbent wound composite sponges containing carboxymethyl cellulose/gelatin/Cu-doped ZnO nanoparticles. *Colloids Surf. B Biointerfaces* **2022**, *218*, 112729. [CrossRef]
63. Godoy-Gallardo, M.; Eckhard, U.; Delgado, L.M.; De Roo Puente, Y.J.; Hoyos-Nogués, M.; Gil, F.J.; Perez, R.A. Antibacterial approaches in tissue engineering using metal ions and nanoparticles: From mechanisms to applications. *Bioact. Mater.* **2021**, *6*, 4470–4490. [CrossRef]

Article

Local Three-Dimensional Characterization of Nonlinear Grain Boundary Length within Bulk ZnO Using Nanorobot in SEM

Feiling Shen [1], Ning Cao [2,*], Hengyu Li [1,*], Zhizheng Wu [1], Shaorong Xie [1] and Jun Luo [1]

1. School of Mechatronic Engineering and Automation, Shanghai University, Shanghai 200444, China
2. College of Mechanical and Electrical Engineering, Zhengzhou University of Light Industry, Zhengzhou 450002, China
* Correspondence: ncao@zzuli.edu.cn (N.C.); lihengyu@shu.edu.cn (H.L.)

Abstract: Aiming at the problems of lack of data on the nonlinear morphology to divide uneven grain boundary in bulk ceramics, a unique approach of nanorobot-based characterization of three-dimensional nonlinear structure length can be creatively proposed under scanning electron microscope to quantify the actual morphology of local micro-area grain boundary in bulk ZnO. Contour shapes of the targeted grain boundaries in plane X-Y can be imaged using SEM. Z-directional relative height differences at different positions can be sequentially probed by nanorobot. Experiments demonstrate that it is effective to characterize three-dimensional length structures of nonlinear grain boundaries in bulk materials. By quantifying Z-directional relative height differences, it can be verified to show that irregular characteristics exist in three-dimensional grain boundary length, which can extend the depth effect on nonlinear bulk conductance. Furthermore, this method can also obtain nonlinear quantitative topographies to divide grain boundaries to uneven structure in the analysis of bulk polycrystalline materials.

Keywords: nonlinear grain boundary length; nanorobot-based three-dimensional characterization; structure reconstruction; bulk ZnO; SEM

1. Introduction

Bulk ZnO ceramics possessing excellent nonlinear voltage-sensitive characteristics can be applied in some typical uses, such as power electronics, varistors, and electronic self-protection devices [1]. Nonlinear macroscopic electrical characteristics in bulk materials can be ultimately determined by the Schottky barrier at the inner grain boundary interface [2,3]. In fact, they can be also essentially attributed to the integrated electrical conductivity effects of multiple nonlinear and irregular grain boundary structures [4].

At present, most studies on bulk electrical property measurements can lead to the lack of electrical conductivity property expressions of grain boundaries in bulk ceramics [5,6]. Moreover, it is usual to assume that the isometric structure in bulk materials can be utilized to divide the grain boundary. This cannot produce the effect of differentiation on a single grain boundary. Thus, electrical property measurements of inner grain boundaries can be developed under optical microscope to accomplish micrometer characterization [7–10]. The author's team has also developed in scanning electron microscope (SEM) to conduct nanometer morphology measurement of the grain boundary [11,12]. However, existing electrical measurements may lack the three-dimensional topography data of a single grain boundary. Although the author has also developed a structure characterization of inner invisible grain boundary using nanorobot in SEM [13], it is still short of nonlinear quantitative topographies. Thus, it is significant to pursue an attempt to address a difficulty regarding to how to characterize three-dimensional grain boundary length.

Due to lack of data on the nonlinear morphology to divide uneven grain boundary in bulk ceramics, an approach of three-dimensional length characterization of nonlinear single grain boundary can be proposed under SEM using nanorobot, to accomplish local nonlinear structure characterization of micro-area grain boundaries morphology in bulk polycrystalline ZnO. Furthermore, it can provide a feasible way for the characterization of the three-dimensional structure of an irregular grain boundary. With respect to potential applications, it can try to provide a better way to enhance the positive effect on polycrystalline materials synthesis and structure characterization.

2. Experiments

A polished bulk polycrystalline ZnO sample was mounted onto a nanorobot embedded into SEM (SU3500, Hitachi, Tokyo, Japan). The nanorobot, consisting of a macro piezoelectric motor and micro positioning table, can be utilized to conduct Z-directional heights probing of different positions in plane X-Y. Nanorobots with end probe tip 500 nm can accomplish 10 mm macro motion with resolution of 20 μm and 20 μm micro motion with a resolution of 1 nm, respectively.

In order to accomplish the three-dimensional characterization of nonlinear grain boundary interface lengths in the micro-area of bulk ZnO, two-dimensional profile structure lengths of the targeted grain boundaries can be imaged in plane X-Y. Next, Z-directional heights in plane X-Z and plane Y-Z can be obtained by nanorobot probing at the equidistant positions. As illustrated in Figure 1, it is a schematic diagram accounting for three-dimensional characterization of nonlinear structure lengths of micro-area grain boundary interfaces in bulk ZnO ceramics. After two-dimensional structures of the targeted grain boundary had been imaged to quantify the profile lengths in plane X-Y. When Z-directional relative differences had been vertically probed between different positions, the nanorobot end tip needed to be detected to accomplish the real contact to the upper surface in plane X-Y. At this point, taking Z-directional height as the initial reference occurred.

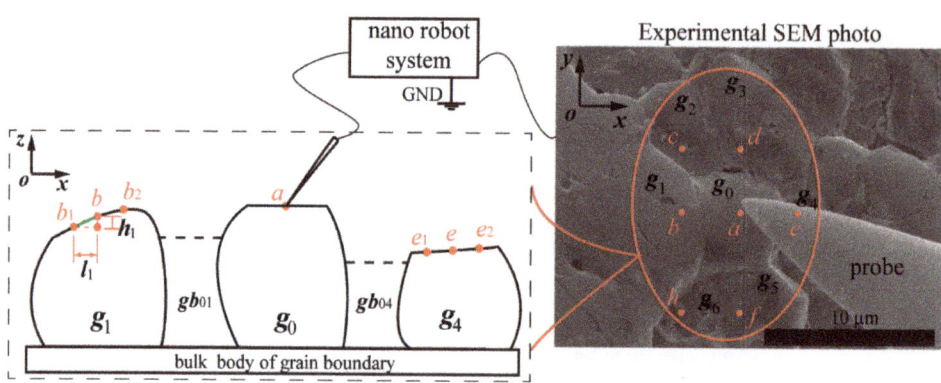

Figure 1. Schematic diagram accounting for three-dimensional characterization of nonlinear structure lengths of micro-area grain boundary interfaces in bulk ZnO ceramics.

In order to verify the rationality of nanorobot-based probing of Z-dimensional heights, six positions at micro-area grain boundaries consisted of the targeted grains g_1, g_2, g_3, g_4, g_5, and g_6 were taken as the Z-directional height probing points. Here, the nanorobot-based Z-directional heights probing of grain boundary g_1-g_0-g_4 in plane X-Z was regarded as an example, seen from the dashed box in details. The probing positions b_1, b, b_2 and position e_1, e, and e_2 were individually selected as the sample probing positions on the adjacent grain g_1 and grain g_4 along the X direction, with a horizontal equidistance of 1 μm. Assuming that the distance l_1 = 1 μm, Z-directional relative vertical height difference h_1 can be acquired by probing position b and b_1 in sequence. As a result, it can be deduced to obtain the hypotenuse length as a part of grain boundary using the trigonometric geometric relation assumed by magnification. Similarly, after the position b_2 has been probed by the nanorobot, partial nonlinear grain boundary length, which can be regarded as a projected length with oblique angle of grain boundary interfacial layer, can be drawn by Z-dimensional heights at position b_1, b, and b_2. It is then reasonable to demonstrate that the probing process of Z-directional heights probed at g_2-g_0-g_6 and g_3-g_0-g_5 in plane Y-Z are similar with that in plane X-Z.

After the grain boundary profile data in plane X-Y were obtained, a model of irregular length can be visually reconstructed by software to realize a virtual stereoscopic intuition expression for the nonlinear length of inner grain boundary.

3. Results and discussion

In order to establish three-dimensional structure morphologies of nonlinear grain boundaries, two-dimensional structures in plane X-Y need to be acquired first. After the targeted grain boundaries have been imaged by SEM, two-dimensional profile structure projections of multiple interface layers can be obtained in plane X-Y, as shown in Figure 2. It can be easily observed that grain boundary length structures projections in plane X-Y can be constructed by the targeted six grains, i.e., g_1, g_2, g_3, g_4, g_5, and g_6. Thus, relative positions away from the central grain g_0 were uniquely identified at the two-dimensional plane. It can be apparent to demonstrate that profile structures of grain boundaries in the two-dimensional plane present the characteristics of irregular shapes and unequal lengths. However, two-dimensional profile structures may be simplified to different shapes composed by straight lines and arc shapes.

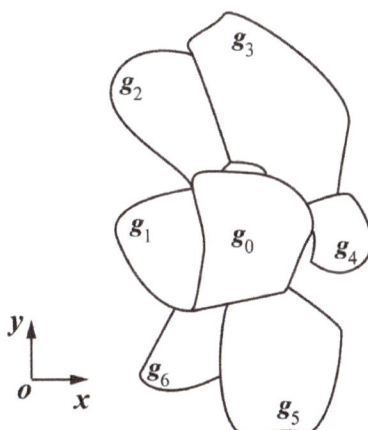

Figure 2. Two-dimensional projections of the contour structure length of the grain boundary interface layer in plane X-Y.

It is worthy of note that it is an assumption to take the contour line of each grain to be considered as the relative uniform thickness of the grain boundary interface. In fact, the thickness of the grain boundary interface layer may vary in different orientations.

It can be further confirmed that grain boundary profile structures can be regarded as the nonlinear expression existing in a two-dimensional plane with irregular and unequal lengths. Furthermore, it can be envisioned that the grain boundary interface layer may exist nonlinearly in three dimensions.

In order to quantify the micromorphology of a grain boundary in plane X-Y, Z-directional height probing of the targeted grain positions along plane X-Z and plane Y-Z profiles can be performed using a nanorobot end probe.

As shown in Figure 3, Z-directional relative heights of three positions on adjacent grains can be individually probed by a nanorobot along plane X-Z and plane Y-Z profiles at equal distances. Figure 3a illustrates the projection drawing of Z-direction relative height differences of g_1 and g_4 relative to g_0 along the X direction. Probing positions on g_1 are individually represented as b_1, b, and b_2, and $L_{b1b} = L_{bb2} = 1$ µm along the X direction. Additionally, probing positions on g_4 are individually represented as e_1, e, and e_2, and $L_{e1e} = L_{ee2} = 1$ µm along the X direction. Correspondingly, Figure 3b illustrates the projection drawing of Z-direction relative height differences of g_2 and g_6 relative to g_0 along the Y direction. Probing positions on g_2 are c_1, c, and c_2, where $L_{c1c} = L_{cc2} = 1$ µm along the Y direction. Additionally, probing positions on g_6 are h_1, h, and h_2, and where $L_{h1h} = L_{hh2} = 1$ µm along the Y direction. Figure 3c illustrates the projection drawing of Z-direction relative height differences of g_3 and g_5 relative to g_0 along the Y direction. Probing positions on g_3 are d_1, d, and d_2, where $L_{d1d} = L_{dd2} = 1$ µm along the Y direction. Additionally, probing positions on g_5 are f_1, f, and f_2, where $L_{f1f} = L_{ff2} = 1$ µm along the Y direction.

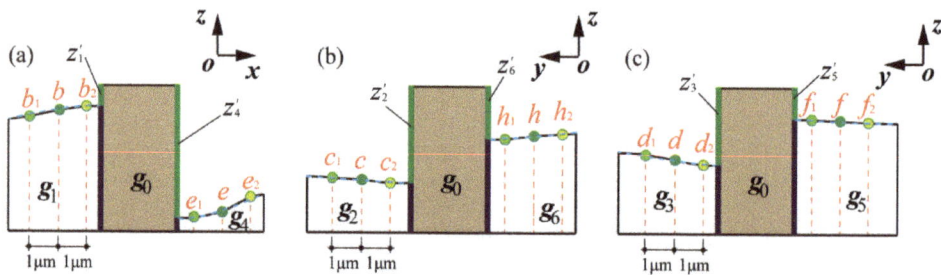

Figure 3. Z-directional relative heights of three positions on adjacent grains individually probed by nanorobot along plane X-Z and plane Y-Z profile at equal distances. (**a**) A projection drawing of Z-direction relative height differences of g_1 and g_4 relative to g_0 along the X direction. (**b**) Projection drawing of Z-direction relative height differences of g_2 and g_6 relative to g_0 along the Y direction. (**c**) Projection drawing of Z-direction relative height differences of g_3 and g_5 relative to g_0 along the Y direction.

After local micro-area Z-directional micromorphology of the targeted grains in plane X-Y have been probed by nanorobot, Z-dimensional relative height differences at different probing positions can be obtained by comparing with grain g_0. Specifically, viewed from Figure 3a, Z-directional relative height differences of three equidistant probing positions b_1, b, b_2 on g_1 and e_1, e, e_2 on g_4 relative to g_0 are −2.3 µm, −2.1 µm, −2.0 µm, and −8.8 µm, −8.7 µm, −8.5 µm, respectively. Viewed from Figure 3b, Z-directional relative height differences of three equidistant probing positions c_1, c, c_2 on g_2 and h_1, h, h_2 on g_6 relative to g_0 are −6.8 µm, −6.9 µm, −7.0 µm and −3.56 µm, −3.5 µm, −3.44 µm, respectively. Viewed from Figure 3c, Z-directional relative height differences of three equidistant probing positions d_1, d, d_2 on g_3 and f_1, f, f_2 on g_5 relative to g_0 are −5.0 µm, −5.2 µm, −5.4 µm and −2.86 µm, −2.9 µm, and −2.93 µm, respectively. Consequently, Z-directional relative height differences at the local grain boundary interface in plane X-Y can be drawn by nanorobot probing at the same equidistant positions on each grain surface.

Viewed from height data obtained by nanorobot Z-directional probing of different positions in plane X-Y, grain boundary lengths can present irregular and non-flat conditions, possessing unequal expressions along three-dimensional directions. Grain boundary lengths in plane X-Y, constructed by grains g_1, g_4, and g_6 away from the targeted grain g_0, present a trend of relatively gradually increasing behavior. On the contrary, grain boundary lengths constructed by grains g_2, g_3, and g_5 present a trend of relatively gradually decreasing behavior. By performing Z-directional nano-probing of adjacent grain surfaces, it is demonstrated that grain shapes can be regarded as irregular characteristics. As a result, three-dimensional nonlinear grain boundaries can also be verified. This method proves that it is effective to characterize three-dimensional length structures of nonlinear polycrystalline grain boundaries using a nanorobot in SEM. It can also make up for the nonlinear morphologies of three-dimensional grain boundaries in bulk ZnO.

Furthermore, SEM imaging combined with robotic Z-directional nano-probing can be utilized to exhibit different internal topographical features of three-dimensional grain boundaries in a micro area of bulk ZnO. Furthermore, it can be used to guide the improvement of composition proportion and synthesis process of bulk ZnO so as to achieve the purpose of improving the macroscopic properties of bulk ZnO by constructing three-dimensional grain boundary morphology.

Correspondingly, a map of local nonlinear lengths of different grain boundary interface layers can be virtually reconstructed, as illustrated in Figure 4. Obviously, the grain boundary interface layer length of each grain possesses a different counter shape. Therefore, it can be deduced that the irregular nonlinear grain boundary length may play a key role in the expression of macro conductance characteristics, which can extend the influences of grain boundary thickness along the depth direction. It is further verified from a nonlinear virtual reconstruction map that an approach of SEM coupled with nanorobots for characterization of micro-area grain boundary lengths is feasible. A virtual reconstruction of the local micro-area of three-dimensional grain boundary length structures can be adopted to interpret the nonlinear electrical characteristics expression of macro bulk polycrystalline ZnO.

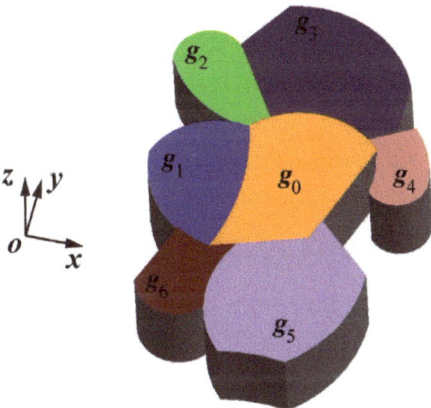

Figure 4. Virtual reconstruction map of local nonlinear lengths of grain boundary interface layers.

It is worth emphasizing that relative heights probed individually along plane X-Z and plane Y-Z can be further utilized to illustrate nonlinear height differences of grain boundary interface lengths. It is further envisioned that if all of the real relative heights of grain boundary interfaces could be quantitatively probed using a nanorobot in plane X-Y, nonlinear grain boundary lengths may be theoretically virtually reconstructed in details. Furthermore, the method can be advanced to accomplish three-dimensional lengths division of grain boundaries, which can potentially provide a unique way to realize the nonlinear mapping characterization of grain boundary length and electrical properties.

Additionally, this proposed method can be utilized to try to address the technological difficulties using isometric brick divisions of grain boundaries. Furthermore, the method can also obtain nonlinear quantitative topographies to divide grain boundaries to uneven structure in property analysis of bulk polycrystalline materials.

As for potential applications, this proposed characterization method can be further utilized to guide a technical improvement of polycrystalline materials synthesis, which can have positive effects on the material composition ratio and processing technology.

4. Conclusions

Aiming at the problem of lack of three-dimensional morphology for nonlinear characterization of a grain boundary to address isometric brick divisions, an approach of three-dimensional length characterization of grain boundary morphology can be proposed under SEM using a nanorobot to accomplish local micro-area nonlinear structure characterization in bulk polycrystalline ZnO. Experiments demonstrate that it is feasible for structure characterization and the reconstruction of the nonlinear length of three-dimensional grain boundaries in bulk materials. It can be further confirmed that grain boundary profile structures can be regarded as the nonlinear expression existing in a two-dimensional plane with irregular and unequal lengths. Nanorobot-based micro-area Z-dimensional relative vertical differences can be utilized to verify that nonlinear and irregular structure characteristics exist in bulk materials. They can also make up for the nonlinear morphologies of three-dimensional grain boundaries in bulk ZnO. It can be deduced that the irregular nonlinear grain boundary length may play a key role in the expression of macro conductance characteristics, which can extend the influences of grain boundary thickness along the depth direction. Virtual reconstruction of local micro-area grain boundary length structures can be adopted to interpret the nonlinear electrical characteristics expression of macro bulk polycrystalline ZnO. Furthermore, it can be advanced to accomplish three-dimensional lengths division of grain boundaries, which can potentially provide a unique way to realize the nonlinear mapping characterization of grain boundary length and electrical properties.

Author Contributions: Conceptualization, N.C., H.L., Z.W., S.X. and J.L.; Data curation, F.S. and N.C.; Investigation, Methodology, F.S., N.C., H.L. and S.X.; Supervision, H.L. and S.X.; Writing—original draft F.S.; Visualization, Writing—review & editing, N.C., H.L. and J.L. All authors have read and agreed to the published version of the manuscript.

Funding: This research was funded by the National Natural Science Foundation of China, grant number 61991415, 61827812, 62073209 and the Key Science and Technology Research Project of Henan Province, China, grant number 222102220032, 212102210359.

Data Availability Statement: Not applicable.

Conflicts of Interest: The authors declare that they have no conflict of interest.

References

1. Kaufmann, B.; Billovits, T.; Supancic, P. Observation of an electrical breakdown at ZnO schottky contacts in varistors. *J. Eur. Ceram. Soc.* **2021**, *41*, 1969–1974. [CrossRef]
2. Bueno, P.R.; Varela, J.A.; Longo, E. Admittance and dielectric spectroscopy of polycrystalline semiconductors. *J. Eur. Ceram. Soc.* **2007**, *27*, 4313–4320. [CrossRef]
3. Kaufmann, B.; Raidl, N.; Supancic, P. Investigation of schottky barriers at Pd-ZnO junctions in varistors. *J. Eur. Ceram. Soc.* **2020**, *40*, 3771–3775. [CrossRef]
4. Bremecker, D.; Keil, P.; Gehringer, M.; Isaia, D.; Rödel, J.; Frömling, T. Mechanically tuned conductivity at individual grain boundaries in polycrystalline ZnO varistor ceramics. *J. Appl. Phys.* **2020**, *127*, 034101. [CrossRef]
5. Ivetić, T.B.; Sekulić, D.L.; Papan, J.; Gúth, I.O.; Petrović, D.M.; Lukić-Petrović, S.R. Niobium and zinc doped titanium-tin-oxide solid-solution ceramics: Synthesis, structure and electrical characterization. *Ceram. Int.* **2018**, *44*, 18987–18995. [CrossRef]
6. Buono, C.; Uriz, A.J.; Aldao, C.M. Effects of intergranular capacitance and resistance dispersion on polycrystalline semiconductor impedance. *Solid State Ion.* **2019**, *343*, 115076. [CrossRef]
7. Nevosad, A.; Hofstätter, M.; Supancic, P.; Danzer, R.; Teichert, C. Micro four-point probe investigation of individual ZnO grain boundaries in a varistor ceramic. *J. Eur. Ceram. Soc.* **2014**, *34*, 1963–1970. [CrossRef]
8. Fleig, J.; Rodewald, S.; Maier, J. Microcontact impedance measurements of individual highly resistive grain boundaries: General aspects and application to acceptor-doped $SrTiO_3$. *J. Appl. Phys.* **2000**, *87*, 2372–2381. [CrossRef]
9. Rodewald, S.; Fleig, J.; Maier, J. Microcontact impedance spectroscopy at single grain boundaries in Fe-Doped $SrTiO_3$ polycrystals. *J. Am. Ceram. Soc.* **2001**, *84*, 521–530. [CrossRef]
10. Shao, R.; Kalinin, S.V.; Bonnell, D.A. Local impedance imaging and spectroscopy of polycrystalline ZnO using contact atomic force microscopy. *Appl. Phys. Lett.* **2003**, *82*, 1869–1871. [CrossRef]
11. Cao, N.; Xie, S.R.; Li, H.Y.; Yang, Y.; Liu, N.; Liu, M.; Pu, H.Y.; Luo, J.; Gong, Z.B. Micro-nano manipulator based localized micro-area electrical impedance measurement for polycrystalline ZnO in scanning electron microscope. *Mater. Lett.* **2018**, *219*, 273–275. [CrossRef]
12. Cao, N.; Shen, F.L. In-situ mapping characterization of structure and electrical property of grain boundary in polycrystalline ZnO using nanorobot in SEM. *Ceram. Int.* **2022**, in press. [CrossRef]
13. Shen, F.L.; Cao, N.; Li, H.Y.; Xie, S.R. Structure characterization of interior invisible single grain boundary by nanorobot pick-and-place grain in bulk ZnO under SEM. *Mater. Lett.* **2022**, *324*, 132777. [CrossRef]

Article

Enhanced Broadband Metamaterial Absorber Using Plasmonic Nanorods and Muti-Dielectric Layers Based on ZnO Substrate in the Frequency Range from 100 GHz to 1000 GHz

Ahmed Emara [1,2,*], Amr Yousef [1,2], Basma ElZein [1,3], Ghassan Jabbour [4] and Ali Elrashidi [1,2]

1 Department of Electrical Engineering, University of Business and Technology, Jeddah 21432, Saudi Arabia
2 Department of Engineering Physics, Alexandria University, Alexandria 21544, Egypt
3 Sustainable Development, Global Council for Tolerance and Peace, VLT1011 Valletta, Malta
4 Advanced Materials and Devices Laboratories, University of Ottawa, 75 Laurier Ave. E, Ottawa, ON K1N 6N5, Canada
* Correspondence: a.emara@ubt.edu.sa

Abstract: A broadband thin film plasmonic metamaterial absorber nanostructure that operates in the frequency range from 100 GHz to 1000 GHz is introduced and analyzed in this paper. The structure consists of three layers: a 200 nm thick gold layer that represents the ground plate (back reflector), a dielectric substrate, and an array of metallic nanorods. A parametric study is conducted to optimize the structure based on its absorption property using different materials, gold (Au), aluminum (Al), and combined Au, and Al for the nanorods. The effect of different dielectric substrates on the absorption is examined using silicon dioxide (SiO_2), aluminum oxide (Al_2O_3), titanium dioxide (TiO_2), and a combination of these three materials. This was followed by the analysis of the effect of the distribution of Al, and Au nanorods and their dimensions on the absorption. The zinc oxide (ZnO) layer is added as a substrate on top of the Au layer to enhance the absorption in the microwave range. The optimized structure achieved more than 80% absorption in the ranges 100–280 GHz, 530–740 GHz and 800–1000 GHz. The minimum optimized absorption is more than 65% in the range 100 GHz to 1000 GHz.

Keywords: electromagnetic absorbers; metamaterial absorbers; SiO_2; Al_2O_3; TiO_2; ZnO; microwave absorbers; plasmonic metamaterial absorbers; absorption spectrum; FDTD

1. Introduction

Recently, electromagnetic wave absorbers (EMAs) have attracted researchers' intertest due to their varied applications in the field of energy harvesting, avion stealth, sensing as well as suppressing the increasing electromagnetic radiations from electronic devices everywhere around us.

Classical EMAs depend on multireflection and interference of electromagnetic waves and could be divided into three types: Salisbury absorbers, Jaumann absorbers, and circuit analog absorbers. Salisbury absorbers consist of a metal plate separated from a resistive sheet by a dielectric material of a quarter wavelength thickness [1–3]. The interference between the reflected wave from the bottom metal plate and the upper resistive sheet is destructive and hence, the EM wave is trapped, and the energy is dissipated in the resistive sheet. Jaumann absorbers use the same concept as the Salisbury absorbers, but Jaumann absorbers use more than one resistive sheet to broaden the absorption bandwidth (BW) of the device [4,5]. The analog circuit absorbers' design is the same as Salisbury absorbers' design but with the top resistive sheet replaced by a periodic top metallic reactive surface that makes the analog circuit absorbers frequency selective absorbers [6,7]. The main disadvantage of the three types of classical absorbers is the need of a dielectric material of quarter wavelength thickness that makes the absorbing device bulky, with limited design flexibility which in turn limits its applicability.

To overcome the drawbacks of classical EMAs, researchers focused on developing new absorbers with thin thickness, light weight, and tunable absorption. In 2008, the first perfect metamaterial absorber (MMA) was proposed by Landy. The unit cell of Landy's absorber consists of two standard split ring resonators connected by an inductive ring parallel to the split wire with a thin dielectric material between them [8]. Generally, MMAs consist of three main layers: ground metal plane, dielectric substrate, and a top metallic periodic patch. The thickness of dielectric layer can be tailored to be much less than the wavelength of the incident electromagnetic wave. Research in the field of EMAs has been accelerated after Landy's perfect absorber due to the prospective applications of the EMAs such as solar energy harvesting, stealth, biological sensors, refractive index sensors, photodetection, photovoltaic devices, and optical switches [9–17].

EMAs can be structured to absorb EM waves in the spectrum range from microwave to visible region. Y. Cheng et al. have reported a metamaterial absorber composed of a single closed-meander-wire resonator structure placed over a metal ground plane separated by a dielectric substrate. They obtained about 90% absorption at different resonance frequencies in the range from 4 to 12 GHz, but this MMA is not suitable for wide bandwidth applications [18]. Using a periodic array of indium tin oxide (ITO) film sandwiched between two polyvinyl chloride layers, Q. Zhou et al. have designed a metamaterial microwave absorber that achieved 90% absorption in the narrow range from 8 to 18 GHz [19]. S. Lai et al. have proposed an MMA with ITO as the top resonance structure array layer, glass as the medium layer, and another ITO as the bottom ground layer. They achieved more than 80% absorption from 15.6 to 39 GHz [20]. J. Ning et al. designed an MMA that operates in the range from 0.4 to 1 GHz with 90% absorption as well, using magnetic nanomaterial and a varactor [21]. Md. Hossain has reported 99.7% to 99.9% EM wave absorption at triple frequencies 5.37, 10.32, and 12.25 GHz using MMA consisting of two split ring copper resonators separated by a dielectric layer [22]. All of these microwave MMAs have a very narrow bandwidth and operate over a narrow range of frequencies.

Recently, thin film MMAs use localized surface plasmon polaritons (LSPPs) to realize small size thin absorber in visible, infrared, and terahertz ranges of the spectrum. W. guo et al. reported infrared MMA using two gold (Au) metallic layers and ZnS dielectric layer sandwiched between them. They got absorption that exceeded 90% in the range from 7.8 to 12.1 µm. The absorption range is enlarged (5.3 to 13.7 µm) by using double Au-ZnS-Au layers on the top of each other but with only 80% absorption [23]. Terahertz plasmonic MMA is reported by Y. Kang et al. using an Au substrate followed by a dielectric material of a dielectric constant of 1.96, with an Au cylinder on the top of the dielectric layer. They obtained two 99% absorption peaks at 275 and 440 THz [24]. Using a ground Au plate, and an array of Au resonators on top of silicon dioxide (SiO_2) substrate, D. Katrodiya et al. have achieved metamaterial broadband solar absorber with an average of 89.79% absorption in the frequency range from 155 to 1595 THz [25]. These reported plasmonic MMAs showed high performance in the infrared and terahertz ranges but suffer from degraded performance in the microwave range.

There is an interest in ZnO driven by its prospects in optoelectronics applications owing to its unique properties such as direct wide band gap $E_g{\sim}3.3$ eV at 300 K. It has been widely known as a versatile material for its different applications in the production of green, blue-ultraviolet, and white light-emitting devices, electronics, and optoelectronics devices. Furthermore, ZnO is known for its strong luminescence in the green–white region of the spectrum, strong sensitivity of surface conductivity to the presence of adsorbed species, and high thermal conductivity. The n-type conductivity of ZnO makes it appropriate for different applications such as metamaterial absorbers [26,27]. On the other hand, SiO_2 is a material of considerable technological importance due to its wide applications in electronics and optoelectronics devices, with a very wide bandgap of 9.6 eV [28]. TiO_2, has been widely investigated in environmental and energy research, due to its wide bandgap of 3.2 eV, which allows it to absorb the UV light [29]. The combination of these materials will play a vital role in affecting the impedance of the proposed metamaterial absorber.

In this work, we introduced an enhanced broadband thin film plasmonic MMA that operates in the frequency range from 100 to 1000 GHz. An absorption above 80% is obtained in the range from 700 to 1000 GHz, while it fluctuates between 60% and 80% in the range below 700 GHz. This proposed absorber is found to be insensitive to light polarization and the direction of incident light. The paper is organized as follows: first, the proposed design of the plasmonic MMA is presented; then the simulation results are illustrated and discussed which yields to an optimized MMA design. Finally, the conclusion summarizes the process and the results of this research.

2. Proposed Structure Design and Its Operation Principle

The typical thin film MMA structure has been adopted in this research, which consists of three layers: a metallic Au ground square plate of 200 nm thickness and cross section area of 1×10^4 nm^2, a 4×4 array of metallic equally spaced nanorods of height h = 50 nm and radius r = 60 nm, and a dielectric substrate of height h_1 = 60 nm sandwiched between the ground plate and the metallic rods. The schematic diagram of the adopted structure is shown in Figure 1. The gold ground plate acts as a back reflector layer that is used to enhance light trapping and reflects the transmitted light to the structure for more light absorption [30,31]. The spacing (X) between any two successive nanorods could be calculated as:

$$X = \frac{L}{4} - 2r \qquad (1)$$

where L is the side length of the square ground metallic plate, and r is the radius of the nanorod. The distance between the center of the outer rod and the edge of the unit cell is assumed X/2.

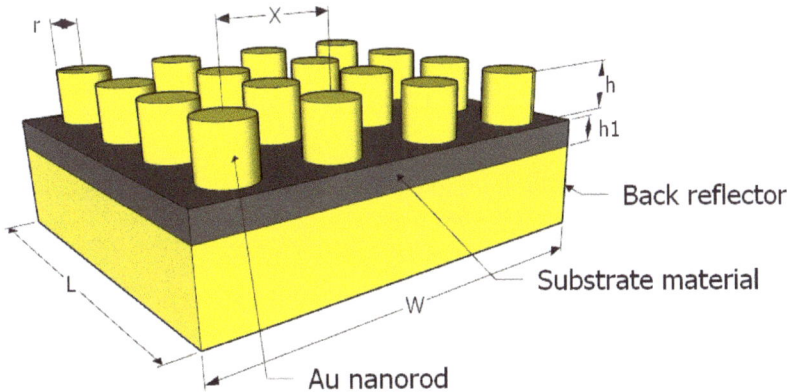

Figure 1. Schematic diagram of the adopted typical MMA structure.

The principle of operation of MMAs depends on resonance. When an electromagnetic wave at a resonance frequency coincides on the MMA, a pair of anti-parallel oscillating currents are induced in the ground metallic layer and the upper metallic nanorods so, a magnetic resonance is established. Moreover, local surface plasmons are generated at the resonance wavelength, and electric resonance is established between the ground metallic layer and the nanorods. Absorption is a result of this resonance, as the electromagnetic wave will be confined in the MMA unit cell and electromagnetic power at the resonance frequency is consumed due to losses in the metallic layer and dielectric layer [8,32,33]. Absorptance (A(f)) of the MMA could be calculated from the relation [34]:

$$A(f) = 1 - R(f) - T(f) \qquad (2)$$

where R(f) and T(f) are the reflectance and transmittance of the absorber, respectively. The reflectance and transmittance could be calculated from the reflection coefficient (S_{11}), and transmission coefficient (S_{21}):

$$R(f) = |S_{11}|^2 \qquad (3)$$

$$T(f) = |S_{21}|^2 \qquad (4)$$

Due to the back reflector metallic ground layer, the transmission coefficient is zero and hence, the absorptance A(f) is given by:

$$A(f) = 1 - |S_{11}|^2 \qquad (5)$$

The absorptance depends on the input impedance of the MMA structure, and is given by [20]:

$$A(f) = 1 - \left| \frac{Z_{in}(f) - Z_o}{Z_{in}(f) + Z_o} \right| \qquad (6)$$

where Z_{in} is the input impedance of the MMA structure, and $Z_0 = 377\ \Omega$ is the free space impedance. At resonance, the input impedance is matched to the free space impedance, so perfect absorption occurs at the resonance frequency.

Plasmonic nanorods distributed on the substrate layer change the absorbed optical power inside the proposed structure, the absorption depends on the maximum reflectivity. Nanorod shape and size are the main parameters that affect the absorbed optical power in addition to the relative permittivity of the plasmonic nanorods and dielectric constant of the surrounding medium [35]. The maximum absorption occurred at the wavelength known as λ_{max}, the maximum peak of the wavelength, which can be calculated using Equation (7).

$$\lambda_{max} = \frac{P}{n} \left(\frac{\varepsilon_n \varepsilon_m (\lambda_{max})}{\varepsilon_m + \varepsilon_n (\lambda_{max})} \right)^{1/2} \qquad (7)$$

where (ε_m) is the permittivity of the surrounding medium, (ε_n) is the plasmonic nanorod dielectric constant at corresponding (λ_{max}), (n) is an integer and (P) is the periodicity of the structural.

Hence, the plasmonic nanorod dielectric permittivity can be calculated using a multi-oscillator Drude-Lorentz model [35] as shown in Equation (8):

$$\varepsilon_n = \varepsilon_\infty - \frac{\omega_D^2}{\omega^2 + j\omega\gamma_D} - \sum_{Y=1}^{6} \frac{\delta_k \omega_k^2}{\omega^2 - \omega_k^2 + 2j\omega\gamma_k} \qquad (8)$$

where (ε_∞) is the nanorod high-frequency dielectric permittivity, (ω_D) is the plasma frequency of the free electrons, (γ_D) is the collision frequency of the free electrons, (δ_k) is the amplitude of Lorentz oscillator, (ω_k) is the resonance angular frequencies and (γ_k) is the damping constants for (Y) value from 1 to 6.

To calculate the absorbed power, the refractive indexes of all used material are given as follows; TiO_2 follows the Devore model [36], however, the value for silicon dioxide is a function of the wavelength and follows Aspnes and Studna model [37], and the refractive index of zinc oxide is considered as given by Kaur et al. [38]. On the other hand, the refractive index of different plasmonic materials is summarized using Equation (8) in Table 1 [39]. The dielectric constants of the used materials are shown in Figures A1–A4 in Appendix A.

Table 1. Plasmonic parameters which are used for the metallic materials.

Material	Term	Strength	Plasma Frequency	Resonant Frequency	Damping Frequency
Au	0	0.7600	0.137188×10^{17}	0.000000×10^{0}	0.805202×10^{14}
	1	0.0240	0.137188×10^{17}	0.630488×10^{15}	0.366139×10^{15}
	2	0.0100	0.137188×10^{17}	0.126098×10^{16}	0.524141×10^{15}
	3	0.0710	0.137188×10^{17}	0.451065×10^{16}	0.132175×10^{16}
	4	0.6010	0.137188×10^{17}	0.653885×10^{16}	0.378901×10^{16}
	5	4.3840	0.137188×10^{17}	0.202364×10^{17}	0.336362×10^{16}
Al	0	0.5230	0.227583×10^{17}	0.000000×10^{0}	0.714047×10^{14}
	1	0.2270	0.227583×10^{17}	0.246118×10^{15}	0.505910×10^{15}
	2	0.0500	0.227583×10^{17}	0.234572×10^{16}	0.474006×10^{15}
	3	0.1660	0.227583×10^{17}	0.274680×10^{16}	0.205251×10^{16}
	4	0.0300	0.227583×10^{17}	0.527635×10^{16}	0.513810×10^{16}

The proposed structure is analyzed and optimized using an electromagnetic wave solver, Lumerical Finite Difference Time Domain (FDTD) solutions software. In the simulation of a unit cell, the boundary conditions are considered as a periodic structure in x and y directions, and the layers are perfectly matched in z-direction. A plane wave source with a frequency band 100–1000 GHz is used as a light source, and the minimum mesh size is 0.5 nm in all directions with an offset time of 7.5 fs is used for the light source. The absorption of the structure is measured at different frequencies.

3. Results and Discussion

In this section, the adopted MMA structure depicted in Figure 1 is optimized to maximize the absorption of the device over a broad spectral band from 100 to 1000 GHz. First, the absorption of the MMA is measured using Lumerical FDTD solution software for different rod materials, then the dielectric substrate material is optimized, and finally, the geometric dimensions of the rod are optimized.

3.1. Effect of the Rod Material on the Absorption of the MMA

The MMA is simulated for different nanorod materials to elect the material that maximizes the absorption. Figure 2 shows the absorption of the MMA, with SiO_2 dielectric substrate, measured when all nanorods are made from gold, and Aluminum. Furthermore, the absorption is measured when the nanorods are arranged such that aluminum and gold rows are alternating as shown in Figure 3. The height and radius of the nanorods are h = 50 nm and radius r = 60 nm, respectively, and the height of the dielectric substrate is chosen as h_1 = 60 nm. It is clear from Figure 2 that the absorber with Au nanorods has better absorption than the absorber with Al nanorods in high frequency range from 600 to 1000 GHz, while the opposite behavior is observed in the lower frequency range from 300 to 600 GHz. Using alternating rows of Au, and Al, the absorption is somewhere between that of the two cases.

The absorption of the MMA depends on the material of the nanorods as the resonance frequency depends on the current generated in the metallic nanorods and the generated plasmons which depend on the material. Additionally, according to the RLC model of the MMA, absorption depends on the losses in the dielectric material and ohmic losses of the nanorods [40]. The average absorptions in the three cases are calculated by finding the area under each curve divided by the frequency span (1000–100 GHz). The obtained average absorption values are 72.09%, 69.94%, and 73.56% in case of Au nanorods, Al nanorods, and alternating rows of Au and Al nanorods, respectively. Hence, the design of the typical structure is modified to that shown in Figure 3 with alternating rows of Au, and Al nanorods.

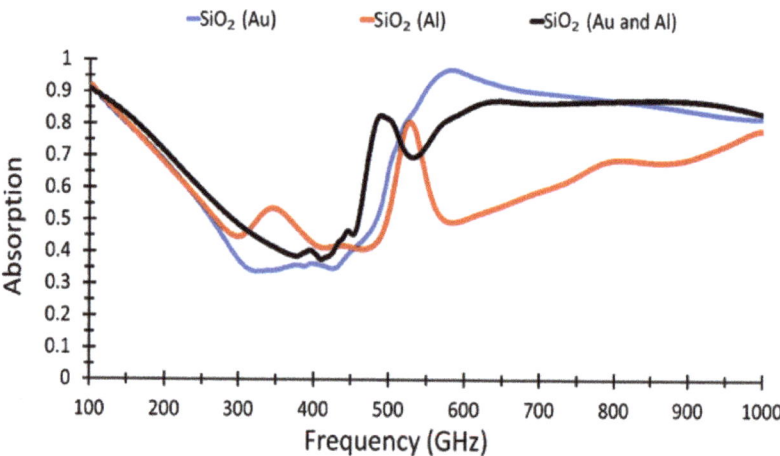

Figure 2. Absorption of the MMA structure.

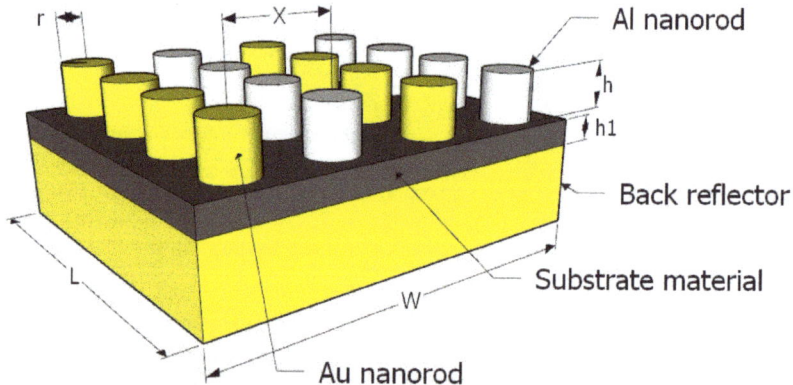

Figure 3. Schematic diagram of the MMA structure with alternating rows of Au, and Al.

3.2. Effect of the Dielectric Substrate Material on the Absorption of the MMA

The absorption of the modified MMA structure shown in Figure 3 is investigated with different dielectric substrate materials of 60 nm fixed thickness. The dimensions of each nanorod in the alternating rows of Au, Al are fixed to h = 50 nm and radius r = 60 nm. Figure 4 shows the absorption of the MMA structure, with Al_2O_3, SiO_2, and TiO_2 substrates. The absorption of the MMA with SiO_2 substrate is the highest in the longer frequency ranging from 600 to 1000 GHz, while the MMA absorber with TiO_2 substrate has the worst absorption over this frequency range. The maximum absorption using SiO_2 substrate exceeds 95%, but the MMA with TiO_2 substrate has the best absorption over the frequency range below 450 GHz with 96% absorption peak at 363 GHz, and another 92% peak at 423 GHz. According to Equation (6), the absorption of the MMA depends on the input impedance of the absorber, which depends on the permittivity, permeability, and refractive index of the dielectric substrate [21].

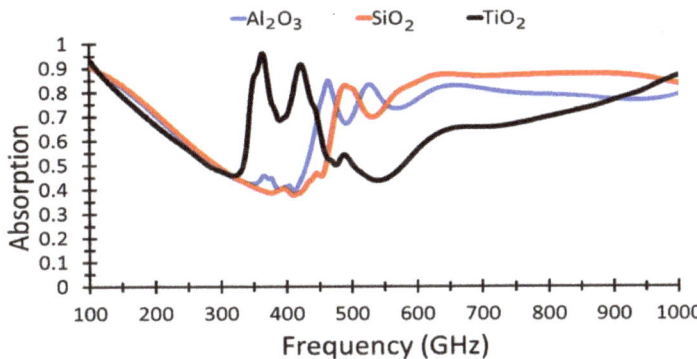

Figure 4. Absorption of the modified structure using different dielectric substrate materials.

The average power absorbed by the MMA is calculated from the obtained absorption spectrum and the absorbed power is 73.6%, 71%, and 70% for SiO$_2$, Al$_2$O$_3$, and TiO$_2$ dielectric substrates, respectively.

As MMAs with SiO$_2$, or Al$_2$O$_3$, give a high absorption in the frequency range above 600 GHz, and MMA with TiO$_2$ substrate has its peak absorption in the frequency range below 450 GHz, in the next stage, the absorption of the absorber is investigated with multi-dielectric layers substrate, as shown in Figure 5.

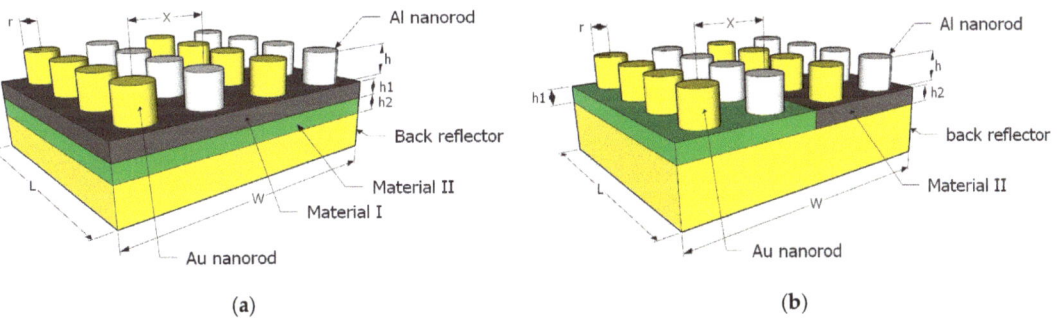

Figure 5. Modified MMA substrate with multi-dielectric layers substrate: (**a**) Two substrates on top of each other; (**b**) two side-by-side substrates.

The performance of the absorber is investigated in two cases. The first case is shown in Figure 5a, where the two substrates are on top of each other, while in the second case shown in Figure 5b, the dielectric substrates are side-by-side, each sharing 50% of the ground plate area. All combinations of SiO$_2$, Al$_2$O$_3$, and TiO$_2$ dielectric substrates are tested to elect the substrate that will give better absorption over a wider band. The absorption of the MMA for the first case is shown in Figure 6a, and for second case is shown in Figure 6b. The difference in the absorption of the absorber between case 1 and case 2 is because of the different overall equivalent capacitance of the absorber which in turn changes the input impedance of the absorber and affects its absorption. The minimum absorption, maximum absorption and average absorption are listed in Table 2.

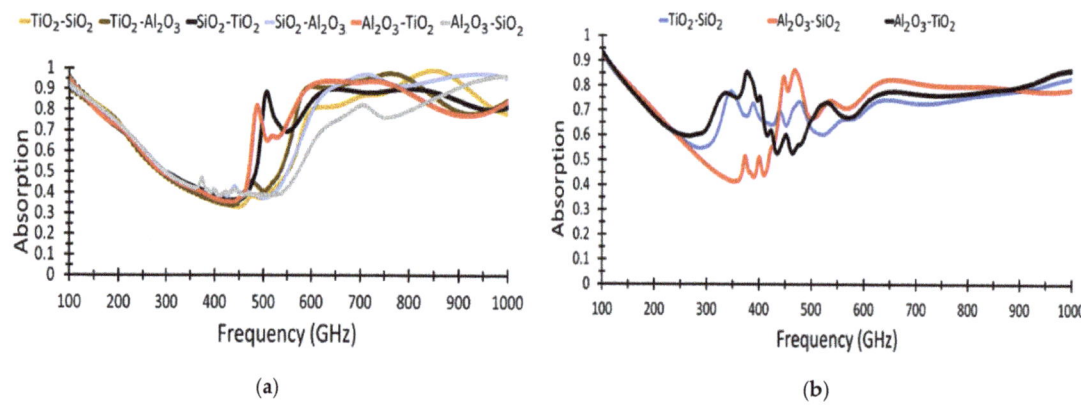

Figure 6. Absorption of the multi-dielectric layers MMA structure: (a) Two dielectric layers on top of each other; (b) Two side-by-side dielectric layers.

Table 2. Absorption for different dielectric substrates.

	Dielectric Substrate Material	Minimum Absorption	Maximum Absorption	Average Absorption
Single dielectric layer	SiO_2	37.6%	90.8%	73.6%
	Al_2O_3	39.4%	91.4%	71.0%
	TiO_2	43.7%	96.1%	70.0%
Multi-dielectric layers on top of each other (The first one is the upper layer)	TiO_2-SiO_2	33.2%	99.0%	69.8%
	TiO_2-Al_2O_3	33.7%	97.7%	71.0%
	SiO_2-TiO_2	36.3%	93.5%	72.4%
	SiO_2-Al_2O_3	37.3%	97.7%	72.7%
	Al_2O_3-TiO_2	35.6%	94.7%	72.7%
	Al_2O_3-SiO_2	38.5%	96.4%	67.4%
Side-by-side multi-dielectric layers	TiO_2-SiO_2	54.8%	92.7%	71.6%
	Al_2O_3-SiO_2	41.5%	92.7%	71.5%
	TiO_2-Al_2O_3	52.6%	93.3%	73.4%

The calculated data presented in Table 1 shows that the absorption of the device is affected by the position of the dielectric layer, and the highest absorption is achieved when the multi-dielectric layers are TiO_2-SiO_2 with SiO_2 is the bottom layer. It's reported that when the top layer is a strong absorber, then the overall absorption of the whole device increases. Small reflection coefficient (S_{11}) of the top layer is a crucial requirement [41]. Furthermore, the minimum absorption is enhanced from 37.6%, and 43.7% in case of single SiO_2, and single TiO_2 layer, respectively, to 54.8% when side-by-side TiO_2-SiO_2 layers are used. This means that the minimum absorption increases by a factor between 25.4% and 46%. Hence, the TiO_2-SiO_2 side-by-side multi-dielectric layers structure with 71.6% average absorption is elected for further enhancement.

3.3. Effect of the Au, Al Array Distribution and Rod Dimensions on the Absorption of the MMA

The structure is modified for further investigation, the nanorod distribution is changed from the alternating rows of nanorods shown in Figure 5b to alternating Au, Al nanorods, so each nanorod is surrounded by four nanorods of the other material as shown in Figure 7.

The absorption of the structure is shown in Figure 8. As an effect of the new nanorods distribution, average power absorption is elevated to 75.7%, the maximum absorption is increased to 94.2%, while the minimum absorption becomes 59.3%.

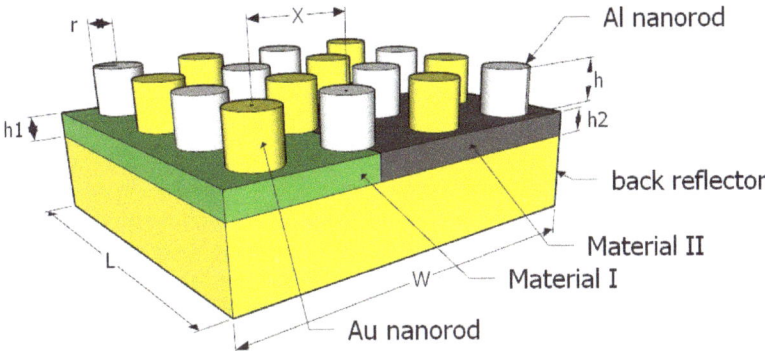

Figure 7. MMA side-by-side multi-dielectric layers structure with alternating nanorods distribution.

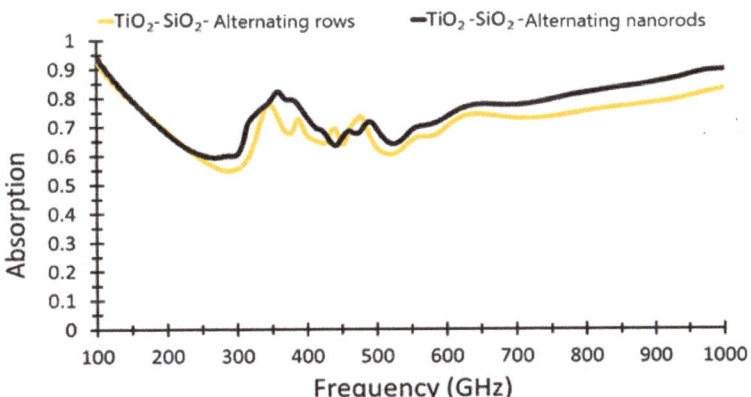

Figure 8. Absorption of the side-by-side multi-dielectric layers MMA structure with alternating rows of nanorods and alternating nanorods.

The distribution of the nanorods changes the distribution of the local surface plasmons induced which affects the electric field in the dielectric material. The absorption of the electromagnetic wave in the dielectric material depends on the magnitude squared ($|E|^2$) of the electric field, which is strongly affected by the design of the structure.

The effect of the radius of the nanorods on the absorption of the MMA is investigated. The radius of the alternating Au, Al nanorods is changed from 40 to 70 nm, while the rod height is fixed to 50 nm, and the absorption is measured in each case. The obtained results are shown in Figure 9a. The effect of the radius is neglected as the maximum absorption is about 94% for all cases and the minimum absorption is between 59.3% and 60%. Moreover, there is a small variation in average absorbed power between 76% and 77%. In addition, the nanorod height is changed from 50 to 80 nm to be optimized. The absorption decreased to almost zero around 400 GHz with nanorods of height 70 nm and 80 nm and increased at the same frequency to 95% with nanorods of height 50 nm then decreased to 55% at 300 GHz, as illustrated in Figure 9b. The optimum nanorod radius and height are 50 nm and 60 nm, respectively.

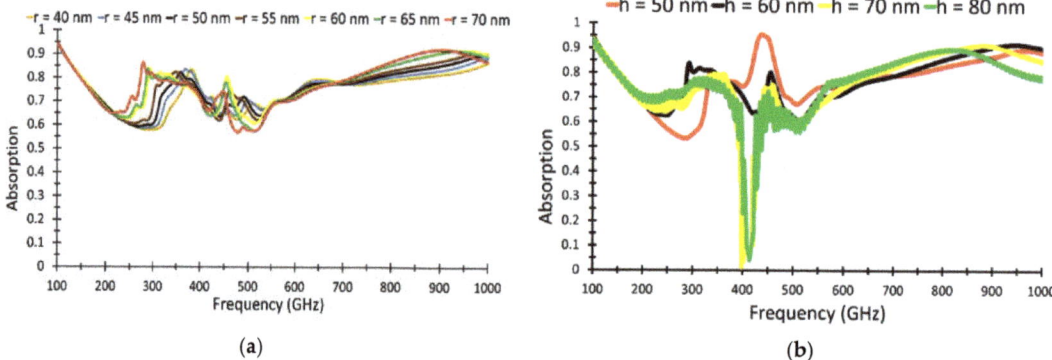

Figure 9. Absorption of the side-by-side multi-dielectric layers MMA structure with alternating nanorods with: (**a**) varying radius from 40 to 70 nm; (**b**) varying heights from 50 to 80 nm.

Finally, a ZnO layer of different thicknesses, 40–70nm, is added on top of the Au ground plate. Recently, ZnO is used to increase the interaction of incident electromagnetic waves and the substrate dielectric layer and thus increase the absorption of the MMA over the operating frequency range [42]. The measured absorption without the ZnO layer and with the ZnO layer of different thicknesses is shown in Figure 10. Adding ZnO layer slightly increases the maximum absorption as illustrated in Table 3. According to the following optimization steps, the optimized structure is shown in Figure 11, where a ZnO layer of thickness 60 nm is added on top of the back reflector. Hence, the optimum design gives an average absorption of 84%, minimum absorption of 65.9% and maximum absorption of 100%.

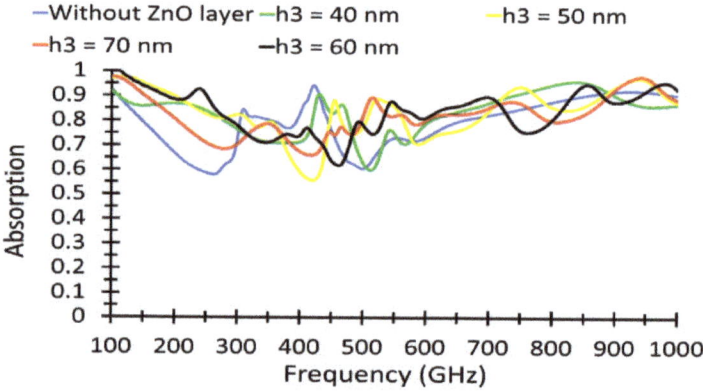

Figure 10. Effect of adding ZnO layer with different thicknesses, 40–70 nm, on the absorption of the absorber.

The enhanced absorption is illustrated in Figure 12 when the ZnO layer is grown on top of the Au layer and used as a base for TiO_2-SiO_2 materials. The absorption is more than 80% in three different regions, 100–280 GHz, 530–740 GHz and 800–1000 GHz, which represent almost 95% of the band. On the other hand, the obtained absorption is more than 65% in the range from 100 GHz to 1000 GHz.

Table 3. Absorption for different dielectric substrates.

ZnO Layer Thickness	Minimum Absorption	Maximum Absorption	Average Absorption
Without ZnO layer	59.3%	94.0%	77.0%
$h_3 = 40$ nm	60.0%	95.7%	83.3%
$h_3 = 50$ nm	55.5%	94.9%	83.5%
$h_3 = 60$ nm	65.9%	100%	84.0%
$h_3 = 70$ nm	62.4%	95.1%	82.1%

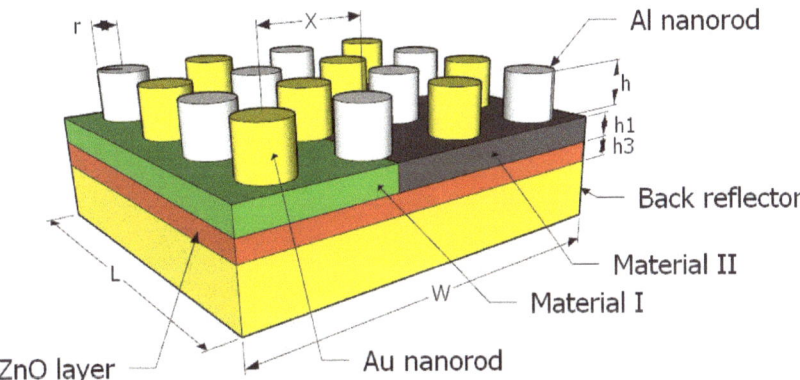

Figure 11. MMA side-by-side multi-dielectric layers structure with alternating nanorods distributed on ZnO substrate.

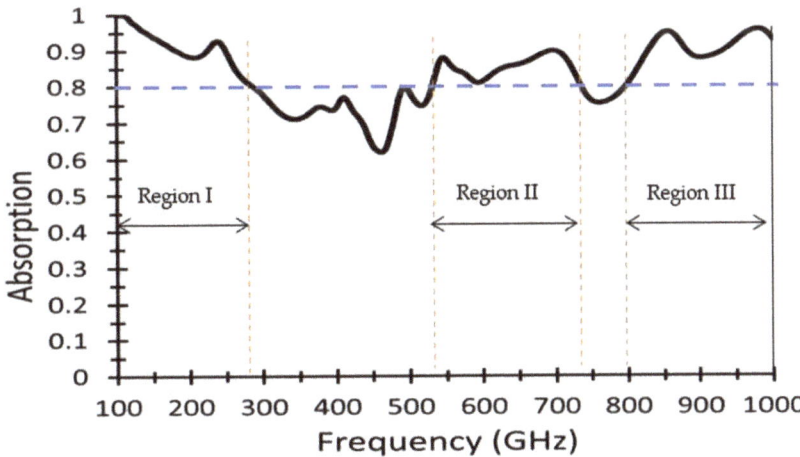

Figure 12. The optimum absorption is more than 65% in the range 100–1000 GHz, and with 65.5% of the band over 80% absorption.

The electric field and magnetic field distributions and absorbed optical power are shown in Figure 13 at three different frequencies (230, 450, and 700 GHz). At 450 GHz, there is an electric and magnetic resonance in the TiO$_2$ layer where the maximum power is absorbed. On the other hand, at 700 GHz, the electric and magnetic field resonance occurs in the SiO$_2$ and the ZnO layer where the maximum power absorption takes place. Moreover, some power is absorbed by the plasmonic nanorods. The effect the ZnO layer is clear at 700 GHz, as the absorbed power increased due the power absorbed in the ZnO layer. At

450 GHz, where one minimum absorption occurs, Figure 13 shows that no resonance occurs at this frequency which leads to minimum power absorption shown at this frequency.

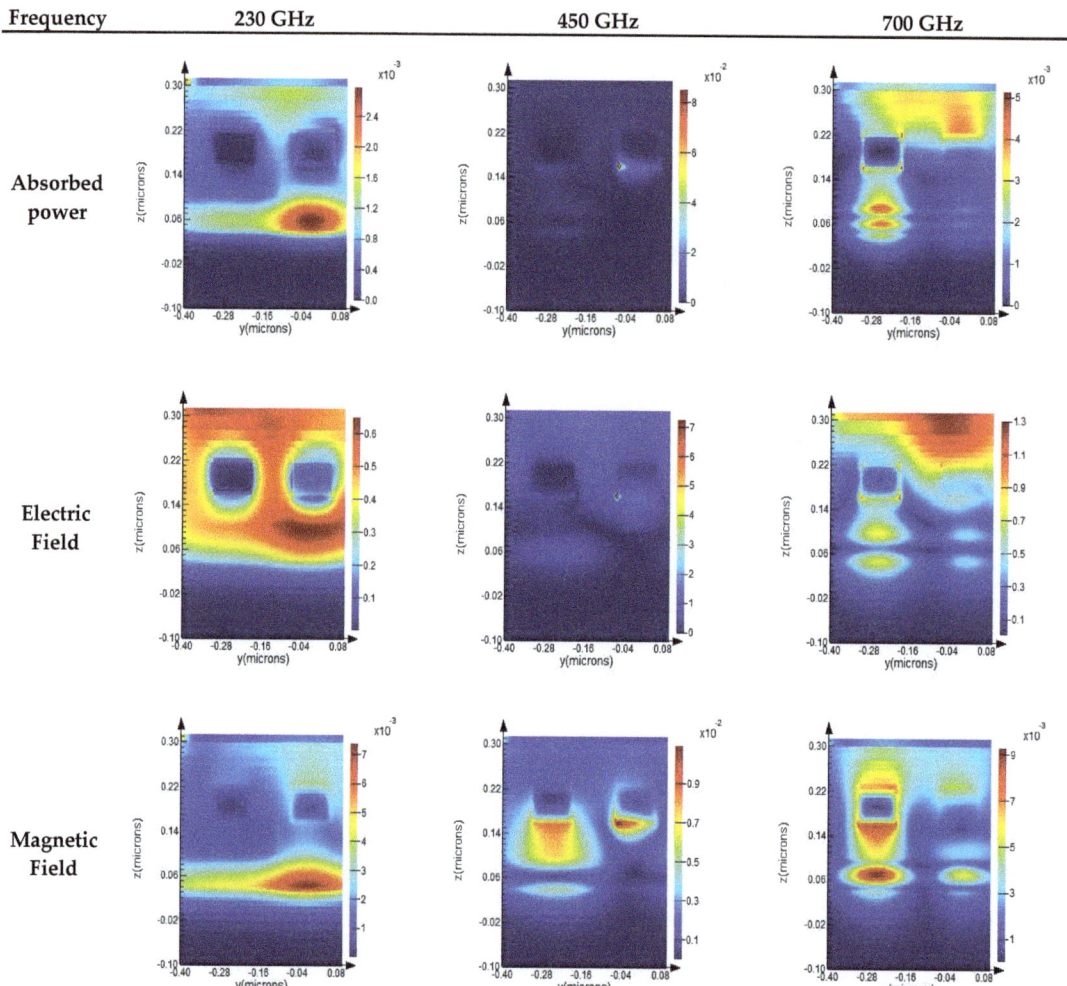

Figure 13. Absorbed optical power, electric field distribution, and magnetic field distribution in the proposed MMA structure at different frequencies.

3.4. Effect of the Incidence Angle and Light Polarization on the Absorption of the MMA

The direction of the incident light and its polarization play an important role in the performance the MMA, so this effect is investigated in this section. Figure 14 shows the absorption of the MMA for different incident angles ranging from 0° (normal incidence) to 70° in a step of 10°. Changing the incidence angle, slightly alters the performance of the absorber with some ripples are observed in the absorbed power. The minimum, maximum, and average power absorbed at different angles are shown in Table 4. The maximum absorbed power changes over a range from 100% to 98.6% which represents a 1.4% decrease in maximum absorption. While the average absorption changes from 84% to 93.3% (about 11% increase), the minimum absorption increases from 65.9% to 78.4% at 50°. Practically, the incident wave is far away from the object and angular stability over a 30° range is enough to ensure absorber stability [20].

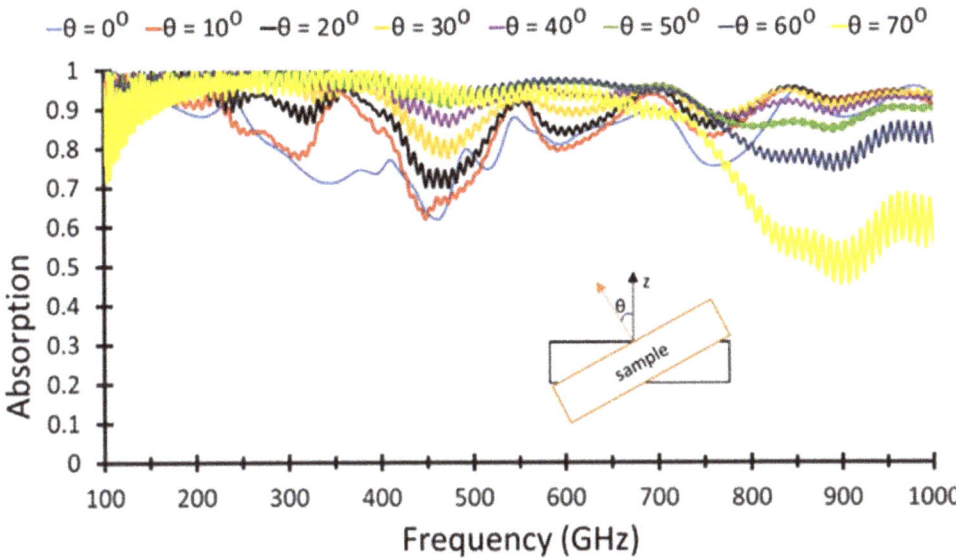

Figure 14. Absorption of the proposed optimized MMA structure for different incident angles.

Table 4. Absorption for different values of incident angle.

θ = 0°	Minimum Absorption	Maximum Absorption	Average Absorption
Direct	65.9%	100%	84%
10°	61.8%	99.1%	86.9%
20°	70.3%	98.8%	89.9%
30°	72.1%	98.6%	92.1%
40°	77.5%	98.8%	93.3%
50°	78.4%	99.1%	93.3%
60°	63.4%	99.1%	91.5%
70°	45%	98.8%	84.5%

The effect of light polarization is investigated by changing the direction of light polarization angle from 0° to 90° in a step of 15°, as shown in Figure 15. Due to the symmetry of the proposed structure, light polarization has no effect on the absorbed power. The effect of light polarization is then investigated for oblique incidence case where the incident angle is 30° and the obtained absorption is shown in Figure 16. It is clear from Figure 16 that the proposed structure is insensitive to light polarization in oblique incidence as well.

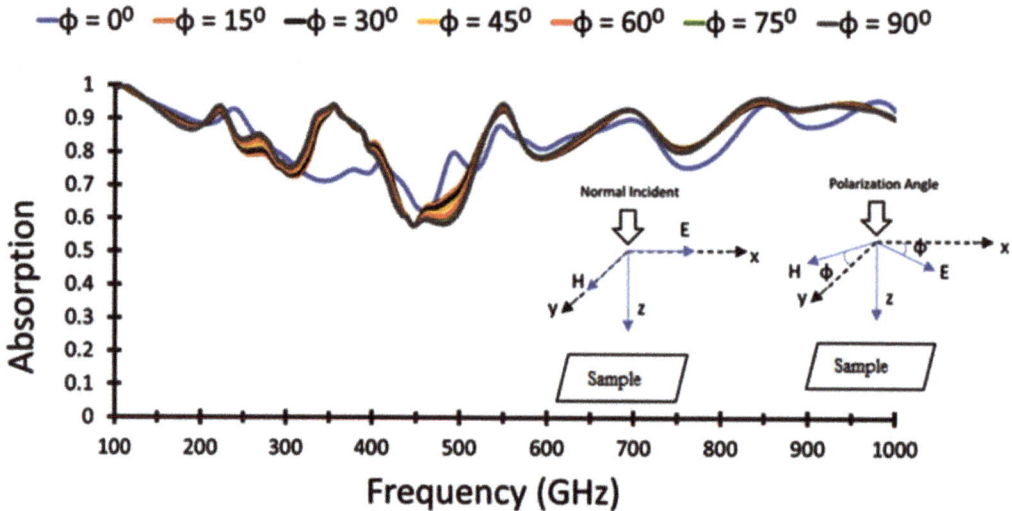

Figure 15. Absorption of the proposed optimized MMA structure for different light polarization at normal incidence.

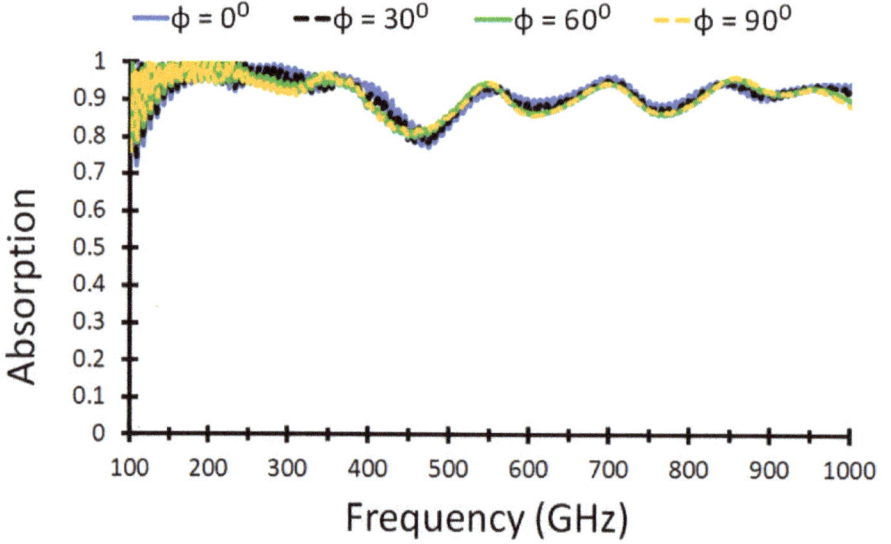

Figure 16. Absorption of the proposed optimized MMA structure for different light polarization at oblique incidence (θ = 30°).

The development of a broadband MMA operating in the wide range of the spectrum has been challenging until now, but comparing our results with the recently reported MMAs shows that the proposed MMA in this work has larger broadband, from 100 to 1000 GHz, with high maximum absorption 100%. The comparison of the absorber's performance is listed in Table 5.

Table 5. Comparison of the MMA performance with recently reported MMAs.

Related Work	Operating Frequency Range	Maximum Absorption	Technique
Ref. [43]	25–37.5 THz	87%	Ti/Ge/Si$_3$N$_4$/Ti metamaterial structure
Ref. [44]	6–16 GHZ	Exceeds 80%	Metallic strips fabricated with lumped resistors on a FR-4 substrate
Ref. [45]	0.79–20.9 GHz and 25.1–40 GHz	90%	Magnetic absorbing material and a multi-layered meta-structure
Ref. [46]	4.2–7.4 THz	98.21%	Split gold and graphene rings over a dielectric and gold plate.
Ref. [47]	7.22–8.84 GHz	90%	Asymmetric section resonator structure with different sizes.
Ref. [48]	10–17 GHz	90%	Array of alternating copper, and FR-4 disks to form a conical frustum
Proposed structure	100–1000 GHz	100%	Au nanorods/TiO$_2$-SiO$_2$/Au ground plate metamaterial structure

4. Conclusions

In this work, a metamaterial absorber structure with multi-dielectric layer is introduced. The structure is optimized to maximize the absorption of the MMA and enhance the minimum absorption of it using SiO$_2$-TiO$_2$ side-by-side multi-dielectric layer on top of a ground Au plate, and an alternating Au, Al nanorods on the dielectric substrate. The ZnO layer is added as a substrate on the top of Au back reflector to enhance the absorption. The MMA has an average absorption of 84%%, a maximum absorption of 100%, and a minimum absorption of about 65.9%. The optimized MMA is shown to have good angular stability as the effect of the incident angle of the electromagnetic wave on the MMA absorption is so small and the absorber is insensitive to polarization for both normal and oblique incidence conditions.

Author Contributions: Methodology, B.E., A.E. (Ahmed Emara) and A.Y.; software, A.E. (Ali Elrashidi) and G.J.; validation, B.E., A.Y. and G.J.; formal analysis, A.E. (Ali Elrashidi) and A.E. (Ahmed Emara); investigation, B.E. and G.J.; resources, A.E. (Ahmed Emara); writing—original draft preparation, A.E. (Ali Elrashidi) and A.Y.; writing—review and editing, G.J.; supervision, A.E. (Ahmed Emara) and G.J. All authors have read and agreed to the published version of the manuscript.

Funding: This research received no external funding.

Data Availability Statement: Not applicable.

Conflicts of Interest: The authors declare no conflict of interest.

Appendix A

The real part and imaginary part of the dielectric constant of the used materials are in the material database of the Lumerical software. These dielectric constant values are shown in Figures A1–A4 for Al$_2$O$_3$, SiO$_2$, TiO$_2$, and ZnO, respectively.

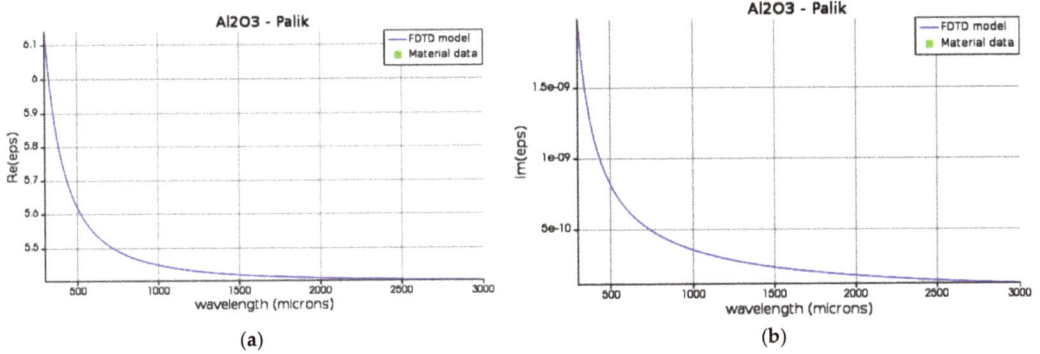

Figure A1. Dielectric constant of Al$_2$O$_3$: (**a**) The real part of the dielectric constant; (**b**) The imaginary part of the dielectric constant.

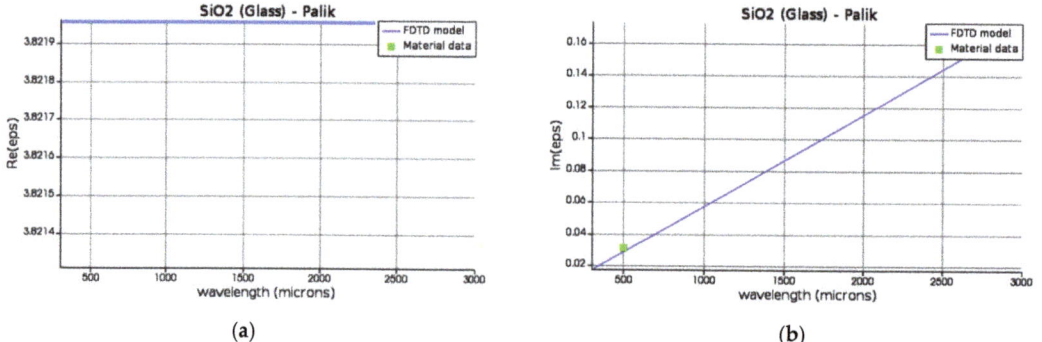

Figure A2. Dielectric constant of SiO_2: (**a**) The real part of the dielectric constant; (**b**) The imaginary part of the dielectric constant.

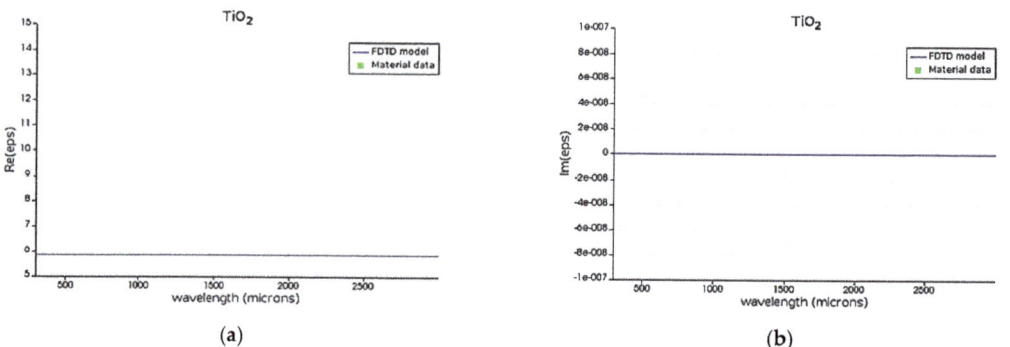

Figure A3. Dielectric constant of TiO_2: (**a**) The real part of the dielectric constant; (**b**) The imaginary part of the dielectric constant.

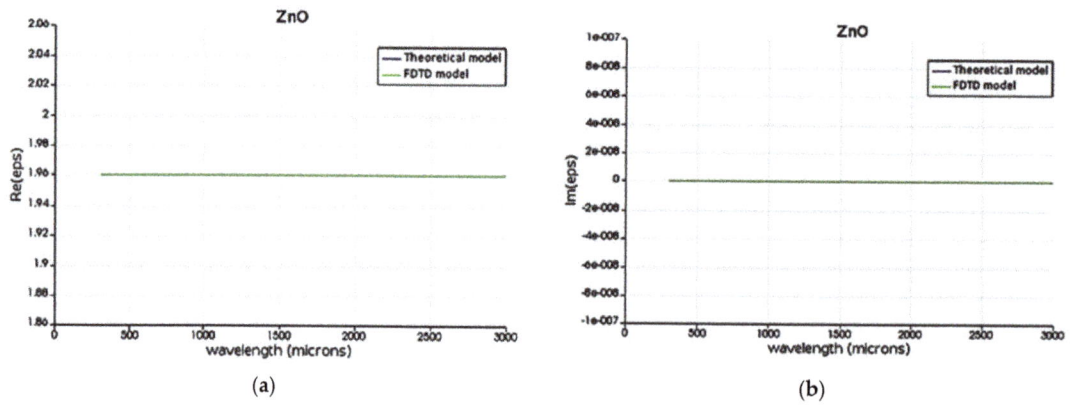

Figure A4. Dielectric constant of ZnO: (**a**) The real part of the dielectric constant; (**b**) The imaginary part of the dielectric constant.

References

1. Monti, A.; Toscano, A.; Bilotti, F. Exploiting the Surface Dispersion of Nanoparticles to Design Optical-Resistive Sheets and Salisbury Absorbers. *Opt. Lett.* **2016**, *41*, 3383–3386. [CrossRef] [PubMed]

2. Medvedev, V.V.; Novikova, N.N.; Zoethout, E. Salisbury Screen with Lossy Nonconducting Materials: Way to Increase Spectral Selectivity of Absorption. *Thin Solid Film.* **2022**, *751*, 139232. [CrossRef]
3. Wulan, Q.; He, D.; Zhang, T.; Peng, H.; Liu, L.; Medvedev, V.V.; Liu, Z. Salisbury Screen Absorbers Using Epsilon-near-Zero Substrate. *Mater. Res. Express* **2021**, *8*, 16406. [CrossRef]
4. Fang, X.; Zhao, C.Y.; Bao, H. Design and Analysis of Salisbury Screens and Jaumann Absorbers for Solar Radiation Absorption. *Front. Energy* **2018**, *12*, 158–168. [CrossRef]
5. Oh, J.; Choi, J. *Adaptable Compressed Jaumann Absorber for Harsh and Dynamic Electromagnetic Environments*; Syracuse Univ Ny Syracuse: Syracuse, NY, USA, 2020.
6. Ma, Z.; Jiang, C.; Cao, W.; Li, J.; Huang, X. An Ultrawideband and High-Absorption Circuit-Analog Absorber with Incident Angle-Insensitive Performance. *IEEE Trans. Antennas Propag.* **2022**. [CrossRef]
7. Yao, Z.; Xiao, S.; Li, Y.; Wang, B. Wide-Angle, Ultra-Wideband, Polarization-Independent Circuit Analog Absorbers. *IEEE Trans. Antennas Propag.* **2022**, *70*, 7276–7281. [CrossRef]
8. Landy, N.I.; Sajuyigbe, S.; Mock, J.J.; Smith, D.R.; Padilla, W.J. Perfect Metamaterial Absorber. *Phys. Rev. Lett.* **2008**, *100*, 207402. [CrossRef]
9. Bagmanci, M.; Karaaslan, M.; Unal, E.; Akgol, O.; Bakır, M.; Sabah, C. Solar Energy Harvesting with Ultra-Broadband Metamaterial Absorber. *Int. J. Mod. Phys. B* **2019**, *33*, 1950056. [CrossRef]
10. Shuvo, M.M.K.; Hossain, M.I.; Mahmud, S.; Rahman, S.; Topu, M.T.H.; Hoque, A.; Islam, S.S.; Soliman, M.S.; Almalki, S.H.A.; Islam, M. Polarization and Angular Insensitive Bendable Metamaterial Absorber for UV to NIR Range. *Sci. Rep.* **2022**, *12*, 1–15. [CrossRef]
11. Wu, J.; Sun, Y.; Wu, B.; Sun, C.; Wu, X. Perfect Metamaterial Absorber for Solar Energy Utilization. *Int. J. Therm. Sci.* **2022**, *179*, 107638. [CrossRef]
12. Hossain, M.J.; Faruque, M.R.I.; Islam, M.T. Perfect Metamaterial Absorber with High Fractional Bandwidth for Solar Energy Harvesting. *PLoS ONE* **2018**, *13*, e0207314. [CrossRef] [PubMed]
13. Abdulkarim, Y.I.; Deng, L.; Altıntaş, O.; Ünal, E.; Karaaslan, M. Metamaterial Absorber Sensor Design by Incorporating Swastika Shaped Resonator to Determination of the Liquid Chemicals Depending on Electrical Characteristics. *Phys. E Low-Dimens. Syst. Nanostruct.* **2019**, *114*, 113593. [CrossRef]
14. Gao, Z.; Fan, Q.; Tian, X.; Xu, C.; Meng, Z.; Huang, S.; Xiao, T.; Tian, C. An Optically Transparent Broadband Metamaterial Absorber for Radar-Infrared Bi-Stealth. *Opt. Mater.* **2021**, *112*, 110793. [CrossRef]
15. Chen, X.; Tian, C.; Che, Z.; Chen, T. Selective Metamaterial Perfect Absorber for Infrared and 1.54 Mm Laser Compatible Stealth Technology. *Optik* **2018**, *172*, 840–846. [CrossRef]
16. Naveed, M.A.; Bilal, R.M.H.; Baqir, M.A.; Bashir, M.M.; Ali, M.M.; Rahim, A.A. Ultrawideband Fractal Metamaterial Absorber Made of Nickel Operating in the UV to IR Spectrum. *Opt. Express* **2021**, *29*, 42911–42923. [CrossRef]
17. Telha, S.; Nouho, A.A.; Ibrahim, I.A.; Achaoui, Y.; Bouaaddi, A.; Jakjoud, H.; Baida, F.I. Annular Hole Array Design as a High Efficiency Absorber for Photovoltaic Applications. *Optik* **2022**, *268*, 169735. [CrossRef]
18. Cheng, Y.; Zou, Y.; Luo, H.; Chen, F.; Mao, X. Compact Ultra-Thin Seven-Band Microwave Metamaterial Absorber Based on a Single Resonator Structure. *J. Electron. Mater.* **2019**, *48*, 3939–3946. [CrossRef]
19. Zhou, Q.; Yin, X.; Ye, F.; Mo, R.; Tang, Z.; Fan, X.; Cheng, L.; Zhang, L. Optically Transparent and Flexible Broadband Microwave Metamaterial Absorber with Sandwich Structure. *Appl. Phys. A* **2019**, *125*, 1–8. [CrossRef]
20. Lai, S.; Wu, Y.; Zhu, X.; Gu, W.; Wu, W. An Optically Transparent Ultrabroadband Microwave Absorber. *IEEE Photonics J.* **2017**, *9*, 1–10. [CrossRef]
21. Ning, J.; Chen, K.; Zhao, W.; Zhao, J.; Jiang, T.; Feng, Y. An Ultrathin Tunable Metamaterial Absorber for Lower Microwave Band Based on Magnetic Nanomaterial. *Nanomaterials* **2022**, *12*, 2135. [CrossRef]
22. Hossain, M.B.; Faruque, M.R.I.; Islam, M.T.; Singh, M.; Jusoh, M. Triple Band Microwave Metamaterial Absorber Based on Double E-Shaped Symmetric Split Ring Resonators for EMI Shielding and Stealth Applications. *J. Mater. Res. Technol.* **2022**, *18*, 1653–1668. [CrossRef]
23. Guo, W.; Liu, Y.; Han, T. Ultra-Broadband Infrared Metasurface Absorber. *Opt. Express* **2016**, *24*, 20586–20592. [CrossRef] [PubMed]
24. Kang, Y.; Gao, P.; Liu, H.; Gao, L. A Polarization-Insensitive Dual-Band Plasmonic Metamaterial Absorber for a Sensor Application. *Phys. Scr.* **2021**, *96*, 65804. [CrossRef]
25. Katrodiya, D.; Jani, C.; Sorathiya, V.; Patel, S.K. Metasurface Based Broadband Solar Absorber. *Opt. Mater.* **2019**, *89*, 34–41. [CrossRef]
26. Xu, W.Z.; Ye, Z.Z.; Zeng, Y.J.; Zhu, L.P.; Zhao, B.H.; Jiang, L.; Lu, J.G.; He, H.P.; Zhang, S.B. ZnO light-emitting diode grown by plasma-assisted metal organic chemical vapor deposition. *Appl. Phys. Lett.* **2006**, *88*, 173506. [CrossRef]
27. Chu, S.; Olmedo, M.; Yang, Z.; Kong, J.; Liu, J. Electrically pumped ultraviolet ZnO diode lasers on Si. *Appl. Phys. Lett.* **2008**, *93*, 181106. [CrossRef]
28. Robertson, J. High dielectric constant oxides. *Eur. Phys. J.-Appl. Phys.* **2004**, *28*, 265–291. [CrossRef]
29. Dette, C.; Pérez-Osorio, M.A.; Kley, C.S.; Punke, P.; Patrick, C.E.; Jacobson, P.; Giustino, F.; Jung, S.J.; Kern, K. TiO_2 anatase with a bandgap in the visible region. *Nano Lett.* **2014**, *14*, 6533–6538. [CrossRef]

30. Li, Z.; Rusli, E.; Foldyna, M.; Wang, J.; Chen, W.; Prakoso, A.B.; Lu, C.; i Cabarrocas, P.R. Nanostructured Back Reflectors Produced Using Polystyrene Assisted Lithography for Enhanced Light Trapping in Silicon Thin Film Solar Cells. *Sol. Energy* **2018**, *167*, 108–115. [CrossRef]
31. Hsu, C.; Burkhard, G.F.; McGehee, M.D.; Cui, Y. Effects of Nanostructured Back Reflectors on the External Quantum Efficiency in Thin Film Solar Cells. *Nano Res.* **2011**, *4*, 153–158. [CrossRef]
32. Li, H.; Yuan, L.H.; Zhou, B.; Shen, X.P.; Cheng, Q.; Cui, T.J. Ultrathin Multiband Gigahertz Metamaterial Absorbers. *J. Appl. Phys.* **2011**, *110*, 14909. [CrossRef]
33. Ogawa, S.; Kimata, M. Metal-Insulator-Metal-Based Plasmonic Metamaterial Absorbers at Visible and Infrared Wavelengths: A Review. *Materials* **2018**, *11*, 458. [CrossRef] [PubMed]
34. Zhang, M.; Song, Z. Switchable Terahertz Metamaterial Absorber with Broadband Absorption and Multiband Absorption. *Opt. Express* **2021**, *29*, 21551–21561. [CrossRef] [PubMed]
35. ElZein, B.; Elrashidi, A.; Dogheche, E.; Jabbour, G. Analyzing the Mechanism of Zinc Oxide Nanowires Bending and Bundling Induced by Electron Beam under Scanning Electron Microscope Using Numerical and Simulation Analysis. *Materials* **2022**, *15*, 5358. [CrossRef] [PubMed]
36. DeVore, J.R. Refractive Indices of Rutile and Sphalerite. *JOSA* **1951**, *41*, 416–419. [CrossRef]
37. Aspnes, D.E.; Studna, A.A. Dielectric Functions and Optical Parameters of Si, Ge, Gap, Gaas, Gasb, Inp, Inas, and Insb from 1.5 to 6.0 Ev. *Phys. Rev. B* **1983**, *27*, 985. [CrossRef]
38. Kaur, D.; Bharti, A.; Sharma, T.; Mahu, C. Dielectric Properties of ZnO-Based Nanocomposites and Their Potential Applications. *Int. J. Opt.* **2021**, *2021*, 995020. [CrossRef]
39. Elrashidi, A. Light Harvesting in Silicon Nanowires Solar Cells by Using Graphene Layer and Plasmonic Nanoparticles. *Appl. Sci.* **2022**, *12*, 2519. [CrossRef]
40. O'Hara, J.F.; Smirnova, E.; Azad, A.K.; Chen, H.-T.; Taylor, A.J. Effects of Microstructure Variations on Macroscopic Terahertz Metafilm Properties. *Act. Passiv. Electron. Compon.* **2007**, *2007*, 049691. [CrossRef]
41. Li, W.; Wu, T.; Wang, W.; Guan, J.; Zhai, P. Integrating Non-Planar Metamaterials with Magnetic Absorbing Materials to Yield Ultra-Broadband Microwave Hybrid Absorbers. *Appl. Phys. Lett.* **2014**, *104*, 22903. [CrossRef]
42. Elrashidi, A. Highly Sensitive Silicon Nitride Biomedical Sensor Using Plasmonic Grating and ZnO Layer. *Mater. Res. Express* **2020**, *7*, 75001. [CrossRef]
43. Zhou, Y.; Qin, Z.; Liang, Z.; Meng, D.; Xu, H.; Smith, D.R.; Liu, Y. Ultra-Broadband Metamaterial Absorbers from Long to Very Long Infrared Regime. *Light Sci. Appl.* **2021**, *10*, 1–12. [CrossRef] [PubMed]
44. Shi, Y.; Li, Y.C.; Hao, T.; Li, L.; Liang, C.-H. A Design of Ultra-Broadband Metamaterial Absorber. *Waves Random Complex Media* **2017**, *27*, 381–391. [CrossRef]
45. Zhang, C.; Yin, S.; Long, C.; Dong, B.W.; He, D.; Cheng, Q. Hybrid Metamaterial Absorber for Ultra-Low and Dual-Broadband Absorption. *Opt. Express* **2021**, *29*, 14078–14086. [CrossRef]
46. Liu, W.; Tian, J.; Yang, R.; Pei, W. Design of a Type of Broadband Metamaterial Absorber Based on Metal and Graphene. *Curr. Appl. Phys.* **2021**, *31*, 122–131. [CrossRef]
47. Cheng, Y.; Luo, H.; Chen, F. Broadband Metamaterial Microwave Absorber Based on Asymmetric Sectional Resonator Structures. *J. Appl. Phys.* **2020**, *127*, 214902. [CrossRef]
48. Thi Quynh Hoa, N.; Huu Lam, P.; Duy Tung, P. Wide-angle and Polarization-independent Broadband Microwave Metamaterial Absorber. *Microw. Opt. Technol. Lett.* **2017**, *59*, 1157–1161. [CrossRef]

Article

Influence of ZnO Morphology on the Functionalization Efficiency of Nanostructured Arrays with Hemoglobin for CO_2 Capture

Alberto Mendoza-Sánchez [1], Francisco J. Cano [2], Mariela Hernández-Rodríguez [1] and Oscar Cigarroa-Mayorga [1,*]

[1] Department Advanced Technologies, UPIITA-Instituto Politécnico Nacional, Av. IPN No. 2580, Mexico City C.P. 07340, Mexico; amendozas1501@alumno.ipn.mx (A.M.-S.); ymhernandez@ipn.mx (M.H.-R.)
[2] Institute of Molecular and Materials of Le Mans e UMR-CNRS 6283, Le Mans Université, 70285 Le Mans, France; francisco.gomez_cano.etu@univ-lemans.fr
* Correspondence: ocigarroam@ipn.mx

Abstract: In this study, nanostructured ZnO arrays were synthesized by an accessible thermal oxidation (TO) methodology. The Zn films were chemically etched with nitric acid (HNO_3) and then oxidized in a furnace at 500 °C for 5 h. Two different morphologies were achieved by modifying the HNO_3 concentration in the etching process: (a) ZnO grass-like nanostructures and (b) rod-like nanostructures, with an etching process in HNO_3 solution at 2 and 8 M concentration, respectively. The physical and chemical properties of the samples were analyzed by X-ray diffraction (XRD), scanning (SEM) and transmission electron microscopy (TEM), energy dispersive X-ray spectroscopy (EDS), and Raman spectroscopy. Both morphologies were functionalized with hemoglobin, and a difference was found in the efficiency of functionalization, which was monitored by UV–Vis spectroscopy. The sample with the highest efficiency was the ZnO grass-like nanostructures. Afterward, the capture of carbon dioxide was evaluated by monitoring a sodium carbonate solution interacting with the as-functionalized samples. The evaluation was analyzed by UV–Vis spectroscopy and the results showed a CO_2 capture of 98.3% and 54% in 180 min for the ZnO grass-like and rod-like nanostructures, respectively.

Keywords: ZnO cane-like nanostructures; ZnO grass-like nanostructures; CO_2 physisorption; artificial photosynthesis; thermal oxidation; carbon dioxide

Citation: Mendoza-Sánchez, A.; Cano, F.J.; Hernández-Rodríguez, M.; Cigarroa-Mayorga, O. Influence of ZnO Morphology on the Functionalization Efficiency of Nanostructured Arrays with Hemoglobin for CO_2 Capture. *Crystals* **2022**, *12*, 1086. https://doi.org/10.3390/cryst12081086

Academic Editors: Yamin Leprince-Wang, Guangyin Jing and Basma El Zein

Received: 30 June 2022
Accepted: 30 July 2022
Published: 3 August 2022

Publisher's Note: MDPI stays neutral with regard to jurisdictional claims in published maps and institutional affiliations.

Copyright: © 2022 by the authors. Licensee MDPI, Basel, Switzerland. This article is an open access article distributed under the terms and conditions of the Creative Commons Attribution (CC BY) license (https://creativecommons.org/licenses/by/4.0/).

1. Introduction

Today, energy production is still dominated by fossil-fuel-based power plants such as thermoelectric stations, coal plants, or gas plants [1]. As a consequence, the energy industry produces around 30 gigatons of carbon dioxide (CO_2) emissions every year [2]. This has led to resource depletion, an increase in costs, and a severe negative environmental impact as global temperatures increasingly rise, ice poles melt, and the number of forest fires increases [3]. CO_2 is considered the main cause of the greenhouse effect. In just one century, it has managed to increase the planet's temperature by 0.6 K. This temperature increase has contributed to the most significant changes in the last two decades [4]. As the Earth's average temperature has increased, many consequences have appeared, such as the rising sea level, promotion of ocean acidification, imbalance of ecosystems, and threats to the life of the species that inhabit them [5]. In addition, great efforts have been made by the scientific community to reduce and control CO_2 emissions. Among the recent proposals, CO_2 capture has appeared as an alternative that takes advantage of the chemical nature of this molecule to trap it on the surface of an active material [6]. CO_2 capture constitutes one step of the artificial photosynthesis (AP) process, which transforms the CO_2 molecule into products such as alcohols and light hydrocarbons, which constitute the current fuels employed by humans in small systems such as gas stoves and water heaters [7]. The AP process comprises systems to convert solar energy into chemical energy [7]. This can be achieved through different nanostructured materials that allow the absorption of light, transportation of electrons, and capture and modification of CO_2 molecules [8]. Thus, proposals for the development of each step that conforms to the process of AP (such as CO_2 capture) are important, and each effort contributes to solving the problems that the surplus of CO_2 in the atmosphere has brought.

Zinc oxide (ZnO) is a widely studied semiconductor material. The ZnO has been obtained as an array of micro- and nanostructures with modified chemical and physical properties. These characteristics make it useful for applications such as gas sensors [9], pressure sensors [10], photonic sensors [11], photodegradation of water pollutants [12,13], optical material for high-power laser applications [14], a highly efficient catalyst for various organic reactions [15], and for corrosion protection and self-cleaning for intelligent surface development [16] or CO_2 capture [17]. These previously mentioned applications are related to the surface properties of ZnO. The electronic properties of ZnO allow the capture of molecules such as CO_2 [17]. Among the methodologies for obtaining ZnO, thermal oxidation (TO) is an accessible technique that promotes modification of the surface due to defects in induction- and atomic-restructuration-related processes [18,19]. Recently, the chemical modification of the Zn surface before TO has become a useful tool to obtain different ZnO morphologies with variations in chemical and physical surface properties in a very accessible way [12,19].

Hemoglobin (Hb) is a molecule capable of transporting mainly O_2 and CO_2. This action depends on factors such as the hydrogen potential (pH), partial pressure of oxygen (pO_2), and partial pressure of carbon dioxide (pCO_2) [20]. At a high pH, Hb mainly transports oxygen; however, at a low pH, Hb can transport 15% to 20% of the undissociated CO_2 in the blood. This transportation is achieved by binding the molecules to hemoglobin at the N terminals of the chains through a reaction [10,21]. Thus, the Hb molecule can be combined with semiconductor materials to promote CO_2 capture. As ZnO has interesting surface properties, which drive the many applications mentioned before, it is a great candidate to serve as a base to be linked with Hb [22]. Even though ZnO has been functionalized with many molecules in other studies that focused on CO_2 capture (such as benzimidazole-based ionic liquid [23], carbon nitride [24], and multiwalled carbon nanotubes and SiO_2 [25]), finding accessible methodologies for achieving a commercial device is a current need [26]. Furthermore, understanding the relationship between the morphology in an inorganic material and the functionalization efficiency with an organic molecule is a topic of great interest for many applications [27].

This paper presents a contribution to the development of accessible CO_2 capture materials by understanding the relationship between the surface properties of ZnO nanostructures and the efficiency of functionalization with Hb. In this study, two kinds of ZnO nanostructures were synthesized by etching treatment in nitric acid, followed by TO. Afterward, the ZnO nanostructures were functionalized with a Hb molecule. The efficiency of the Hb functionalization process was studied in each morphology by UV–Vis spectroscopy. The CO_2 capture was evaluated by monitoring the spectroscopic changes in carbonate dilution across time. The influence of morphology on the surface properties of the ZnO nanostructures was determined. In addition, the influence of ZnO morphology on the CO_2 capture effectiveness was discussed.

2. Materials and Methods

2.1. Chemicals and Reagents

In this study, zinc foil (Zn, SKU: GF00595391, SIGMA-ALDRICH, USA) was used as a substrate for the growth of ZnO. Nitric acid (HNO_3, SKU: 225711-475ML, SIGMA-ALDRICH, USA), hydrogen chloride (HCl, SKU: 17935-250ML, SIGMA-ALDRICH, USA), sodium carbonate (Na_2CO_3, SKU: 223484-500G, SIGMA-ALDRICH, USA), ethanol (CH_3CH_2OH, SKU: 1070179026, SIGMA-ALDRICH, USA), deionized water (Resistivity of 18 MΩ), ammonia

(NH₃, SKU: 294993, SIGMA-ALDRICH, USA), tetraethyl orthosilicate (TEOS, SKU: 333859, SIGMA-ALDRICH, USA), and chromatographic nitrogen (N$_2$) were commercially acquired and used in the experiments without further purification processes.

2.2. Synthesis of the Nanostructured ZnO Array

The nanostructured ZnO arrays were obtained by the known thermal oxidation (TO) methodology [12,19]. Two morphologies were selected from our past studies [19,28]: (a) ZnO grass-like nanostructures and (b) cane-like nanostructures. To create the ZnO grass-like nanostructures, the Zn film was cleaned in a HCl solution for 10 min to remove organic pollutants attached to the surface of the Zn foil. Then, the Zn substrate was etched into a 2 M HNO$_3$ solution for 1 min. To create the ZnO rod-like nanostructures, the Zn foil was cleaned in an HCl solution for 10 min, and was subsequently etched into an 8 M HNO$_3$ solution for 1 min. In both cases, the samples were cleaned in deionized water after each step and were immediately afterward dried by a N$_2$ flux of 20 L/min. Afterward, the samples were oxidized under a 20% of oxygen (weight) in the atmosphere at 500 °C. The oxidation process was continued for 5 h by an isothermal process in a horizontal furnace. After the oxidation process, the samples were rinsed using deionized water and dried under a N$_2$ flux.

2.3. Characterization of Physic and Chemical Properties

Once the samples were synthesized, the chemical and physical properties of the samples were compared. The structural features were analyzed by X-ray diffraction (XRD) by a Bruker D8 Eco Advance diffractometer. The excitation source was a copper tube with Kα radiation (λ = 1.5418 Å). The achieved morphologies in the ZnO arrays were studied by field-emission scanning electron microscopy (FESEM). The electron microscopy was a JEOL 7401F model equipped with an energy dispersive spectroscopy (EDS) system, which was employed to determine the chemical composition of samples. The evolution of atoms in the individual nanostructures of the ZnO arrays was analyzed across the TO process by transmission electron microscopy (TEM) in high-resolution (HRTEM) mode, employing a transmission electron microscope (JEOL ARM200F model). The vibrational modes exhibited by the samples were analyzed by Raman spectroscopy in INTEGRA Spectra equipment with a green laser as the excitation source (wavelength of 532 nm).

2.4. Functionalization of Nanostructured ZnO Array with Hemoglobin

For Hb functionalization, Hb was diluted in deionized water to obtain a 2 mM aqueous solution. The ZnO structures were preconditioned for interacting with the Hb molecule. Thus, 500 mL of ethanol, 500 mL of deionized water, and 400 mL of ammonia were mixed and placed in an ultrasonic bath for 10 min. Afterward, the ZnO nanostructures and 500 mL of the TEOS were added to the previously mentioned mixture. The mixture was agitated for 8 h at room temperature (RT). Then, the ZnO structures were removed from the reaction and immersed in deionized water to remove the remanent molecules. The preconditioned ZnO structures were immersed in the aqueous dispersion of Hb at a temperature of 80 °C for 1 h and then suddenly cooled down at 0 °C. Afterward, the samples were held at this temperature for 180 min. The remanent liquid was analyzed every 30 min by measuring the absorbance spectra with a UV–Vis spectrometer (i3 UV-VIS SPECTROPHOTOMETER, Hanon Instruments) to determine the percentage of Hb functionalized with the ZnO structure.

2.5. Evaluation of CO$_2$ Capture

The capability of the samples (Hb-functionalized ZnO structures) for CO$_2$ capture was studied. The samples were exposed to CO$_2$ (gas) coming from the following reaction:

$$Na_2CO_3 + 2HCl \rightarrow 2NaCl + H_2O + CO_2 \tag{1}$$

The samples' interaction with CO_2 was performed in a hermetic quartz cell to determine the functionality of the immobilized Hb for CO_2 capture. Thus, a Na_2CO_3 solution (186 mM concentration) was employed for the test. The CO_2 adsorption was indirectly measured from the absorbance signal measured from the remanent solution across time with a UV–Vis spectrometer (i3 UV-VIS SPECTROPHOTOMETER, Hanon Instruments). Absorbance spectra were recorded at room temperature every 15 min for 180 min.

3. Results and Discussion

3.1. Morphological and Structural Characterization of ZnO Arrays

A comparison between the XRD pattern of the Zn foil after the HNO_3 etching process under a solution with 2 and 8 M concentrations is shown in Figure 1. The XRD pattern was measured before and after the TO process for both etching conditions. When the Zn foil was etched with the 2 M HNO_3 solution, three peaks at 43.22, 70.09, and 70.29 degrees were found to be in agreement with the diffraction exhibited by the (101), (103), and (110) planes in the hexagonal Zn phase (PDF # 04–0831). For the Zn foil etched with the 8 M HNO_3 solution, three peaks were located at 43.38, 54.40, and 70.08 degrees. These peaks were in agreement with the (101), (102), and (103) planes of the hexagonal Zn phase (PDF # 04–0831). In the Zn foil etched with the lower concentration, a predominant intensity of the plane (103) was observed, while the sample etched with the higher concentration promoted the prevalence of (102) plane exposition. These observations suggest that the concentration of HNO_3 has an important effect on the plane exposition in a similar way that was reported in the grain observation process for metallurgical practices [29]. After the oxidation process was completed, the peaks located at 47.67, 56.60, 62.91, 68.06, and 69.19 degrees were found in the sample treated with 2 M HNO_3 solution, which is labeled as ZnO-g in Figure 1. The peaks located at 47.59, 56.64, 62.91, 67.98, and 69.04 were found in the sample etched with 8 M HNO_3. In both cases, the peaks corresponded with the diffraction of the (102), (110), (103), (112), and (201) planes for the wurtzite hexagonal structure of ZnO (PDF # 036–1451). Both samples exhibited the diffraction of the (101) and (103) planes for Zn, which indicated that the substrate was not totally oxidized during the TO process.

Figure 2 shows a comparison between the obtained morphology from the Zn foil after the TO process across the explored etching conditions. As can be seen, both conditions led to nanostructures growing on the entire surface of the substrate. The sample etched with the 2 M HNO_3 solution was obtained with grass-like morphology (see Figure 2a), while the sample etched with the 2 M HNO_3 solution exhibited cane-like morphology (see Figure 2b). A close examination of both kinds of structures showed that ZnO grass-like structures (ZnO-g) were formed by flat structures with an average area of $1 \pm 0.01\ \mu m^2$ and a thickness of 10 ± 0.2 nm (see Figure 2c). The ZnO cane-like structures (ZnO-c) were $100 \pm 0\ 0.5$ nm in length and had a diameter of 10 nm $\pm$ 0.1 nm. In addition, EDS was performed on each sample after the TO process, finding Zn and O as unique elements. This suggested that this methodology offers a great level of clean control for obtaining ZnO. For the ZnO-g (see Figure 2e), the amount of Zn was less than that found in the ZnO-c (see Figure 2f). This suggested that the etching with a higher concentration of HNO_3 promotes a reduction in oxidation in the Zn substrates, showing the signal of Zn that was not fully oxidized.

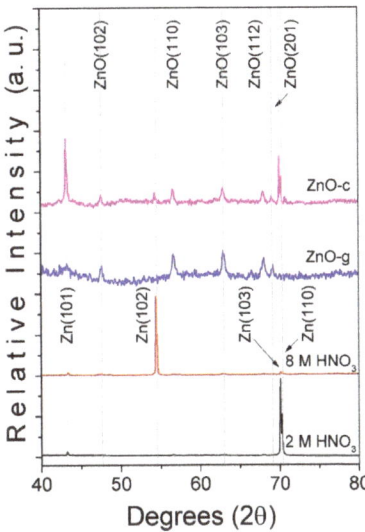

Figure 1. Comparison between XRD patterns of Zn foil etched with HNO$_3$ (2 and 8 M concentration) before and after TO at 500 °C. Note that samples obtained after TO for 2 and 8 M were the grass-like (ZnO-g) and cane-like (ZnO-c) ZnO nanostructures.

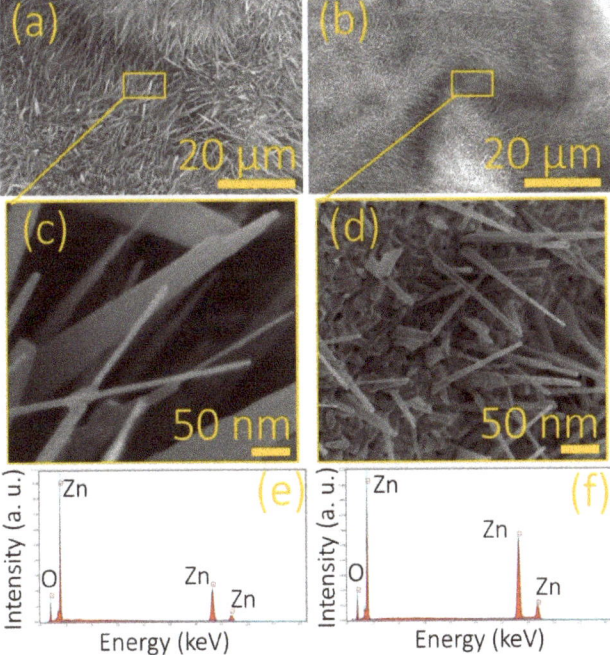

Figure 2. Zn foils etched with HNO$_3$ at (**a**) 2 M and (**b**) 8 M imaged by FESEM after the TO processing. (**c,d**) Close views of the samples in (**a,b**), respectively. Recorded EDS spectra from the (**e**) ZnO grass-like and (**f**) ZnO cane-like nanostructures.

3.2. Growing Path of ZnO Arrays

The individual structures were analyzed by electron microscopy to identify the differences in the interfaces of ZnO grass-like structures (see Figure 3a) and ZnO cane-like structures (see Figure 3c). Once the samples were imaged by HRTEM, the interplanar distance was measured, and we an average distance of 2.85 Å for the ZnO-g and 1.95 Å for the ZnO-c structures. Thus, according to the measured data, the prevalence of plane (010) in the ZnO grass-like structures was revealed (see Figure 3b), while the plane (012) was found in high density on the ZnO cane-like structures (see Figure 3d). The difference in the exposed planes in each kind of structure explained the resulting morphology for each substrate, as was seen in other studies [12].

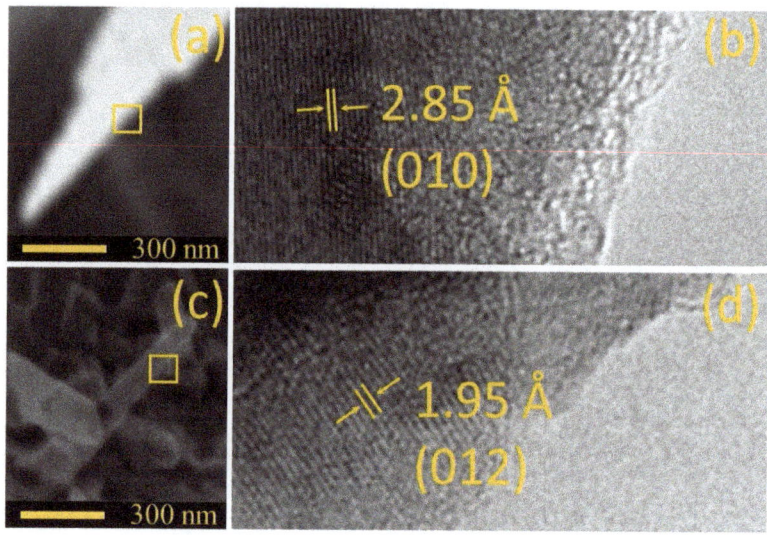

Figure 3. Comparison between (a) a single ZnO grass-like nanostructure and (b) ZnO cane-like nanostructure imaged by SEM. (c,d) Close view of the interface imaged by HRTEM of samples (a,b), respectively.

According to the results, the Zn etching made with HNO_3 is an important factor in producing morphological changes on the surface of the substrates that influence the final characteristics of the ZnO samples. It was reported [30] that oxide preferential growth occurs in Zn zones affected by etching along grain boundaries and scratches on nanoscaled dimensions. These nanostructured oxide regions [31] act as nuclei for further ZnO surface array deposition, as shown in Figure 2. Yu et al. [31] explained that ZnO nanowire deposition occurs at temperatures between the Zn melting and boiling point by means of two mechanisms: first, the development of ZnO nuclei sites formed by vapor-solid action; second, the nanowire formation controlled by the vapor–liquid mechanism. The oxidation temperature (T) in the experiments carried out in this study was 600 °C, which is a temperature between the Zn melting and boiling points (420 °C < T < 907 °C) [31]. The ZnO nanofeatures' growth can be explained by the liquid–solid growth mechanism with preferential nucleation on the oxidized Zn nuclei first produced by acid etching [30] and then by the initial thermal process [30,31]. A scheme showing the influence of HNO_3 etching on the growth mechanism is depicted in Figure 4. The process starts with the formation of surface features by acid etching; according to the results, a nitric acid concentration of 2 M induces scratches on the surface, and the 8 M concentration promotes spherical craters; scratches and craters are marked in red in the scheme in Figure 4a. On the etched surface, nanostructured oxide formation takes place (Figure 4b), mainly on the features promoted by etching (preferential sites for oxidized nuclei formation). Finally, when the temperature reaches the isothermal point, at 500 °C, which is a temperature between Zn melting and boiling points [31], the known liquid–solid growth mechanism controls the arrays' growth, with surface features acting as nucleation sites (Figure 4c), allowing the formation of ZnO surface nanoarrays on Zn foils.

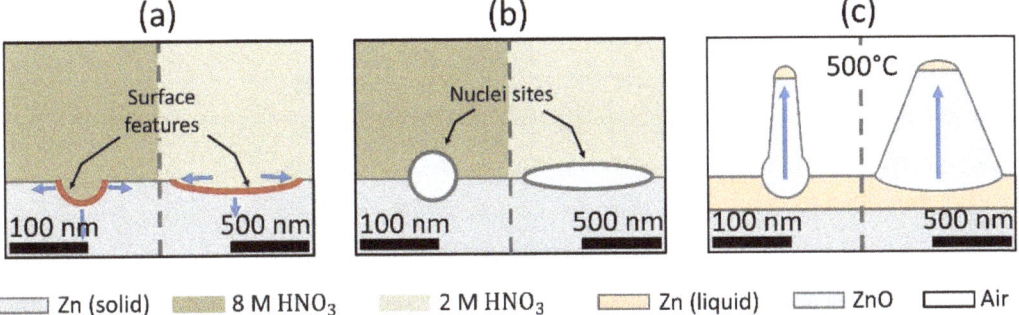

Figure 4. Schematic representation for the growth path of nanostructured ZnO arrays: (a) the pattern on Zn foil promoted by the acid etching process, (b) the phase transition given by the nucleation mechanism, and (c) growth of nanostructured ZnO arrays by the solid–liquid mechanism on nucleation sites into the TO process.

The Raman spectra were recorded to understand the structural differences between the ZnO grass-like and ZnO cane-like nanostructures. Figure 5 shows a comparison between the spectra recorded from the Raman spectroscopy from the Zn foils etched with the 2 and 8 M HNO$_3$ solutions before and after TO. As shown, the Raman shift was completely modified after the TO process. A Lorentzian fit was applied to the spectra recorded from the ZnO grass-like (see Figure 5b) and ZnO cane-like (see Figure 5c) nanostructures. In both samples, six vibrational modes were identified (see Table 1). The E_2 (H)–E_2 (L), A_1 (TO), E_2 (H), and E_1 (LO) vibrational modes corresponded to the ZnO phase and were localized in all samples [19,32–35]. Although the same vibrational modes were found in both kinds of structures, the relationship between the area under the curve of the E_2 (H) mode with respect to the others was different in the ZnO-g and ZnO-c. These results suggested a difference in the defect concentration from one sample to another, which could have a significant impact on the functionalization efficiency of the Hb molecule.

Table 1. Comparison of Raman shift values for vibrational modes between the as-oxidized ZnO structures previously etched with 2 M (ZnO-g) and 8 M (ZnO-c) HNO$_3$ solutions. Note that * point a vibrational mode not reported for ZnO.

Vibrational Mode	ZnO-g (cm^{-1})	ZnO-c (cm^{-1})	Label in Figure 5	ZnO Bulk (cm^{-1}) [33,34]	Nanotubes ZnO (cm^{-1}) [32]
E_2 (H)–E_2 (L)	321.97	331.03	(a)	321	331
A_1 (TO)	387.51	392.98	(b)	384	383
E_2 (H)	437.29	438.80	(c)	438	437
*	486.72	502.26	(d)	-	-
E_1 (LO)	567.23	566.22	(e)	580	578

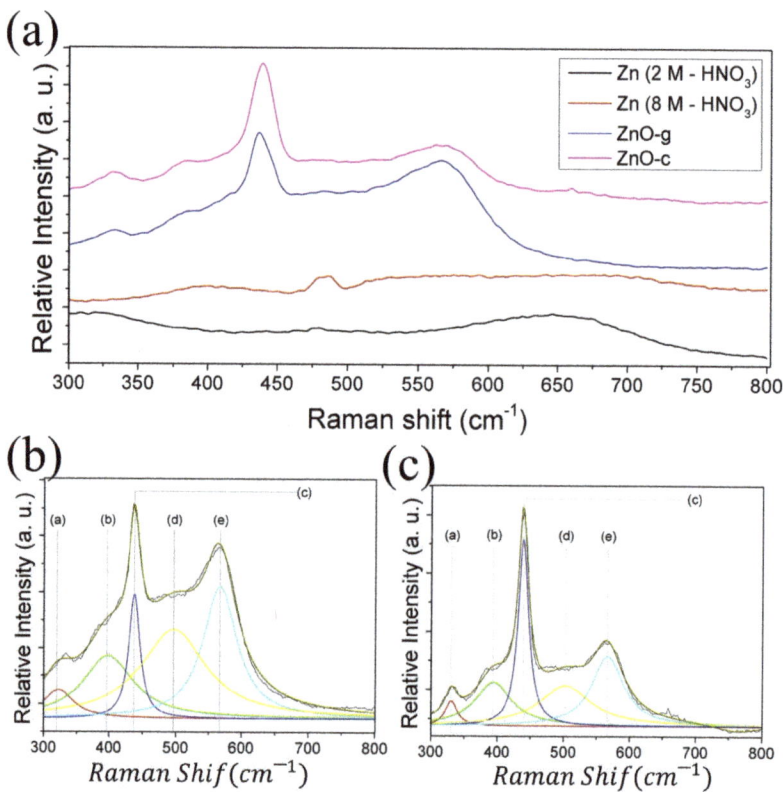

Figure 5. (a) Comparison of the recorded Raman shift between the Zn foil etched with HNO3 2 M (Black line) and 8 M (red line), before and after TO (blue and magenta line, respectively). Lorentzian fit applied on the Raman shift recorded from (b) ZnO grass-like and (c) ZnO cane-like structures.

3.3. Hb Functionalization of ZnO Nanostructures and CO_2 Capture

The supernatant from the Hb-functionalization reaction was analyzed over time with UV–Vis spectroscopy (see Section 2 for details). Figure 6a shows the remnant from the interaction of Hb with the ZnO grass-like structures, while Figure 6b shows the results from the interaction with the ZnO cane-like structures. By comparing these figures, it is clear that the ZnO grass-like structures exhibited a high efficiency for attaching to Hb molecules on the surface. This is clearest in Figure 6c, where the percentage of remnant Hb molecule over the reaction time is depicted. The cane-like structures presented the slowest functionalization rate and achieved a 32.1% Hb-functionalization efficiency. In comparison, the grass-like structures exhibited the fastest rate with an efficiency of 99.2%. This difference in Hb-functionalization efficiency could be attributed to the predominance of the (010) plane in the grass structures, along with the high density of defects in ZnO. This difference in functionalization efficiency due to morphology was seen in other systems [27]. The insert in Figure 6c shows the Hb-functionalized ZnO grass-like structures after the whole process, where a complete surface modification can be seen, possibly due to the incorporation of the Hb molecule at the surface of the sample.

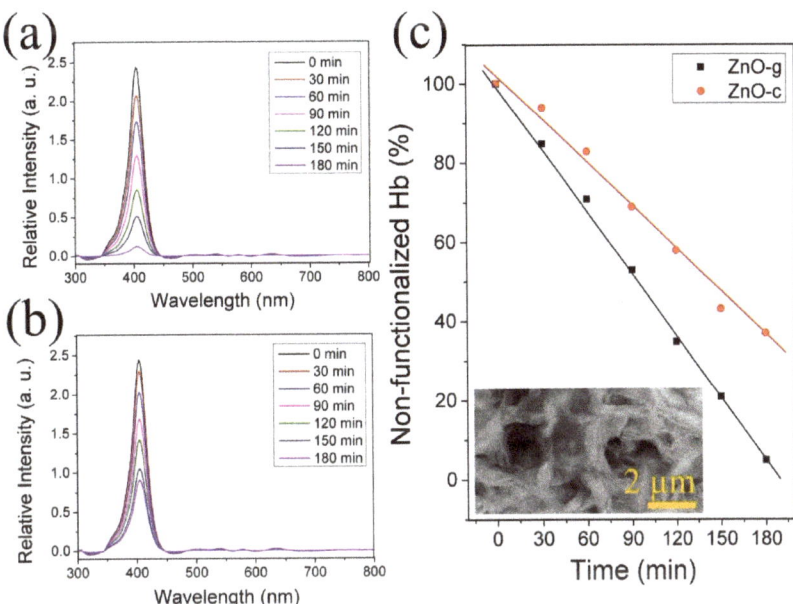

Figure 6. Measured absorbance by UV–Vis of the remanent hemoglobin in the functionalization process over time with (**a**) grass-like and (**b**) cane-like structures. (**c**) Relationship between absorbance signal over time of supernatant in the Hb-functionalization process. Insert in (**c**) shows a Hb-functionalized sample imaged by SEM.

The CO_2 capture due to the Hb-functionalized ZnO structures was indirectly measured by the absorption spectra of sodium carbonate (see details in Section 2). The reaction for obtaining CO_2 presented an absorption peak below 250 nm. Thus, the decrease in this intensity was indicative of CO_2 capture. Note that the shape of the UV–Vis spectrum did not change in the entire process. This could be explained because the molecules in the sample not changing their chemical structure by the interaction with the ZnO structures. The intensity of the UV–Vis spectrum decreased over time because some molecules were trapped by the ZnO; here, this was the CO_2 molecule. As Figure 7a shows, the ZnO grass-like structures allowed significant CO_2 capture after 180 min, with a 98.3% of reduction in the initial intensity. The ZnO cane-like structures exhibited a 54% of reduction in intensity after 180 min of exposure to the Na_2CO_3 solution (see Figure 6b). To compare the rate of CO_2 adsorption, the kinetics of the reaction was computed according to the following equation:

$$\ln\left[\frac{C_O}{C}\right] = kt$$

where C_O is the concentration when time is equal to zero, C is the concentration at time (t), and k is the reaction rate constant [12]. For the sample ZnO-g, the k at 180 min was computed as 0.0118, while that for sample ZnO-c was 0.0040. This provided evidence of a the great difference in CO_2 capture promoted by the morphology in the ZnO arrays. The ZnO can capture CO_2 by itself [17]. According to our results, the morphology of ZnO nanostructures influences the functionalization efficiency with Hb because of the charge distribution on the surface (which is directly related to the structural properties). These observations are in agreement with the theoretical results reported by Usseinov et al. [17]. After the Hb-functionalization process, both kinds of ZnO nanostructures adsorbed the CO_2 molecule by the known interaction between the hemoglobin and CO_2 molecule [36]. Thus, the CO_2 capture difference depends on the Hb efficiency achieved by each ZnO nanostructure.

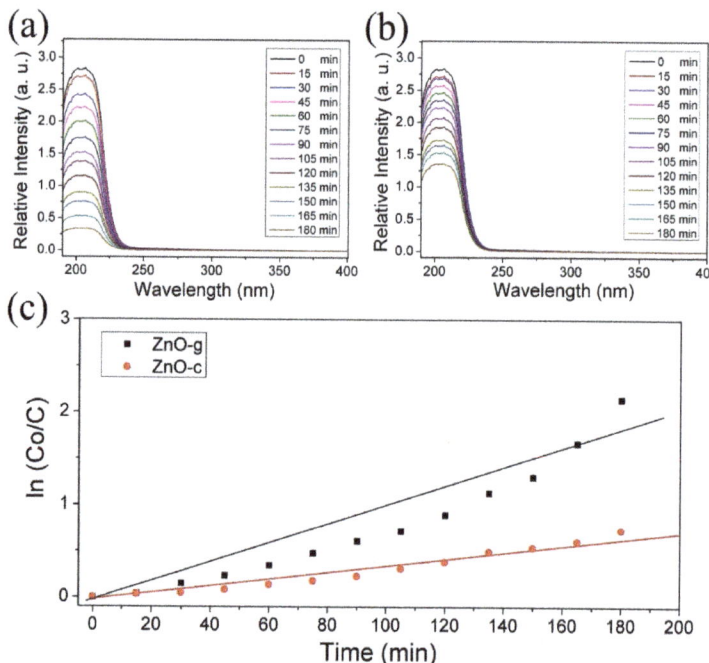

Figure 7. Absorbance spectra of Na_2CO_3 dissolution over time of interaction with (**a**) ZnO grass-like (ZnO-g) and (**b**) ZnO cane-like (ZnO-c) structures. (**c**) The comparison of kinetics in CO_2 adsorption between ZnO-g and ZnO-c.

4. Conclusions

In this study, ZnO grass-like (ZnO-g) and ZnO cane-like (ZnO-c) nanostructures were obtained as an array with an accessible thermal oxidation technique. The ZnO-g structures were mainly constituted by the (010) plane, while the ZnO-c structures were mainly constituted by the (012) plane. The growth and formation of both kinds of ZnO structures explored in this study can be explained by the vapor–liquid mechanism influenced by the HNO_3 etching treatment. The ZnO structures were functionalized with hemoglobin, and we found that the ZnO grass-like structures exhibited the best efficiency, possibly due to the predominance of the (010) plane at the interface. This kind of structure also presented the best efficiency for CO_2 capture, with a 98.3% of initial concentration after 180 min.

Author Contributions: Conceptualization, A.M.-S.; methodology, M.H.-R. and O.C.-M.; software, F.J.C.; validation, M.H.-R. and O.C.-M.; formal analysis, A.M.-S.; investigation, A.M.-S.; resources, M.H.-R. and O.C.-M.; data curation, F.J.C.; writing—original draft preparation, A.M.-S. and F.J.C.; writing—review and editing, M.H.-R. and O.C.-M.; visualization, F.J.C.; supervision, O.C.-M.; project administration, M.H.-R.; funding acquisition, O.C.-M.; All authors have read and agreed to the published version of the manuscript.

Funding: This research was funded by the Secretaría de Investigación y Posgrado of the Instituto Politécnico Nacional, grant number SIP 20220370 and SIP 20221569.

Institutional Review Board Statement: Not applicable.

Informed Consent Statement: Not applicable.

Data Availability Statement: All data included in this study are available upon request by contact with the corresponding author.

Acknowledgments: The authors acknowledge the *Secrtaría de Investigación y Posgrado* of the *Instituto Politécnico Nacional* for the economic support for this research. The authors also thank the *Centro de Nanociencias y Micro y Nanotecnologías* of the *IPN* for the sample characterization facilities.

Conflicts of Interest: The authors declare no conflict of interest.

References

1. Vigoya, M.F.; Mendoza, J.G.; Abril, S.O. International Energy Transition: A Review of its Status on Several Continents. *Int. J. Energy Econ. Policy* **2020**, *10*, 216–224. [CrossRef]
2. Castells-Quintana, D.; Dienesch, E.; Krause, M. Air pollution in an urban world: A global view on density, cities and emissions. *Ecol. Econ.* **2021**, *189*, 107153. [CrossRef]
3. Kim, J.S.; Kug, J.S.; Jeong, S.; Zeng, N.; Hong, J.; Jeong, J.-H.; Zhao, Y.; Chen, X.; Williams, M.; Ichii, K.; et al. Arctic warming-induced cold damage to East Asian terrestrial ecosystems. *Commun. Earth Environ.* **2022**, *3*, 16. [CrossRef]
4. Beusch, L.; Nauels, A.; Gudmundsson, L.; Gütschow, J.; Schleussner, C.-F.; Seneviratne, S.I. Responsibility of major emitters for country-level warming and extreme hot years. *Commun. Earth Environ.* **2022**, *3*, 7. [CrossRef]
5. Lyam, P.T.; Duque-Lazo, J.; Hauenschild, F.; Schnitzler, J.; Muellner-Riehl, A.N.; Greve, M.; Ndangalasi, H.; Myburgh, A.; Durka, W. Climate change will disproportionally affect the most genetically diverse lineages of a widespread African tree species. *Sci. Rep.* **2022**, *12*, 7035. [CrossRef]
6. Alivand, M.S.; Mazaheri, O.; Wu, Y.; Stevens, G.W.; Scholes, C.A.; Mumford, K.A. Catalytic Solvent Regeneration for Energy-Efficient CO_2 Capture. *ACS Sustain. Chem. Eng.* **2020**, *8*, 18755–18788. [CrossRef]
7. Nguyen, V.-H.; Nguyen, B.-S.; Jin, Z.; Shokouhimehr, M.; Jang, H.W.; Hu, C.; Singh, P.; Raizada, P.; Peng, W.; Lam, S.S.; et al. Towards artificial photosynthesis: Sustainable hydrogen utilization for photocatalytic reduction of CO_2 to high-value renewable fuels. *Chem. Eng. J.* **2020**, *402*, 126184. [CrossRef]
8. Liu, Y.; Deng, L.; Sheng, J.; Tang, F.; Zeng, K.; Wang, L.; Liang, K.; Hu, H.; Liu, Y.-N. Photostable core-shell CdS/ZIF-8 composite for enhanced photocatalytic reduction of CO_2. *Appl. Surf. Sci.* **2019**, *498*, 143899. [CrossRef]
9. Zheng, C.; Zhang, C.; He, L.; Zhang, K.; Zhang, J.; Jin, L.; Asiri, A.M.; Alamry, K.A.; Chu, X. $ZnFe_2O_4$/ZnO nanosheets assembled microspheres for high performance trimethylamine gas sensing. *J. Appl. Phys.* **2020**, *849*, 156461. [CrossRef]
10. Park, J.; Ghosh, R.; Song, M.S.; Hwang, Y.; Tchoe, Y.; Saroj, R.K.; Ali, A.; Guha, P.; Kim, B.; Kim, S.-W.; et al. Individually addressable and flexible pressure sensor matrixes with ZnO nanotube arrays on graphene. *NPG Asia Mater.* **2022**, *14*, 40. [CrossRef]
11. Oh, H.; Tchoe, Y.; Kim, H.; Yun, J.; Park, M.; Kim, S.; Lim, Y.-S.; Kim, H.; Jang, W.; Hwang, J.; et al. Large-scale, single-oriented ZnO nanostructure on h-BN films for flexible inorganic UV sensors. *J. Appl. Phys.* **2021**, *130*, 223105. [CrossRef]
12. Rojas-Chávez, H.; Miralrio, A.; Hernández-Rodríguez, Y.M.; Cruz-Martínez, H.; Pérez-Pérez, R.; Cigarroa-Mayorga, O.E. Needle- and cross-linked ZnO microstructures and their photocatalytic activity using experimental and DFT approach. *Mater. Lett.* **2021**, *291*, 129474. [CrossRef]
13. Lin, Y.-P.; Polyakov, B.; Butanovs, E.; Popov, A.A.; Sokolov, M.; Bocharov, D.; Piskunov, S. Excited States Calculations of MoS_2@ZnO and WS_2@ZnO Two-Dimensional Nanocomposites for Water-Splitting Applications. *Energies* **2022**, *15*, 150. [CrossRef]
14. Uklein, A.V.; Multian, V.V.; Kuz'micheva, G.M.; Linnik, R.P.; Lisnyak, V.V.; Popov, A.I.; Gayvoronsky, V.Y. Nonlinear optical response of bulk ZnO crystals with different content of intrinsic defects. *Opt. Mater.* **2018**, *84*, 738–747. [CrossRef]
15. Khaliullin, S.M.; Zhuravlev, V.D.; Ermakova, L.V.; Buldakova, L.Y.; Yanchenko, M.Y.; Porotnikova, N.M. Solution Combustion Synthesis of ZnO Using Binary Fuel (Glycine + Citric Acid). *Int. J. Self-Propag. High-Temp. Synth.* **2019**, *28*, 226–232. [CrossRef]
16. Zheng, H.; Liu, W.; He, S.; Wang, R.; Zhu, J.; Guo, X.; Liu, N.; Guo, R.; Mo, Z. A superhydrophobic polyphenylene sulfide composite coating with anti-corrosion and self-cleaning properties for metal protection. *Colloids Surf. A: Physicochem. Eng. Asp.* **2022**, *648*, 129152. [CrossRef]
17. Usseinov, B.; Akilbekov, A.T.; Kotomin, E.A.; Popov, A.I.; Seitov, D.D.; Nekrasov, K.A.; Giniyatova, S.G.; Karipbayev, Z.T. The first principles calculations of CO_2 adsorption on (10–10) ZnO surface. *AIP Conf. Proc.* **2019**, *2174*, 020181. [CrossRef]
18. Tang, C.; Sun, F.; Chen, Z.; Chen, D.; Liu, Z. Improved thermal oxidation growth of non-flaking CuO nanorod arrays on Si substrate from Cu film and their nanoscale electrical properties for electronic devices. *Ceram. Int.* **2019**, *45*, 14562–14567. [CrossRef]
19. Rojas-Chávez, H.; Cruz-Martínez, H.; Montejo-Alvaro, F.; Farías, R.; Hernández-Rodríguez, Y.M.; Guillen-Cervantes, A.; Ávila-García, A.; Cayetano-Castro, N.; Medina, D.I.; Cigarroa-Mayorga, O.E. The formation of ZnO structures using thermal oxidation: How a previous chemical etching favors either needle-like or cross-linked structures. *Mater. Sci. Semicond. Process.* **2020**, *108*, 104888. [CrossRef]
20. Gibson, J.S.; Cossins, A.R.; Ellory, J.C. Oxygen-sensitive membrane transporters in vertebrate red cells. *J. Exp. Biol.* **2000**, *203 Pt 9*, 1395–1407. [CrossRef]
21. Chu, H.; McKenna, M.M.; Krump, N.A.; Zheng, S.; Mendelsohn, L.; Thein, S.L.; Garrett, L.J.; Bodine, D.M.; Low, P.S. Reversible binding of hemoglobin to band 3 constitutes the molecular switch that mediates O_2 regulation of erythrocyte properties. *Blood* **2016**, *128*, 2708–2716. [CrossRef]
22. Gibson, J.S. Mechanism of O_2-sensitive red cell properties. *Blood* **2016**, *128*, 2593–2595. [CrossRef]

23. Rojas, L.M.G.; Huerta-Aguilar, C.A.; Orta-Ledesma, M.T.; Sosa-Echeverria, R.; Thangarasu, P. Zinc oxide nanoparticles coated with benzimidazole based ionic liquid performing as an efficient CO_2 capture: Experimental and Theoretical studies. *J. Mol. Struct.* **2022**, *1265*, 133466. [CrossRef]
24. Anuar, S.A.; Ahmad, K.N.; Al-Amiery, A.; Masdar, M.S.; Isahak, W.N.R.W. Facile Preparation of Carbon Nitride-ZnO Hybrid Adsorbent for CO_2 Capture: The Significant Role of Amine Source to Metal Oxide Ratio. *Catalysts* **2021**, *11*, 1253. [CrossRef]
25. Jena, K.K.; Panda, A.P.; Verma, S.; Mani, G.K.; Swain, S.K.; Alhassan, S.M. MWCNTs-ZnO-SiO_2 mesoporous nano-hybrid materials for CO_2 capture. *J. Alloys Compd.* **2019**, *800*, 279–285. [CrossRef]
26. Barzagli, F.; Giorgi, C.; Mani, F.; Peruzzini, M. Screening Study of Different Amine-Based Solutions as Sorbents for Direct CO_2 Capture from Air. *ACS Sustain. Chem. Eng.* **2020**, *8*, 14013–14021. [CrossRef]
27. Cigarroa-Mayorga, O.E. Tuning the size stability of $MnFe_2O_4$ nanoparticles: Controlling the morphology and tailoring of surface properties under the hydrothermal synthesis for functionalization with myricetin. *Ceram. Int.* **2021**, *47*, 32397–32406. [CrossRef]
28. Mendoza-Sánchez, R.; Hernández-Rodríguez, Y.M.; Cigarroa-Mayorga, O.E. Degradation of ZnO Nanostructures Interface by Effect of Photocatalytic Activity in Methylene Blue. In Proceedings of the 2021 IEEE International Summer Power Meeting/International Meeting on Communications and Computing (RVP-AI/ROC&C), Acapulco, Mexico, 14–18 November 2021; pp. 1–5. [CrossRef]
29. Tan, W.K.; Razak, K.A.; Ibrahim, K.; Lockman, Z. Oxidation of etched Zn Foil for the formation of ZnO nanostructure. *J. Alloys Compd.* **2011**, *509*, 6806. [CrossRef]
30. Yuan, L.; Wang, C.; Cai, R.; Wang, Y.; Zhou, G. Temperature-dependent growth mechanism and microstructure of ZnO nanostructures grown from the thermal oxidation of zinc. *J. Crystal Growth* **2014**, *390*, 101. [CrossRef]
31. Xiang, K.; Chai, L.; Wang, Y.; Wang, H.; Guo, N.; Ma, Y.; Murty, K.L. Microstructural characteristics and hardness of CoNiTi medium-entropy alloy coating on pure Ti substrate prepared by pulsed laser cladding. *J. Alloys Compd.* **2020**, *849*, 156704. [CrossRef]
32. Xing, Y.J.; Xi, Z.H.; Xue, Z.Q.; Zhang, X.D.; Song, J.H.; Wang, R.M.; Xu, J.; Song, Y.; Zhang, S.L.; Yu, D.P. Optical properties of the ZnO nanotubes synthesized via vapor phase growth. *Appl. Phys. Lett.* **2003**, *83*, 1689–1691. [CrossRef]
33. Wang, M.; Luo, Q.; Hussain, S.; Liu, G.; Qiao, G.; Kim, E.J. Sharply-precipitated spherical assembly of ZnO nanosheets for low temperature H2S gas sensing performances. *Mater. Sci. Semicond. Process.* **2019**, *100*, 283–289. [CrossRef]
34. Souissi, A.; Sartel, C.; Sayari, A.; Meftah, A.; Lusson, A.; Galtier, P.; Sallet, V.; Oueslati, M. Zn- and O-polar surface effects on Raman mode activation in homoepitaxial ZnO thin films. *Solid State Commun.* **2012**, *152*, 794–797. [CrossRef]
35. Souissi, A.; Sartel, C.; Amiri, G.; Meftah, A.; Lusson, A.; Galtier, P.; Sallet, V.; Oueslati, M. Raman study of activated quasi-modes due to misorientation of ZnO nanowires. *Solid State Commun.* **2012**, *152*, 1729–1733. [CrossRef]
36. Christiansen, J.; Douglas, C.G.; Haldane, J.S. The absorption and dissociation of carbon dioxide by human blood. *J. Physiol.* **1914**, *48*, 244–271. [CrossRef]

Article

Motility Suppression and Trapping Bacteria by ZnO Nanostructures

Ningzhe Yan, Hao Luo *, Yanan Liu, Haiping Yu * and Guangyin Jing

School of Physics, Northwest University, Xi'an 710069, China; yanningzhe@foxmail.com (N.Y.); yanan.liu@nwu.edu.cn (Y.L.); jing@nwu.edu.cn (G.J.)
* Correspondence: luo@nwu.edu.cn (H.L.); yuhp@nwu.edu.cn (H.Y.)

Abstract: Regulating the swimming motility of bacteria near surfaces is essential to suppress or avoid bacterial contamination and infection in catheters and medical devices with wall surfaces. However, the motility of bacteria near walls strongly depends on the combination of the local physicochemical properties of the surfaces. To unravel how nanostructures and their local chemical microenvironment dynamically affect the bacterial motility near surfaces, here, we directly visualize the bacterial swimming and systematically analyze the motility of *Escherichia coli* swimming on ZnO nanoparticle films and nanowire arrays with further ultraviolet irradiation. The results show that the ZnO nanowire arrays reduce the swimming motility, thus significantly enhancing the trapping ability for motile bacteria. Additionally, thanks to the wide bandgap nature of a ZnO semiconductor, the ultraviolet irradiation rapidly reduces the bacteria locomotion due to the hydroxyl and singlet oxygen produced by the photodynamic effects of ZnO nanowire arrays in an aqueous solution. The findings quantitatively reveal how the combination of geometrical nanostructured surfaces and local tuning of the steric microenvironment are able to regulate the motility of swimming bacteria and suggest the efficient inhibition of bacterial translocation and infection by nanostructured coatings.

Keywords: ZnO nanowire arrays; bacteria motility; antibacterial; particle tracking

Citation: Yan, N.; Luo, H.; Liu, Y.; Yu, H.; Jing, G. Motility Suppression and Trapping Bacteria by ZnO Nanostructures. *Crystals* **2022**, *12*, 1027. https://doi.org/10.3390/cryst12081027

Academic Editor: Witold Łojkowski

Received: 31 May 2022
Accepted: 20 July 2022
Published: 23 July 2022

Publisher's Note: MDPI stays neutral with regard to jurisdictional claims in published maps and institutional affiliations.

Copyright: © 2022 by the authors. Licensee MDPI, Basel, Switzerland. This article is an open access article distributed under the terms and conditions of the Creative Commons Attribution (CC BY) license (https://creativecommons.org/licenses/by/4.0/).

1. Introduction

The migration, colonization, and reproduction of bacteria play pivotal roles in human health and the ecological environment. Biofilm formation, for example, is a response to most inflammation and metabolism [1–4], and some bacteria can accelerate the decomposition of minerals and extract some metal materials [5,6]. The living activities of bacteria are almost entirely conducted within the confined walls of their habitat; thus, detailed bacterial movements near the surface have attracted extensive interest [7–9]. The complex geometry of real wall surfaces and the dynamical fluctuation of local environments within the immediate interfacial layer are nevertheless essential to bacterial motion and thus their attachment or detachment behavior. The swimming behaviors of bacteria driven by the flagella connected to the rotating motors on the cell membrane demonstrate fascinating features from the hydrodynamic point of view [10]. Berg and Turned [11] reported the clockwise (CW) circular motion parallel to a rigid surface due to the balance of angular momentum between the counter-rotations of flagella and body, which inevitably increased the trapping probability of bacteria near the surface. The dynamic tracking method has been proposed to directly observe the 3D trajectories of bacteria near solid surfaces under microscopy [12,13], showing that the swimming speed and tumble frequency on surfaces were lowered to a certain extent. It has been elucidated that a solid wall is able to attract bacteria with the mechanism of bacterial circular motion on the surface [14,15]. The near-surface movement of bacteria is determined to be a form of ultra-low flight [16] with an accumulation at a distance of 500 nm away from the surface, which has been further measured by Total Internal Reflection microscopy [17,18].

Except for the hydrodynamic origin of slowing down close to the surface, the physical contact between the flagellum and surface has been proven by directly capturing the collision and scattering processes [19]. It is well accepted that the steric interaction between bacteria and surfaces is also important to the motile behavior near the surfaces [20–25]. Since the bacteria swim the distance to the surface down to sub-hundred nanometers, the physicochemical properties of the surface play vital roles in movement behavior. Hu et al. [26], through their mesoscale hydrodynamic simulations, found that a change in surface roughness at the nanoscale was perceivable by cells. When bacteria swim distances of hundreds of nanometers, there is a critical slip length of about 30 nm, over which bacteria swim straight or even reverse the circular motion. In addition, charge density, wettability, and the stiffness of the surface have unavoidable influences on bacterial motility [27,28]. The motility, a measure of the swimming capability of the bacteria, is represented here by the average speed calculated from a vast number of tracked trajectories. However, most studies of bacterial contamination on surfaces roughly relied on the global adsorption number and reproduction rate of bacteria [29–33]. In contrast to adhesion, dynamical bacterial motility on rough substrates is still lacking [31–36]. Obviously, the motility of bacteria near the surface directly relates to bacterial spreading and colonization. Chang et al. found that when the surface structure size is smaller than a lower limit, the bacteria cannot feel the topography; when the structure size is around a critical value, the bacteria are more likely to travel parallel to the local crystal axis [36]. Quantitative and systematical analysis of the effect of surface topography on bacteria from both the trajectories and velocities of bacteria are rather rare. In addition, for real surfaces, such as real skin, nasal mucosa, and intestinal surfaces, except for the surface topography, the chemical microenvironment near the surface also affects bacterial behavior. As addressed in the review by Kołodziejczak-Radzimska et al. [37], the investigation of bacterial movement subjected to surfaces with complex textures and fluctuated chemical reactions are even more demanding. Particularly, from the point of view of the microscopic bacteriostasis mechanism, nanostructured coatings are excellent for capturing the physicochemical picture of modeling real surfaces, where surface morphology and the chemical microenvironment immediately near the surface (nanoscale) are both present in a natural way.

Here we use ZnO nanowire arrays (NWA) as a model system in vitro to demonstrate how the surface structures and local chemical microenvironment controlled by ultraviolet (UV) light are able to efficiently regulate bacterial swimming and quantitively evaluate the bacteriostasis merits. The tracking protocol and statistical analysis of a vast number of trajectories of individual bacteria are proposed to quantify the swimming behavior on a nanoparticles film (NPF) and NWA. Due to the photocatalytic effect of ZnO, the antibacterial components are generated at the interface under ultraviolet light irradiation and diffuse into the bulk liquid. A great number of detailed trajectories and the swimming speed of *E. coli* on different surfaces with nanostructures and under exposure to UV irradiation are systematically analyzed. It is realized that ZnO nanostructures are successful at suppressing bacterial motility within a quarter of an hour, and the spatial spreading of the bacteria is thus restricted. Interestingly, UV irradiation is found to enhance the slowing down of bacterial swimming on ZnO NWA. Our results help to understand the impact of real surfaces on bacterial motile behavior and provide a potential protocol and a general method to improve the sterilization efficiency for nanostructured materials.

2. Materials and Methods

2.1. Bacteria Culturing

E. coli (strain of RP437) with a plasmid fluorescence of protein marker YFP [38] (EX 489–505 nm, EM 524–546 nm) was used in all of the experiments. The bacterial cells were grown overnight in 15 mL of LB buffer (0.01 g/g NaCl (OXOID, Basingstoke, UK), 0.01 g/g of Tryptone (OXOID), and 0.005 g/g of Yeast (OXOID) and dissolved in DI (deionized water) culture medium, plus antibiotics (Chloramphenicol (Aladdin, Shanghai, China), 1 µL/mL), in the incubator at a temperature of 30 °C, under a shaking speed of 200 rpm. Then the culture solution was diluted by DI water at a ratio of 1:100 into fresh medium (15 mL LB) and was kept for another growth period of 5.5 h until it reached the exponential growth phase (optical density ~0.7 at 600 nm of wavelength). The harvesting time was determined under the microscope when the bacteria presented with uniform body dimensions (about 2 μm in length and 0.5 μm in diameter) and the highest motility (~25 μm/s on average). Then, the bacterial cells were concentrated from the culture media by centrifugation (1500× g, 5 min), followed by the removal of the supernatant, and were then gently redispersed into the motility buffer (100 mM EDTA, 1 mM L-methionine (Sigma-Aldrich, St. Louis, MO, USA), 1 M of sodium lactate (Sigma-Aldrich) and 0.1 M of potassium phosphate buffer (EMSURE), pH = 7, and dissolved in DI water and 0.025 g/mL L-Serine (Sigma-Aldrich). The final bacteria concentration used in the tracking experiment on the glass slides and ZnO nanostructured surfaces was about 0.05 of the optical density, dissolved in motility buffer + L-Serine.

2.2. ZnO Nanostructures Preparation

ZnO NPF and NWA were synthesized using classical protocols [39]. Briefly, the glass substrate was first cleaned by ultrasonication in acetone, ethanol, and DI water for 15 min in sequence. Then, 1.44 g of $Zn(CH_3COO)_2 \cdot 2H_2O$ (Alfa Aesar, Haverhill, MA, USA), 0.75 g of PVA17-88 (Aladdin), and 1 mL of absolute ethyl alcohol (General-Reagent, Merck KGaA, Germany) were dissolved in 10 mL of DI water under constant stirring and heated for 30 min to prepare a seed layer solution. Then, a uniform, thin gel layer of the seed layer solution was spin-coated onto the substrate (10 s at 500 rpm followed by 20 s at 3000 rpm), which was transformed into ZnO NPF after annealing in a Muffle furnace (heating speed 5 °C/min, 450 °C in air, for 120 min). The substrate with ZnO NPF was placed upside down in a Teflon container and immersed in a so-called hydrothermal solution (3.5 mM $Zn(NO_3)_2 \cdot 6H_2O$ (Sigma-Aldrich) and 3.5 mM Hexamethylenetetramine (HMTA, Alfa Aesar) dissolved in 35 mL DI water) for 4 h of growth at 95 °C. After cooling down to room temperature, the ZnO NWAs grown from the seed layers were washed with ethanol and DI water several times. Finally, the wurtzite ZnO NWAs were post-annealed in a Muffle furnace (heating speed 5 °C/min, 450 °C in air, hold for 120 min) to remove the residues. The scanning electron microscopy images (SEM, FEI Apreo S, Thermo Fisher Scientific, Waltham, MA, USA), at 10 kV acceleration voltage and 11 mm of working distance) of the ZnO nanostructures are shown in Figure 1b–e. The nanowires were measured and had dimensions of 140 ± 14 nm in diameter, 8.1 ± 1.3 μm in length, 140 nm in spacing, and 21/μm² in number density. The grain size of the NPF was about 42.6 ± 6.1 nm in diameter and 576/μm² in number density (~50 nanowires or nanoparticles within 100 square microns have been selected for this statistical measurement). The roughness is defined here as the ratio of the actual area to the projected area. Then the roughness of NWA was about 96.6, which is much larger than 2.6 of NPF.

2.3. Observation Setup

A Nikon Eclipse Ti2 inverted microscope combined with a 60X Nikon (Tokyo, Japan) water immersion objective with a 1.20 numerical aperture was used for observation. Fluorescent and bright field image sequences were captured using a Flash 4.0 camera with a field-view size of 225 μm in a square at a resolution of 2048 × 2048 pixels and a framerate of 50 fps at an exposure time of 20 ms. Drops of bacterial suspension with a ~10 μL volume were transferred onto the substrates and enclosed in a 1.0 × 1.0 cm² chamber (Gene Frame, Thermo AB-0576, with a height of 250 μm (Figure 1a). The temperature was fixed at 24 ± 2 °C for all of the experiments. The peak wavelength of the UV light used here was about 398 nm, and the wavelength range at half maximum was 390-406 nm. The energy density of the UV light irradiated on the sample was about 4 mW/cm².

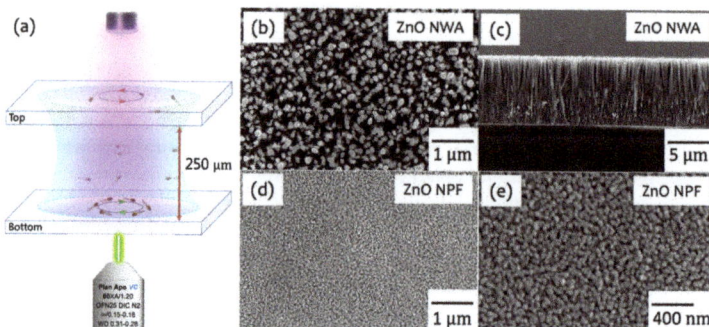

Figure 1. Experimental setup and SEM images of ZnO nanostructures. (**a**) Schematic bacterial swimming within the sandwich confinement. The bacterial suspension is surrounded by four plastic side walls to avoid evaporation. (**b**,**c**) are SEM images of ZnO NWA from the top and side view, respectively. (**d**,**e**) are SEM and zoom-in images of ZnO NPF from the top view at different magnifications, respectively.

2.4. Bacteria Tracking

The swimming bacteria on the surface were recorded by the camera as an image sequence, and the swimming path was then extracted from the linking positions of the tracked individual bacteria at a continuous time sequence. The tracking plugin of TrackMate (Fiji, NIH) was used to detect and analyze the real paths of the swimming bacteria. The displacement and speed were then calculated and plotted with MATLAB coding. Note that the bacteria suspension was rather dilute and that the interaction between each other can be neglected. Meanwhile, roughly tens to hundreds of the total number of independent trajectories were needed for our statistical analysis. To avoid damage or the photobleaching effect of the fluorescence illumination by the laser light source, each image stack was independently captured at different locations (at least 400 μm at the distance with 30% of the total fluorescence intensity).

3. Results and Discussion

3.1. Swimming on ZnO NWA

From the SEM measurements, the ZnO nanowires have a typical diameter of ~140 nm and a height of ~8.1 μm. The body (head) of the *E. coli* has a rod shape, which is about 2 μm in length and 0.5 μm in radius and is connected to 4–6 left-handed flagellum. Each flagellum is about 20 nm in diameter, 200 nm in helical radius, and 5~7 μm in length [40]. Due to the large body size of the bacteria, they are unable to penetrate into gaps between the nanoarrays. However, the rotating flagella are able to sense the ZnO NWA due to its very small diameter and large length. Therefore, the swimming behavior is expected to be different from that on ideal smooth surfaces. The swimming of bacteria on the ZnO NWA was traced with time intervals of 2 min. The typical evolving trajectories of bacterial swimming near the surface are displayed in Figure 2a, with a view field of 225 × 225 μm². As shown in Figure 2b, all of the trajectories are overlapped together with the same start point at the origin of the coordinates. In the first minute, the average moving range of swimming bacteria is about 100 μm, and some of the bacteria could reach 200 μm. While only a few bacteria could move more than 100 μm after 5 min (the majority of the motion range is less than 50 μm). Obviously, the length of the trajectory monotonically decreases over time, as shown in Figure 2a, b, indicating that the spreading of bacteria is significantly inhibited.

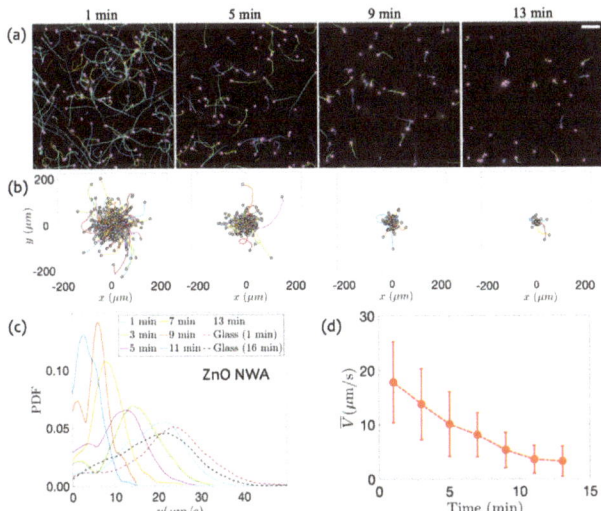

Figure 2. Swimming trajectories and motility variation on ZnO NWA. (**a**) Bacteria trajectory varying with time from 1 min to 13 min. Each trajectory is observed with a duration longer than 2 s at the focal plane. (**b**) All the trajectories are overlapped into a centered start point. (**c**,**d**) Probability density function and mean swimming speed, respectively. The scale bar represents 30 µm.

To quantify the motility of the bacteria, the instantaneous swimming speed at each time step (0.02 s) and probability density function (PDF) during each observation time window are calculated and plotted in Figure 2c. Immediately after being transferred onto the NWA from the motility buffer, the bacteria motions show that the standard variation of the swimming speed is rather wide in the beginning and then significantly shrinks with time. Moreover, the peaks of PDF gradually shift leftwards, i.e., the smaller speed region, indicating the decay of swimming motility. Furthermore, to confirm the motility supersession resulting from the ZnO NWA, the control experiment of bacteria swimming on bared glass is conducted. A glass substrate with a roughness of about 1 nm in magnitude of order measured by atomic force microscopy is a good control with a completely different material nature to ZnO nanostructured substrates. The results are shown in Figure 2c, with the red and black dashed line, respectively, which illustrate the weak decay of bacteria motility. It can also be seen that the swimming speed on the glass surface is almost maintained during the total observation time, which clearly testifies the reduction in motility due to the presence of ZnO NWA. The mean speed with error bar (standard deviation SD = $\sqrt{\frac{\Sigma |V-\overline{V}|^2}{N}}$) over time is also calculated and showed in Figure 2d, presenting the speed dropping from 17.7 µm/s to 3.1 µm/s within 12 min. Therefore, the interactions between the nanoarrays and the swimming bacteria are expected to contribute to the drag forces and the slowing down of bacterial swimming. However, as a complex interface (not ideal flat and chemical homogeneity), ZnO NWA plays a complicated role in the motility reduction in the swimming bacteria when considering the possible steric force, van der Waals force, hydrodynamic attraction with the virtual image swimmer in theoretical modeling, and the dynamical bouncing force due to the direct mechanical contact between flagella and the nanowires.

3.2. Swimming on ZnO NWA in Response to UV Irradiation

Thanks to ZnO being a wide bandgap semiconductor, UV illumination contributes to the generation of electron–hole pairs. The excess electrons are able to activate possible photodynamic effects or increase the charge density on the surfaces of the ZnO NWA.

Therefore, motility suppression and then effective attachment are expected. Then, the bacterial motility on the ZnO NWA under UV irradiation is measured. During the observation of bacterial swimming, the UV light is kept running at all times. At the initial stage, shown in Figure 3a, the bacteria swim with curved trajectories, showing the run-and-tumble process similar to that in the absence of UV light. Thereafter, the extent of the sticking and motility reduction are increasingly enhanced. Within the first minute, the length of the trajectories is almost around 50 µm, much shorter than the displacements in the absence of UV light. After 12 min, almost all of the bacteria stick to the surface and could not escape from the surface again during the observation time. Overall, the expansion range of bacterial motion became significantly smaller than that without UV irradiation, as illustrated in Figure 3c, d with the speed distribution and the mean value. It is possible that bacteria become non-motile after introducing the UV irradiation since the fluorescence of the mutation *E. coli* disappeared.

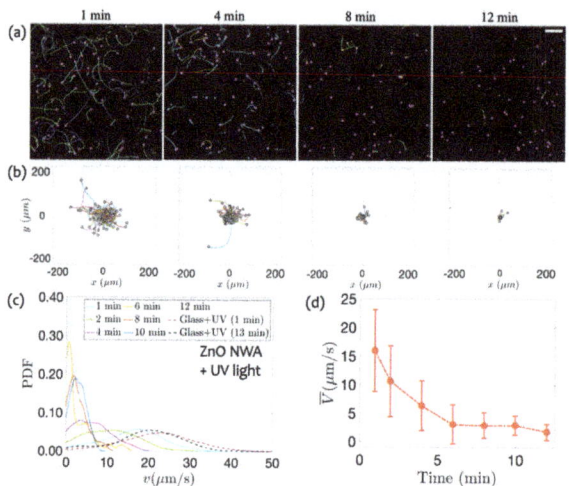

Figure 3. Bacterial swimming under UV irradiation on ZnO NWA. (a,b) Typical trajectories of swimming bacteria on NWA with varying time. (c,d) Speed distribution and mean speed at different time duration, respectively. A control experiment of bacteria swimming on bared glass under UV light is also plotted in (c). Scale bar is 30 µm.

To evaluate the effect of UV illumination on the swimming speed, a control experiment is carried out with UV light on bare glass under the same conditions. Figure 3c shows all the probability density distributions of the swimming speed on both the surfaces of glass and ZnO NWA under UV irradiation. It is confirmed that the UV light played a negligible role for those bacteria swimming on the glass surface without ZnO mediation, which will also be discussed in a later section.

3.3. Swimming on ZnO NPF with and without UV Radiation

To identify the influence of surface morphology on bacterial motility, a nanoparticle film of ZnO is used as a counterpart to the nanowire array. From the SEM measurements (Figure 1d, e), the NPF consisted of nanosized particles with a diameter of ~42.6 nm. Figure 4a shows the typical bacterial trajectories over time on the NPF without UV irradiation. As shown in Figure 4d, the movement range of bacteria during the tracked period can reach 200 µm in 1 min, with a typical value of 150 µm. At 9 min, only a few bacteria can spread around 100 µm. A long time later (37 min), the range of bacterial motion shrinks slightly. The mean swimming speed shown in Figure 4g, reaches a plateau in 13 min, and the mean speed drops to 5.6 µm/s in 37 min from the initial speed of 17.7 µm/s.

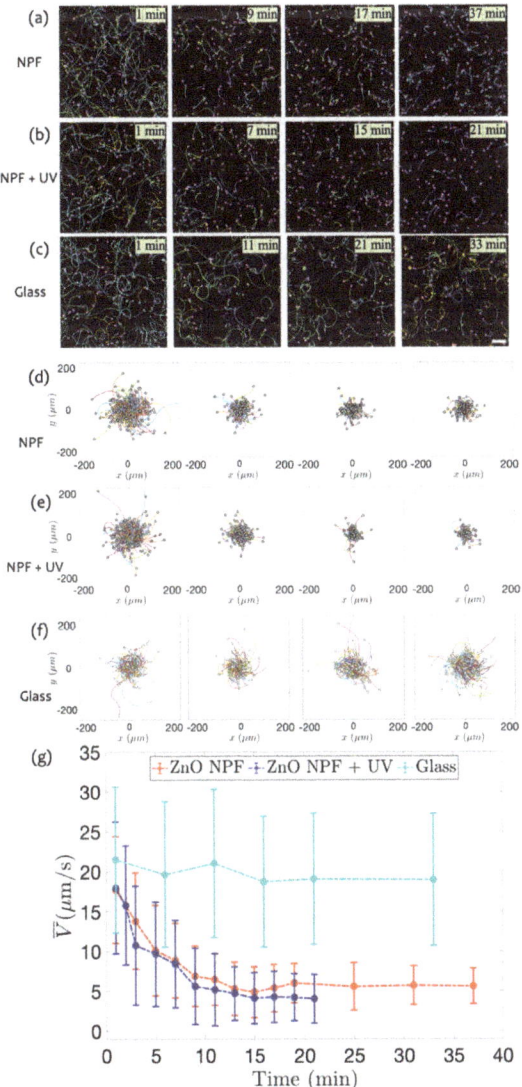

Figure 4. Swimming on nanoparticle film of ZnO. (**a**–**f**) Snapshots and trajectories of tracked bacteria on NPF and glass with and without UV irradiation, respectively. (**g**) Mean swimming speed on NPF with and without UV irradiation over time. The scale bar represents 30 µm.

Then, the bacteria motility on the ZnO NPF with UV irradiation is also measured and shown in Figure 4b,e. After 1 min, the motion range of bacteria during the tracking period can reach 200 µm, very close to that of those swimming on glass and ZnO NWA with and without UV light. Compared with the data of ZnO NWA under UV irradiation, the shift and contraction of peak positions are not sharp. The mean bacteria motility over time with an error bar, calculated by the standard deviation of each trajectory, is shown in Figure 4g, where a negligible decrease occurs, and the mean speed drops from 18.0 to 4.0 µm/s in 21 min. Most of the bacteria on ZnO NPF + UV lose their fluorescence after 21 min, resulting in the number of bacteria not reaching the minimum.

3.4. Comparison of Bacterial Motility

The anti-contamination efficiency of physical structures and chemical reactions due to the coupling of photon–electron and UV illumination is now evaluated. The experiments about the effect of UV irradiation are carried out on three types of substrates, including bared glass, ZnO NPF, and ZnO NWA. As shown in Figure 5a, there are basically no differences in the speed of the bacteria in the two groups of experiments on glass with and without UV illumination marked with the colors cyan and purple, respectively, indicating the negligible effect of the UV light for swimming bacteria on a glass surface. Here, the error bar and data after 15 min are hidden in Figure 5a for clearer alignment. The mean speed on the ZnO NPF shows a substantial reduction, implying a specific contribution from the material nature of ZnO. Furthermore, the motility of bacteria on the ZnO NWA is also slightly lower than that on the ZnO NPF, suggesting that the physical barriers created by roughness have a weak effect on bacterial motility. However, a distinct decay of mean speed appears on ZnO NWA with UV irradiation, i.e., the mean speed dropped to 3.5 µm/s after 6 min, which is faster than that without UV irradiation. The apparent enhancement of bacteriostatic ability illustrates that the roughness of the nanowire arrays plays a weak role from nanoparticle to nanoarray while significantly contributing to motility suppression due to chemical reactions by converting photonic energy to excess electrons.

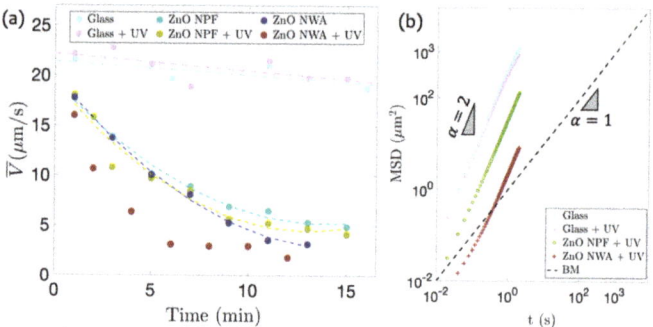

Figure 5. Mean speed and ballistic diffusion of bacterial spreading on various surfaces with and without UV irradiation. (a) Replot of bacterial motility as a function of time on ZnO NWA, ZnO NPF, and bared glass substrates, respectively. (b) Mean squared displacement depends on the lag time, showing the ballistic and super diffusion processes.

MSD (mean square displacement) $\langle |x(t) - x_0|^2 + |y(t) - y_0|^2 \rangle$ is used here to evaluate the spreading efficiency of bacteria on the surface based on the trajectories by the calculation of the time-evolving position $x(t)$ and $y(t)$. MSD is also ascribed to the motility of bacteria. As shown in Figure 5b, the MSD depending on lag time demonstrates a ballistic diffusion feature, i.e., MSD ~ t^2, which is reasonable by recognizing the bacteria as a type of self-propelled particles. As a reference to the Brownian motion of passive particles due to thermal agitation, polystyrene beads with a 1 µm diameter (comparable with the bacterial size) and with a density of 1.04 g/cm^3, which is also similar to that of the bacteria (1.04 g/cm^3), are used to present the spreading capability, as marked with a black dashed line in Figure 5b. In the case of ZnO NWA under UV irradiation, bacterial motility, characterized by an active ballistic diffusion process, is even smaller than that of pure Brownian particles at a short time range (less than 1 s). In other cases, bacteria behave with greater motility than Brownian particles do in the all-time window. The slopes of the four MSD curves are both greater than 1, indicating a super diffusion mode of bacterial swimming [41]. In other words, swimming bacteria, similar to typical, active Brownian particles, prefer straight motion compared to a random walk with a long-time limit.

Briefly, the motility of bacteria on ZnO was significantly inhibited compared to that on glass. This inhibition effect occurs immediately (within 1 min) after the transfer of bacteria onto ZnO nanostructures. The reduction in bacterial speed is supposed to be the reason that the ZnO nanomaterial is a typical polar material [42,43], and the oxygen vacancies on the surface can form a large number of equivalent positive charge centers [44–46], which is more likely to adsorb negatively charged E. coli [47]. In the absence of UV irradiation, the motility of bacteria on the ZnO NWA is slightly lower than that on ZnO NPF. It might be caused by the large undulation of the arrays, collisions, or flagella entanglement when E. coli swims close to the surface with a geometrical landscape. This is consistent with the more random swimming directions indicated by the trajectories of bacteria on the ZnO NWA, as shown in Figure 2.

The ZnO NWA presents a significant antibacterial effect under UV irradiation, and the corresponding enhancement effect is clear compared with results from measurements on the NPF. It is believed that the photodynamic-effect efficiency of the nanowire arrays with a large surface area comes to play an important role. The roughness is defined here as the ratio of the actual area to the projected area. Then the roughness of NWA is about 96.6, which is much larger than 2.6 of NPF. UV irradiation is adopted to generate hydroxyl radicals and singlet oxygen in the aqueous phase in the presence of ZnO, which has been proven to be curtailed [48–51], shown in the sketch of Figure 6. Compared with the insensitivity of the ZnO NPF to UV light, the higher roughness of the ZnO NWA leads to a significant improvement in the bacteriostatic effect.

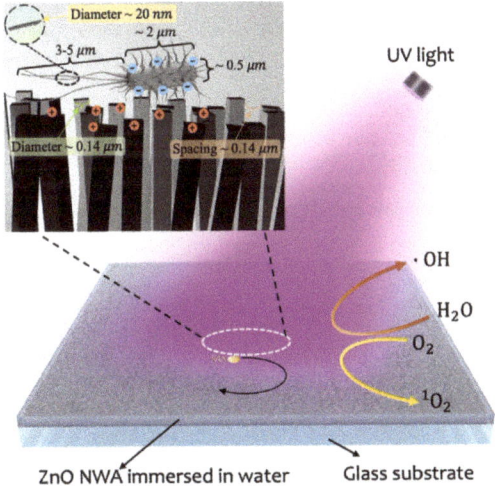

Figure 6. Sketch of bacterial swimming on ZnO NWA. Generation of hydroxyl radical and singlet oxygen are initiated in the presence of UV irradiation.

4. Conclusions

In summary, ZnO nanostructures are used to investigate the interfacial swimming behavior of E. coli by tuning surface textures and chemical reactions by remote undulation towards bacteriostatic capability. By visualizing the swimming trajectories of bacteria and analyzing their locomotion ability, the natural bacteriostatic ability of ZnO is quantitatively determined, showing that the average speed of bacteria can be reduced by 20% immediately after introducing ZnO nanostructures into the bacterial suspension. Increasing the roughness and the exposure of UV irradiation can further improve the antibacterial effect of the ZnO nanomaterials. Interestingly, when large roughness and UV irradiation are combined, the bacteriostatic effect is significantly enhanced. ZnO NWA with UV light irradiation reduces almost 90% of bacterial motility (slowing down to ~2 μm/s) within only 6 min. This highly effective bacteriostatic ability might originate from the combined effect of the polar nature of ZnO crystallization, the structural properties of the nanowire arrays, and the photodynamic effects. The strong trapping ability of ZnO NWA can strongly inhibit the migration of bacteria and prevent bacterial infection. Trapped bacteria become non-motile with possible bacteriostatic substances produced by photodynamic effects from ZnO material nature. The findings here not only demonstrate the detailed motility of bacteria swimming on nanostructured surfaces but also provide means for antibacterial applications of ZnO nanomaterials.

Author Contributions: Conceptualization, N.Y., H.L. and G.J.; Methodology, N.Y., H.L. and G.J.; Software, N.Y. and Y.L.; Investigation, N.Y., H.L. and G.J.; Formal analysis, N.Y., H.L., H.Y. and G.J.; Resources, N.Y.; Data Curation, N.Y.; Writing—Review and Editing, N.Y., H.L., Y.L. and G.J.; Supervision, N.Y., H.L. and G.J.; Project administration, N.Y., H.L., H.Y. and G.J.; Funding acquisition, G.J., Y.L., H.Y. and H.L. All authors have read and agreed to the published version of the manuscript.

Funding: This research received the funding from NSFC (12004308, 12174306, 11804275), and the special Scientific Research Program of Shaanxi Provincial Education Department (No. 203010036).

Institutional Review Board Statement: Not applicable.

Informed Consent Statement: Not applicable.

Data Availability Statement: Not applicable.

Acknowledgments: The authors thanks E. Clément for providing *E. coli* RP437 bacteria, and gratefully acknowledge the help from F. Teng.

Conflicts of Interest: The authors declare no conflict of interest.

References

1. Knights, H.E.; Jorrin, B.; Haskett, T.L.; Poole, P.S. Deciphering bacterial mechanisms of root colonization. *Environ. Microbiol. Rep.* **2021**, *13*, 428–444. [CrossRef] [PubMed]
2. Gensollen, T.; Iyer, S.S.; Kasper, D.L.; Blumberg, R.S. How colonization by microbiota in early life shapes the immune system. *Science* **2016**, *352*, 539–544. [CrossRef]
3. Siegel, S.J.; Weiser, J.N. Mechanisms of bacterial colonization of the respiratory tract. *Annu. Rev. Microbiol.* **2015**, *69*, 425–444. [CrossRef] [PubMed]
4. Ribet, D.; Cossart, P. How bacterial pathogens colonize their hosts and invade deeper tissues. *Microbes Infect.* **2015**, *17*, 173–183. [CrossRef] [PubMed]
5. Bosecker, K. Bioleaching: Metal solubilization by microorganisms. *FEMS Microbiol. Rev.* **1997**, *20*, 591–604. [CrossRef]
6. Kolencik, M.; Vojtkova, H.; Urik, M.; Caplovicova, M.; Pistora, J.; Cada, M.; Babicova, A.; Feng, H.; Qian, Y.; Ramakanth, I. Heterotrophic bacterial leaching of zinc and arsenic from artificial adamite. *Water Air Soil Pollut.* **2017**, *228*, 1–11. [CrossRef]
7. Vaccari, L.; Molaei, M.; Niepa, T.H.; Lee, D.; Leheny, R.L.; Stebe, K.J. Films of bacteria at interfaces. *Adv. Colloid Interface Sci.* **2017**, *247*, 561–572. [CrossRef]
8. Alapan, Y.; Yasa, O.; Yigit, B.; Yasa, I.C.; Erkoc, P.; Sitti, M. Microrobotics and microorganisms: Biohybrid autonomous cellular robots. *Annu. Rev. Control Robot. Auton. Syst.* **2019**, *2*, 205–230. [CrossRef]
9. Raina, J.-B.; Fernandez, V.; Lambert, B.; Stocker, R.; Seymour, J.R. The role of microbial motility and chemotaxis in symbiosis. *Nat. Rev. Microbiol.* **2019**, *17*, 284–294. [CrossRef]
10. Lauga, E. Bacterial hydrodynamics. *Annu. Rev. Fluid Mech.* **2016**, *48*, 105–130. [CrossRef]
11. Berg, H.C.; Turner, L. Chemotaxis of bacteria in glass capillary arrays. Escherichia coli, motility, microchannel plate, and light scattering. *Biophys. J.* **1990**, *58*, 919–930. [CrossRef]
12. Frymier, P.D.; Ford, R.M.; Berg, H.C.; Cummings, P.T. Three-dimensional tracking of motile bacteria near a solid planar surface. *Proc. Natl. Acad. Sci. USA* **1995**, *92*, 6195–6199. [CrossRef] [PubMed]
13. Frymier, P.D.; Ford, R.M. Analysis of bacterial swimming speed approaching a solid–liquid interface. *AIChE J.* **1997**, *43*, 1341–1347. [CrossRef]
14. Lauga, E.; DiLuzio, W.R.; Whitesides, G.M.; Stone, H.A. Swimming in circles: Motion of bacteria near solid boundaries. *Biophys. J.* **2006**, *90*, 400–412. [CrossRef]
15. Lopez, D.; Lauga, E. Dynamics of swimming bacteria at complex interfaces. *Phys. Fluids* **2014**, *26*, 400–412. [CrossRef]
16. Li, G.; Bensson, J.; Nisimova, L.; Munger, D.; Mahautmr, P.; Tang, J.X.; Maxey, M.R.; Brun, Y.V. Accumulation of swimming bacteria near a solid surface. *Phys. Rev. E* **2011**, *84*, 041932. [CrossRef]

17. Vigeant, M.A.-S.; Ford, R.M.; Wagner, M.; Tamm, L.K. Reversible and irreversible adhesion of motile Escherichia coli cells analyzed by total internal reflection aqueous fluorescence microscopy. *Appl. Environ. Microbiol.* **2002**, *68*, 2794–2801. [CrossRef] [PubMed]
18. Li, G.; Tam, L.-K.; Tang, J.X. Amplified effect of Brownian motion in bacterial near-surface swimming. *Proc. Natl. Acad. Sci. USA* **2008**, *105*, 18355–18359. [CrossRef]
19. Kantsler, V.; Dunkel, J.; Polin, M.; Goldstein, R.E. Ciliary contact interactions dominate surface scattering of swimming eukaryotes. *Proc. Natl. Acad. Sci. USA* **2013**, *110*, 1187–1192. [CrossRef]
20. Vigeant, M.; Ford, R.M. Interactions between motile Escherichia coli and glass in media with various ionic strengths, as observed with a three-dimensional-tracking microscope. *Appl. Environ. Microbiol.* **1997**, *63*, 3474–3479. [CrossRef]
21. Bos, R.; Van der Mei, H.C.; Busscher, H.J. Physico-chemistry of initial microbial adhesive interactions–its mechanisms and methods for study. *FEMS Microbiol. Rev.* **1999**, *23*, 179–230. [CrossRef]
22. Hermansson, M. The DLVO theory in microbial adhesion. *Colloids Surf. B Biointerfaces* **1999**, *14*, 105–119. [CrossRef]
23. Zhao, W.; Walker, S.L.; Huang, Q.; Cai, P. Adhesion of bacterial pathogens to soil colloidal particles: Influences of cell type, natural organic matter, and solution chemistry. *Water Res.* **2014**, *53*, 35–46. [CrossRef] [PubMed]
24. Bianchi, S.; Saglimbeni, F.; Frangipane, G.; Dell'Arciprete, D.; Di Leonardo, R. 3D dynamics of bacteria wall entrapment at a water–air interface. *Soft Matter* **2019**, *15*, 3397–3406. [CrossRef]
25. Kim, D.; Kim, Y.; Lim, S. Effects of swimming environment on bacterial motility. *Phys. Fluids* **2022**, *34*, 031907. [CrossRef]
26. Hu, J.; Wysocki, A.; Winkler, R.G.; Gompper, G. Physical sensing of surface properties by microswimmers–directing bacterial motion via wall slip. *Sci. Rep.* **2015**, *5*, 1–7. [CrossRef]
27. Gottenbos, B.; Grijpma, D.W.; van der Mei, H.C.; Feijen, J.; Busscher, H.J. Antimicrobial effects of positively charged surfaces on adhering Gram-positive and Gram-negative bacteria. *J. Antimicrob. Chemother.* **2001**, *48*, 7–13. [CrossRef]
28. Zheng, S.; Bawazir, M.; Dhall, A.; Kim, H.-E.; He, L.; Heo, J.; Hwang, G. Implication of surface properties, bacterial motility, and hydrodynamic conditions on bacterial surface sensing and their initial adhesion. *Front. Bioeng. Biotechnol.* **2021**, *9*, 82. [CrossRef]
29. Schwibbert, K.; Menzel, F.; Epperlein, N.; Bonse, J.; Krüger, J. Bacterial adhesion on femtosecond laser-modified polyethylene. *Materials* **2019**, *12*, 3107. [CrossRef]
30. Chien, H.-W.; Chen, X.-Y.; Tsai, W.-P.; Lee, M. Inhibition of biofilm formation by rough shark skin-patterned surfaces. *Colloids Surf. B Biointerfaces* **2020**, *186*, 110738. [CrossRef]
31. Tang, M.; Chen, C.; Zhu, J.; Allcock, H.R.; Siedlecki, C.A.; Xu, L.-C. Inhibition of bacterial adhesion and biofilm formation by a textured fluorinated alkoxyphosphazene surface. *Bioact. Mater.* **2021**, *6*, 447–459. [CrossRef] [PubMed]
32. Du, C.; Wang, C.; Zhang, T.; Yi, X.; Liang, J.; Wang, H. Reduced bacterial adhesion on zirconium-based bulk metallic glasses by femtosecond laser nanostructuring. *Proc. Inst. Mech. Eng. Part H J. Eng. Med.* **2020**, *234*, 387–397. [CrossRef] [PubMed]
33. Lutey, A.H.; Gemini, L.; Romoli, L.; Lazzini, G.; Fuso, F.; Faucon, M.; Kling, R. Towards laser-textured antibacterial surfaces. *Sci. Rep.* **2018**, *8*, 1–10. [CrossRef] [PubMed]
34. Yang, M.; Ding, Y.; Ge, X.; Leng, Y. Control of bacterial adhesion and growth on honeycomb-like patterned surfaces. *Colloids Surf. B Biointerfaces* **2015**, *135*, 549–555. [CrossRef] [PubMed]
35. Helbig, R.; Günther, D.; Friedrichs, J.; Rößler, F.; Lasagni, A.; Werner, C. The impact of structure dimensions on initial bacterial adhesion. *Biomater. Sci.* **2016**, *4*, 1074–1078. [CrossRef]
36. Chang, Y.-R.; Weeks, E.R.; Ducker, W.A. Surface topography hinders bacterial surface motility. *ACS Appl. Mater. Interfaces* **2018**, *10*, 9225–9234. [CrossRef]
37. Kolodziejczak-Radzimska, A.; Jesionowski, T. Zinc oxide—from synthesis to application: A review. *Materials* **2014**, *7*, 2833–2881. [CrossRef]
38. Hecht, A.; Endy, D.; Salit, M.; Munson, M.S. When wavelengths collide: Bias in cell abundance measurements due to expressed fluorescent proteins. *ACS Synth. Biol.* **2016**, *5*, 1024–1027. [CrossRef]
39. Yin, Z.; Shan, Y.; Yu, M.; Yang, L.; Song, J.; Hu, P.; Teng, F. Enhanced performance of UV photodetector based on ZnO nanorod arrays via TiO_2 as electrons trap layer. *Mater. Sci. Semicond. Processing* **2022**, *148*, 106813. [CrossRef]
40. Son, K.; Brumley, D.R.; Stocker, R. Live from under the lens: Exploring microbial motility with dynamic imaging and microfluidics. *Nat. Rev. Microbiol.* **2015**, *13*, 761–775. [CrossRef]
41. Wu, X.-L.; Libchaber, A. Particle diffusion in a quasi-two-dimensional bacterial bath. *Phys. Rev. Lett.* **2000**, *84*, 3017. [CrossRef] [PubMed]
42. Heiland, G.; Kunstmann, P. Polar surfaces of zinc oxide crystals. *Surf. Sci.* **1969**, *13*, 72–84. [CrossRef]
43. Ching, K.-L.; Li, G.; Ho, Y.-L.; Kwok, H.-S. The role of polarity and surface energy in the growth mechanism of ZnO from nanorods to nanotubes. *CrystEngComm* **2016**, *18*, 779–786. [CrossRef]
44. Luo, H.; Ma, J.; Wang, P.; Bai, J.; Jing, G. Two-step wetting transition on ZnO nanorod arrays. *Appl. Surf. Sci.* **2015**, *347*, 868–874. [CrossRef]
45. Pacchioni, G. Oxygen vacancy: The invisible agent on oxide surfaces. *ChemPhysChem* **2003**, *4*, 1041–1047. [CrossRef]
46. Scorza, E.; Birkenheuer, U.; Pisani, C. The oxygen vacancy at the surface and in bulk MgO: An embedded-cluster study. *J. Chem. Phys.* **1997**, *107*, 9645–9658. [CrossRef]
47. Güler, S.; Oruç, Ç. Comparison of the behavior of negative electrically charged E. coli and E. faecalis bacteria under electric field effect. *Colloids Surf. B Biointerfaces* **2021**, *208*, 112097. [CrossRef]

48. Yi, C.; Yu, Z.; Ren, Q.; Liu, X.; Wang, Y.; Sun, X.; Yin, S.; Pan, J.; Huang, X. Nanoscale ZnO-based photosensitizers for photodynamic therapy. *Photodiagnosis Photodyn. Ther.* **2020**, *30*, 101694. [CrossRef]
49. Jyothi, M.; Nayak, V.; Padaki, M.; Balakrishna, R.G.; Ismail, A. The effect of UV irradiation on PSf/TiO2 mixed matrix membrane for chromium rejection. *Desalination* **2014**, *354*, 189–199. [CrossRef]
50. Bono, N.; Ponti, F.; Punta, C.; Candiani, G. Effect of UV irradiation and TiO2-photocatalysis on airborne bacteria and viruses: An overview. *Materials* **2021**, *14*, 1075. [CrossRef]
51. Fakhar-e-Alam, M.; Kishwar, S.; Willander, M. Photodynamic effects of zinc oxide nanowires in skin cancer and fibroblast. *Lasers Med. Sci.* **2014**, *29*, 1189–1194. [CrossRef] [PubMed]

Article

ZnO Nanowire-Based Piezoelectric Nanogenerator Device Performance Tests

Linda Serairi [1,2] and Yamin Leprince-Wang [1,*]

1. Univ Gustave Eiffel, CNRS, ESYCOM, F-77454 Marne-la-Vallée, France
2. University of Montpellier, Department GEII, F-34000 Montpellier, France
* Correspondence: yamin.leprince@univ-eiffel.fr

Abstract: Over the past two decades, the quick development of wireless sensor networks has required the sensor nodes being self-powered. Pushed by this goal, in this work, we demonstrated a ZnO nanowire-array-based piezoelectric nanogenerator (NG) prototype, which can convert mechanical energy into electricity. High-quality single crystalline ZnO nanowires, having an aspect ratio of about 15, grown on gold-coated silicon substrate, were obtained by using a low-cost and low-temperature hydrothermal method. The NG-device fabrication process has been presented in detail, and the NG's performance has been tested in both compression and vibration modes. Peak power of 1.71 µW was observed across an optimal load resistance of 5 MΩ for the ZnO nanowires-based NG, with an effective area of 0.7 cm^2, which was excited in compression mode, at 9 Hz, corresponding to ~38.47 mW/cm^3 volume-normalized power output. The measured voltage between the top and bottom electrodes was 5.6 V. In vibration mode, at 500 Hz, the same device showed a potential of 1.4 V peak-to-peak value and an instantaneous power of 0.04 µW, corresponding to an output power density of ~0.9 mW/cm^3.

Keywords: ZnO nanowires; piezoelectric nanogenerator; VING nanogenerator; energy harvesting

1. Introduction

Recent advances in the internet of things (IoT) have stimulated the quick development of self-powered sensor nodes with characteristics of low power consumption [1,2]. Due to their disadvantages, such as limited lifespan, complicated maintenance, and environmental unfriendliness of battery materials, the conventional power supply technologies based on electrochemical batteries will not be promising way to sustainable development. This is why more and more research is focusing on energy harvesting systems as self-sustained power sources, by harvesting and transforming ambient energy into electricity, in order to realize the self-autonomous sensors [3,4].

The surrounding energy can be funded in different forms: in mechanical, thermal, and electromagnetic forms. Even though electromagnetic energy, especially solar radiation, represents a vast energy reservoir, mechanical energy sources are also largely presented in our daily life (such as vibrations, shocks, flows, etc.) and electromechanical conversion presents a definite advantage in dark or inaccessible environments, such as in some aeronautical, automotive, and underground applications. Extracting part of this energy from the ambient environment is particularly interesting at small-scales, because the system of reduced energy recovery could be directly integrated into the microsystem devices to make them semi-autonomous or autonomous. However, this energy must have two distinguishing features: (1) the abundance of its sources; (2) an environmentally friendly nature, meaning no nuisance or greenhouse gases will be generated during electricity production. In addition, the mechanical energy has a highly appreciable asset: it can be produced at the place of consumption, so there is no remote energy transfer, or loss due to the transfer. The conversion of mechanical energy into electrical energy can be realized via the piezoelectric properties of some materials, i.e., their ability to generate electricity when mechanical stress is applied to them.

Nowadays, the study and application of piezoelectric nanowires (NWs) attracts a lot of attention because of their interesting properties due to their small size. The advantage of using NWs for energy harvesting is not only related to their small size, which allows the reduction in the device size, but also to their sensitivity to the least mechanical stimulations.

In 2006, Z.L. Wang's research group at Georgia Tech (USA) demonstrated, for the first time, a piezoelectric nanogenerator based on ZnO NWs by recovering the mechanical energy to electricity [5]. The obtained results were very encouraging and created a desire, in researchers and specialists worldwide to develop this field for more than a decade. As a result, ZnO NWs have become one of the most popular materials for designing various NGs, due to their piezoelectric characteristics and biocompatibility, as well as their low-cost synthesis processes, with easy morphology control [6–9].

NGs based on piezoelectric NWs can be classified into two broad categories: NGs based on vertically oriented NWs (VINGs) [4,10], and NGs based on lateral NWs (LINGs) [4,11,12]. In general, in the case of the VINGs, the piezoelectric nanowires are intentionally synthesized vertically, on a conductive substrate, and the device is completed by an upper electrode. In the case of the LINGs, piezoelectric NWs are first transferred from growth substrates to a receiver substrate, and then fabrication of metallic contacts is undertaken, to ensure that the piezo potential can be coupled to an external load, when actuated by a bending moment [13]. Owing to the relative ease of fabrication and apparent effectiveness, vertically integrated nanogenerators (VINGs) are among the most extensively exploited structures in this field.

In this work, we demonstrated the characteristics of a VING: the NG was based on the vertical ZnO nanowire arrays grown on gold-coated silicon substrate. The energy harvesting tests were carried out both in compression mode at low frequency, and in vibrational mode, at a frequency range from 50 Hz to 3 kHz; the choice of frequency range was based on the aeronautic applications, by considering the different airplane vibrations.

2. Materials and Methods

2.1. ZnO Nanowire Synthesis

As part of this work, we used the hydrothermal method for the synthesis of ZnO nanowires. This is a low-cost, low-temperature method, consisting of two main phases. The first phase was the nucleation phase, which can be described in several steps:

Firstly, a metallized silicon substrate (Si/Pt or Si/Au) having a size of 1 cm × 1 cm was placed in an ultrasonic bath, containing a surfactant solution, for 10 min. Then, it was rinsed with distilled (DI) water (milli-Q quality, R > 18 MΩ), followed by drying with a hot air flow, followed by an annealing in the oven at 200 °C for 10 min, in order to remove any solvent residues from the substrate. Finally, the substrate was put in a plasma cleaner for 10 min, for a perfectly-cleaned surface.

Then, the buffer layer was prepared with dehydrated zinc acetate (ZnAc$_2$·2H$_2$O, VWR Norma Pur Analytical Reagent 99%) in solution, in absolute ethanol (EtOH$_{abs}$, VWR AnalaR Norma Pur 99.9%) at 0.01 M. The buffer layer deposition was carried out with the dip-coating method. To do this, we immersed the substrate vertically in a bath containing the buffer layer solution, and then removed it at a controlled rate. Finally, the substrate covered with the buffer layer was placed in an oven at 350 °C for 20 min. under ambient atmosphere. The ethanol evaporated and the zinc acetate decomposed, to create the ZnO-nanocrystallitesto form the nucleation sites. The equation below explains the decomposition process of zinc acetate, to form ZnO seeds, according to the reaction shown in Equation (1).

$$Zn(CH_3COO)_2 + \frac{1}{2}O_2 \rightarrow ZnO + 2CO_2 + 3H_2 \tag{1}$$

For a hydrothermal growth process, zinc nitrate (Zn(NO$_3$)$_2$, Sigma-Aldrich ACS Reagent 98%) and hexamethylenetetramine (HMTA) (C$_6$H$_{12}$N$_4$, Sigma-Aldrich ACS Reagent 99%) were used for ZnO nanowires growth. The growth solution was prepared from two separate solutions: one of 50 mM Zn(NO$_3$)$_2$, and another of 25 mM HMTA. Then both solutions, in equal volumes, were mixed at room temperature, to obtain the growth solution, with concentrations of 25 mM for Zn(NO$_3$)$_2$ and 12.5 mM for HMTA.

The substrate with the buffer layer was then immersed face down, in a Teflon bottle containing the growth solution, which had been preheated to 90 °C in a conventional oven. The different chemical reactions that take place during the process of ZnO nanowire growth can be described by the following equations [9,14]:

$$(CH_2)_6 N_4 + 6H_2O \leftrightarrow 4NH_3 + 6HCHO \quad (2)$$

$$4NH_3 + 4H_2O \leftrightarrow 4NH_4^+ + 4OH^- \quad (3)$$

$$Zn(NO_3)_2 \leftrightarrow Zn^{2+} + 2NO_3^- \quad (4)$$

$$Zn^{2+} + 2OH^- \leftrightarrow Zn(OH)_2 \quad (5)$$

$$Zn(OH)_2 \leftrightarrow ZnO + H_2O \quad (6)$$

Although the exact function of HMTA during the growth of ZnO nanowires is not yet fully understood, it is generally considered a weak base, which will be slowly hydrolyzed in water, to progressively produce OH$^-$. This step is important in the synthesis process of ZnO nanowires, because if the HMTA hydrolyzes too rapidly and produces too much OH$^-$ in a short time, the Zn^{2+} ions in the solution will precipitate too quickly, by forming Zn(OH)$_2$. This precipitation will lead to quick consumption of the Zn^{2+} ions, which will reduce the nanowire growth kinetic. That is why we chose the 2:1 ratio between zinc nitrate and HMTA, instead of equimolar, which can often be seen in the literature [15,16].

After the growth of the ZnO nanowires, the substrate was rinsed by DI water, to remove the possible residues from the intermediate reactions, and was then dried in an oven at 80 °C for 15 min. The morphology of ZnO NWs was characterized using a scanning electron microscope (FEG–SEM, NEON 40 ZEISS, Oberkochen, Germany), operating at 10 kV accelerating voltage. Figure 1a,b shows both planar- and cross-section-view SEM images, from one of our samples exhibiting a typical morphology of our samples: a homogeneous ZnO nanowire array, in which the diameter varies between 35 and 40 nm and the length of the nanowires is approximately 0.9 µm (for 2-h growth time).

The X-ray diffraction (XRD) analysis was performed on a PANalytical X'Pert PRO MPD diffractometer (Almelo, The Netherlands) using Co Kα radiation (λ = 0.17890 nm) in θ-2θ configuration, equipped with a diffracted beam monochromator. Figure 1c shows a typical XRD pattern of a sample of ZnO nanowires on a gilded silicon substrate. We can clearly identify that the synthesized nanowires crystallized according to the compact hexagonal phase B4 (Würtzite type), with a preferential growth towards the c-axis. The high-resolution transmission electron microscopy (HRTEM) observation, performed on a FEI TECNAI G^2 F20 operating at 200 kV (Hillsboro, OR, USA), confirms this preferential directional growth and shows excellent monocrystallinity of our nanowires, obtained from an easy, and low-cost, hydrothermal method (Figure 1d).

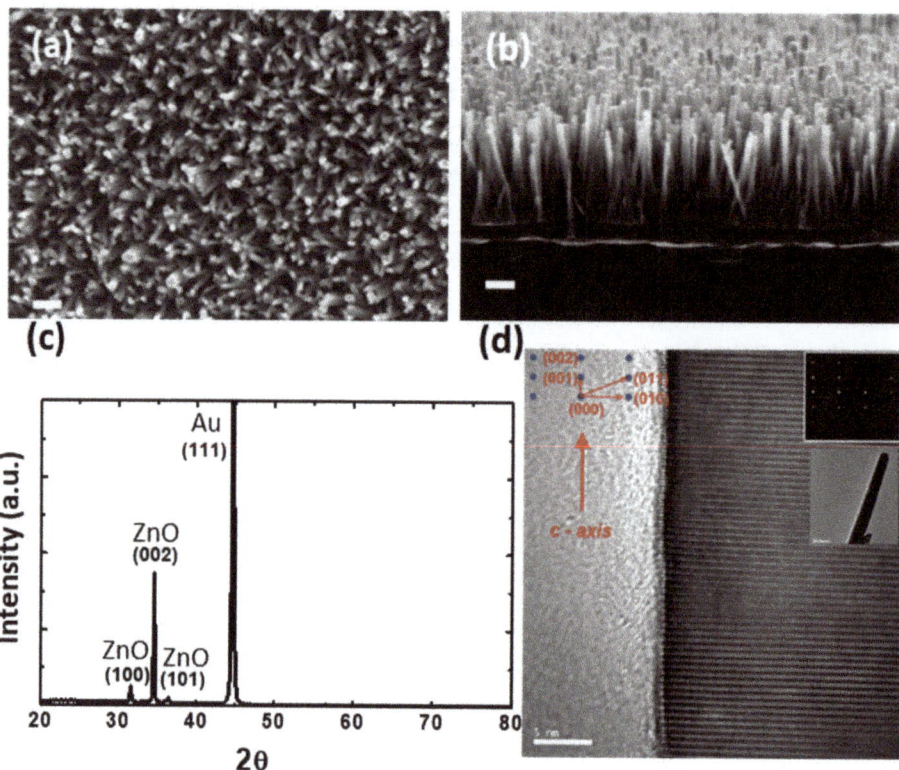

Figure 1. Representative SEM images show: (**a**) top view; (**b**) side view of as-grown vertically orientated ZnO NW arrays on the gilded Si substrates (scale bar: 200 nm); (**c**) XRD analysis of hydrothermally-grown ZnO NWs on gilded Si substrates; and (**d**) HRTEM image of a representative ZnO NW.

2.2. ZnO-Nanowire-Based Nanogenerator Device Fabrication

For ZnO NW-based nanogenerator fabrication, we used the ZnO NW sample with 4 h growth time, in order to obtain suitable nanowires for the NG devices: L ~ 1.1 µm, D ~ 45 nm with aspect ratio α ~ 25. Then, to protect the ZnO NWs during the experimental test phase, and to electrically isolate between the top metallic electrode and the Au layer of substrate (the bottom electrode), the substrate, covered with a vertical network of ZnO NWs, was encapsulated in a matrix of PMMA (poly-methyl methacrylate, Acros Organics, Illkirch, France), thus forming a ZnO/PMMA composite layer. To do this, the PMMA 495 A4 solution (MMA 495K MW in 4% Anisol) was spin-coated on the surface of the nanowires, leading to a penetration of PMMA into the volume of nanowires, as shown in Figure 2. It is worth mentioning that this process is quite difficult to perform, we had to carry out multiple encapsulations to successfully develop an optimized process. In fact, there are no standard spin-coating parameters to apply, because it depends on the ZnO NW's density, even with a fixed PMMA solution viscosity.

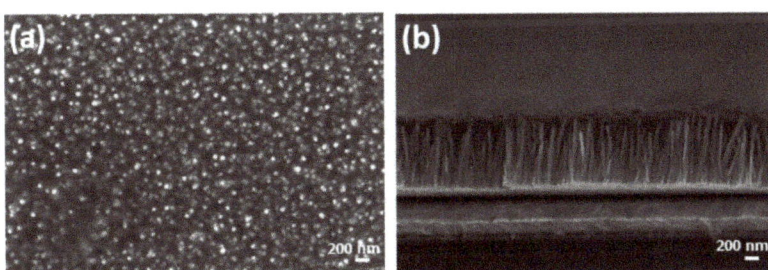

Figure 2. ZnO nanowires after PMMA deposition, (**a**) top view and (**b**) side view of as-grown, vertically orientated ZnO NW arrays.

Figure 3 illustrates the ZnO NWs-based NG device-fabrication method. After PMMA spin-coating (step 2), an aluminum (Al) layer of 700 nm thickness, as the top contact electrode, was evaporated on the upper sample surface using a stainless-steel shadow mask (step 3). After inserting connecting wires (step 4), the NG was encapsulated with a PDMS layer, for protection and robustness during the different tests (step 5). A photograph of the final device has been also presented in Figure 3. It should be emphasized that the obtained device Al/ZnO/Au forms a Schottky contact, as studied both in our previous work [17] and in literature [18,19].

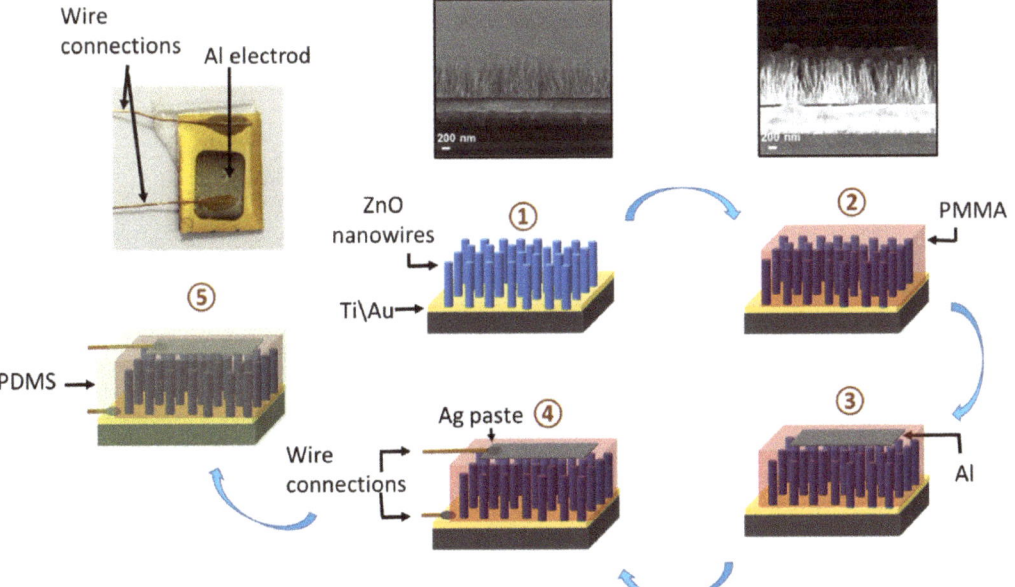

Figure 3. Schematic showing the device and reference sample assembly, along with a photograph of a completed device.

3. Results and Discussions

The measurement apparatus was based on a custom-built test bench, carefully designed to minimize anomalous signals that can occur due to effects other than that of the piezoelectric effect. In the first case, the NG was tested under vibration mode, so to perform this, the NG was placed on a shaker, which acted as an actuator and transmitted the vibrations to the NG, with controlled frequencies and accelerations, through a vibration controller. During the measurement, we used a voltage-follower amplifier, which is characterized by a very large input impedance and a low output impedance, in order to have maximum voltage transmitted and recovered. The test bench was connected to a data acquisition system, to view and record the data.

To assess the impact of resistive loading on the measured NG output characteristics, we carried out a series of experiments, by varying the load resistors (R_L) between 1 MΩ and 1 GΩ, and by applying a fixed vibrational frequency of 500 Hz to the NG. Figure 4a presents the experimental data resulting from the various R_L, in which it can be clearly observed that the output voltages for the understudied device increase with increasing R_L value and saturate at a maximal value of about 720 mV for R_L = 30 MΩ. The generated signal has a peak-to-peak value of about 1.4 V, as shown in Figure 4b. Therefore, we can conclude that 30 MΩ represents the optimal value of the load resistance for our NG.

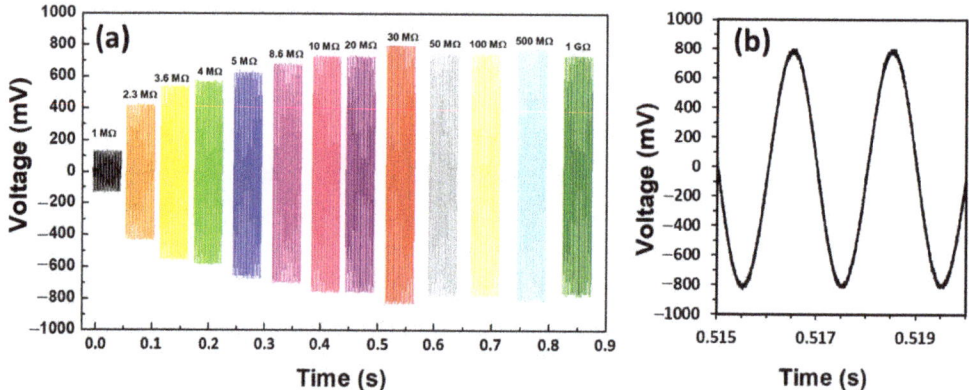

Figure 4. (a) Voltage response according to load resistance R_L, under vibration mode, with f = 500 Hz. (b) The generated signal, with a peak-to-peak value of about 1.4 V.

With these considerations, we can determine the instantaneous power generated by our device, using the following relationships:

$$P_i = \frac{(V_{RMS})^2}{R_L}; \text{ with } V_{RMS} = \frac{V_{max}}{\sqrt{2}} \tag{7}$$

From the measured values of maximal voltage "V_{max}" as a function of load resistance R_L, we can plot the variation of the instantaneous power (P_i) as a function of the R_L (Figure 5). We found that the NG delivered a maximal instantaneous output power of 0.04 µW, corresponding to an R_L value of about 4 MΩ; the corresponding power density value was about 0.9 mW/cm^3. To give a comparison, a similar work on ZnO NWs-based NG, from Dahiya et al. [20], generated an output power density of ~16 µW/cm^3, with a potential peak-to-peak value of 2 V.

Next, we studied the performance of our NG in the frequency range corresponding to aircraft vibration, in order to be coherent with future potential applications of NG as an energy source for self-powered sensors. Figure 6a (black curve) shows the variation of the instantaneous power of the NG as a function of the frequency from 50 Hz to 3 KHz, for a load resistor of 4 MΩ (optimized value). We noticed that the instantaneous power increased exponentially with the vibrational frequency. However, the energy per cycle had a different behavior, as shown in Figure 6b (black curve). In fact, its value increased very quickly with the frequency before 300 Hz and reached the maximum energy between 300 and 1200 Hz, and then it decreased for the frequency above 1200 Hz. From this observation, we can deduce that the optimal range of vibrational frequency for our NG is between 300 and 1200 Hz. It is worth highlighting that the maximal aircraft vibration energy is approximatively located between 500 Hz and 1000 Hz. Thus, our NG response is quite coherent with this application.

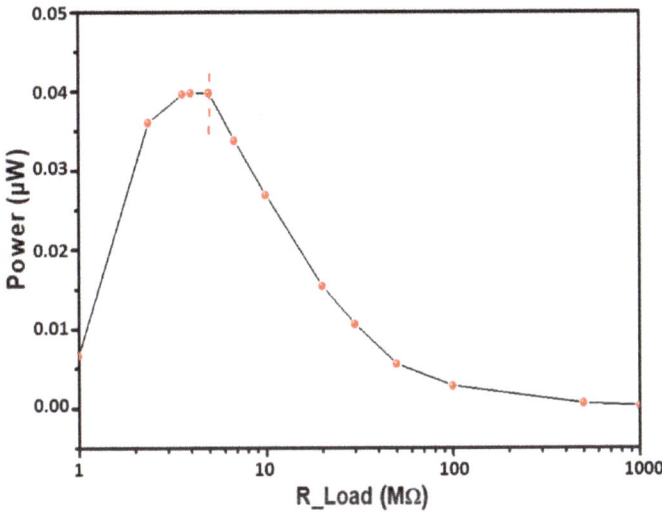

Figure 5. Estimated instantaneous power as a function of load resistance.

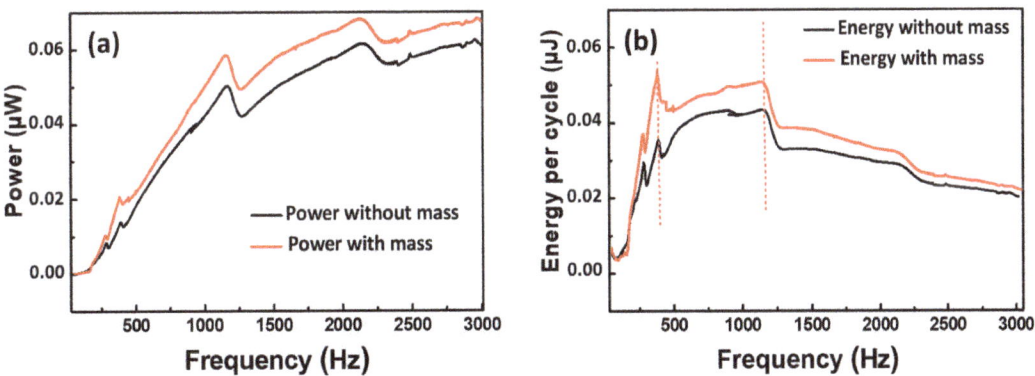

Figure 6. (a) Variation of instantaneous power, with and without mass (2 g). (b) Variation of energy per cycle as function of vibrational frequency, with and without mass (2 g).

After defining the optimal load resistor and vibrational frequency for the optimal working conditions of our ZnO nanowire-based NG, we were interested in improving its performance, by adding a mass of 2 g on the top of the NG. This mass could increase the compressive constraint on the ZnO NWs and decrease the value of the resonance frequency of the device. We knew that we did not want to go beyond 2 g, to avoid making the device cumbersome for later applications, while maintaining the same range of excitation frequency from 50 Hz to 3 KHz. Figure 6a (red curve) illustrates the variation of the instantaneous power as a function of the vibrational frequency. By comparing the power curves before and after the mass addition, we can see that a small improvement has been obtained for the NG output power, but not in a significant way (15%). Similar behavior was also observed in energy per cycle (see Figure 6b) (red curve).

In the second case, the NG was tested under compression mode. The NG was placed in front of a shaker which acted as an actuator (by applying a compression force on the

NG), with a fixed low frequency, for potential applications under low frequency motion, such as body movement. The applied force was measured by a force sensor, linked on the shaker, as shown in Figure 7.

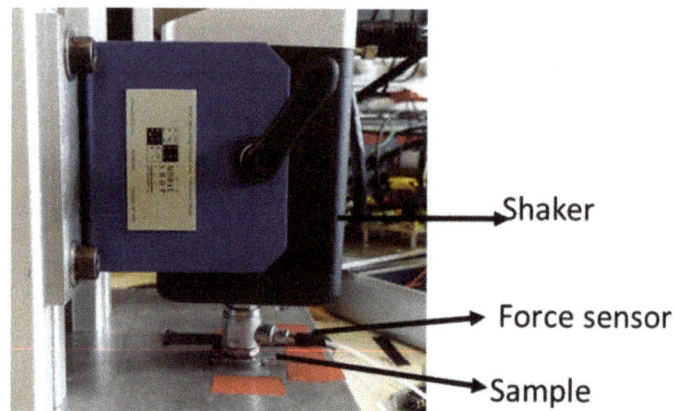

Figure 7. Experimental setup for the NG test in compression mode.

Figure 8a shows the voltage generated by the NG, based on the ZnO nanowire array, synchronized with the force applied by the actuator (Figure 8b) (calculated from the force sensor). The device withstood a large number of cycles, with a frequency of 9 Hz and a maximum force of about 6 N. It is important to note here that these forces, applied by the actuator to deform the NG device, did not induce significant damage to the polymer matrix, the electrical contacts, or the silicon substrate. No cracking or degradation was observed by optical microscopy after numerous cycle tests.

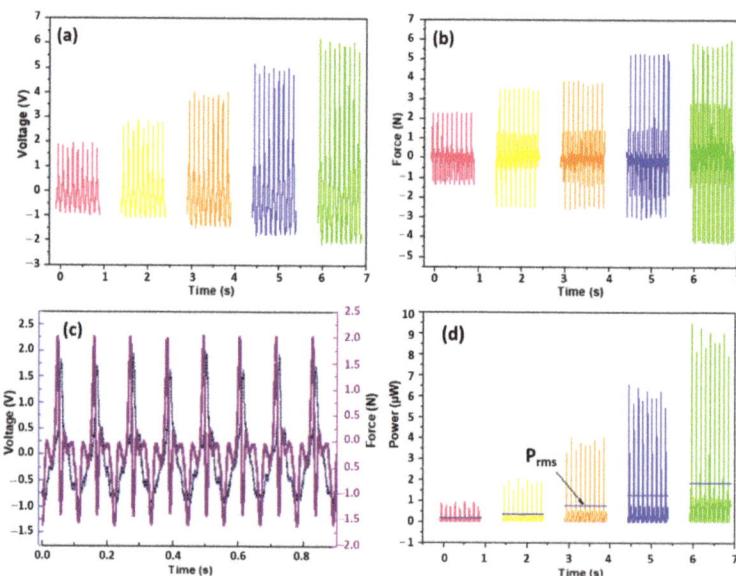

Figure 8. Performance obtained from a ZnO nanowire-array-based piezo-nanogenerator. (**a**) Potential generated under a compressive force. (**b**) Corresponding forces applied on the NG. (**c**) A zoom on the applied force. (**d**) Power obtained from NG under different applied forces.

Note that, under these biased conditions, the nanowires were compressed. Nanowires generate a negative signal when the NG is compressed and a reverse signal when released. The profile of the generated signal perfectly followed the profile of the applied force (Figure 8c). The piezo-nanogenerator delivered an alternating voltage, both during compression of the device and during its relaxation. This behavior is in concordance with the piezo-nanogenerators found in the literature [21–24].

Figure 8d presents the powers generated under different excitation forces. In this case, the NG generated a mean power higher than in the previous case (in vibration mode), with an effective power of 1.71 µW for a compression force of 6 N. This is due to a more important stress having been applied towards the c-axis of the nanowires in the compression–excitation mode.

4. Conclusions

In this work, we have studied the electrical signal produced by a piezoelectric nanogenerator, incorporating vertically-aligned piezoelectric ZnO NWs encapsulated in a polymer matrix. ZnO NWs were hydrothermally synthesized, showing high crystalline quality and homogeneous morphology. PMMA was used as a matrix material, to physically support the ZnO NWs, and as a dielectric layer on top of the NWs. The theory proposed to describe the ZnO NWs-based nano-generator operation is that the dielectric material acts as an effective barrier layer for induced charges in metal electrodes, while providing electrical isolation between adjacent NWs. The devices assembled here employed a ~1.2 µm thick PMMA layer over the top of the NWs, which offered effective electrical isolation for the top electrode. Thinner layers proved challenging when trying to electrically isolate the two conductive layers.

This work was devoted to the design and study of the performance of an NG based on ZnO nanowire arrays (NWs: ~1.1 µm length and ~45 nm diameter). The device was tested in two different modes: firstly, in vibration mode, then in compression mode. In the vibration mode, the experimental tests demonstrated the best working conditions for our NG, including the load resistance R_L at 4 MΩ and the range of vibrational frequency between 300 and 1200 Hz. Our ZnO NWs-based NG produced a potential of 1.4 V peak-to-peak value, and an instantaneous power of 0.04 µW (output power density ~0.9 mW/cm^3). In the compression mode, with the same R_L value, a larger effective power of 1.71 µW (output power density ~38.47 mW/cm^3) was obtained at a low frequency of 9 Hz, with an applied force of 6 N. From these results, it can be concluded that the VINGs are more sensitive to the compression mode than the vibration mode.

Author Contributions: Conceptualization, L.S. and Y.L.-W.; methodology, L.S. and Y.L.-W.; software, L.S.; validation, Y.L.-W.; writing—original draft preparation, L.S. and Y.L.-W.; writing—review and editing, Y.L.-W.; supervision, Y.L.-W.; project administration, Y.L.-W.; funding acquisition, Y.L.-W. All authors have read and agreed to the published version of the manuscript.

Funding: This research was funded by Ile-de-France region within the FUI 15 Program.

Institutional Review Board Statement: Not applicable.

Informed Consent Statement: Not applicable.

Data Availability Statement: Not applicable.

Acknowledgments: The authors would like to thank Philippe BASSET and Yingxian LU for their help on materials used for experiments.

Conflicts of Interest: The authors declare no conflict of interest.

References

1. Wang, Z.L. Nanogenerators, self-powered systems, blue energy, piezotronics and piezo-phototronics—A recall on the original thoughts for coining these fields. *Nano Energy* **2018**, *54*, 477–483. [CrossRef]
2. Li, Z.; Zheng, Q.; Wang, Z.L.; Li, Z. Nanogenerator-Based Self-Powered Sensors for Wearable and Implantable Electronics. *Research* **2020**, *2020*, 1–25. [CrossRef] [PubMed]

3. Hu, Y.; Lin, L.; Zhang, Y.; Wang, Z.L. Replacing a Battery by a Nanogenerator with 20 V Output. *Adv. Mater.* **2011**, *24*, 110–114. [CrossRef] [PubMed]
4. Xu, S.; Qin, Y.; Xu, C.; Wei, Y.; Yang, R.; Wang, Z.L. Self-powered nanowire devices. *Nat. Nanotechnol.* **2010**, *5*, 366–373. [CrossRef]
5. Wang, Z.L.; Song, J. Piezoelectric Nanogenerators Based on Zinc Oxide Nanowire Arrays. *Science* **2006**, *312*, 242–246. [CrossRef]
6. Wang, Z.L. Piezoelectric Nanostructures: From Growth Phenomena to Electric Nanogenerators. *MRS Bull.* **2007**, *32*, 109–116. [CrossRef]
7. Xu, L.; Guo, Y.; Liao, Q.; Zhang, A.J.; Xu, D. Morphological Control of ZnO Nanostructures by Electrodeposition. *J. Phys. Chem. B* **2005**, *109*, 13519–13522. [CrossRef]
8. Mezy, A.; Gerardin, C.; Tichit, D.; Ravot, D.; Suwanboon, S.; Tedenac, J.-C. Morphology control of ZnO nanostructures. *J. Ceram. Soc. Jpn.* **2008**, *116*, 369–373. [CrossRef]
9. Tong, Y.; Liu, Y.; Dong, L.; Zhao, D.; Zhang, J.; Lu, Y.; Shen, A.D.; Fan, X. Growth of ZnO Nanostructures with Different Morphologies by Using Hydrothermal Technique. *J. Phys. Chem. B* **2006**, *110*, 20263–20267. [CrossRef]
10. Wang, X.; Song, J.; Liu, J.; Wang, Z.L. Direct-Current Nanogenerator Driven by Ultrasonic Waves. *Science* **2007**, *316*, 102–105. [CrossRef]
11. Yang, R.; Qin, Y.; Dai, L.; Wang, Z.L. Power generation with laterally packaged piezoelectric fine wires. *Nat. Nanotechnol.* **2008**, *4*, 34–39. [CrossRef] [PubMed]
12. Qin, Y.; Wang, X.; Wang, Z.L. Microfibre–nanowire hybrid structure for energy scavenging. *Nature* **2008**, *451*, 809–813. [CrossRef] [PubMed]
13. Hu, Y.; Zhang, Y.; Xu, C.; Zhu, G.; Wang, Z.L. High-Output Nanogenerator by Rational Unipolar Assembly of Conical Nanowires and its Application for Driving a Small Liquid Crystal Display. *Nano Lett.* **2010**, *10*, 5025–5031. [CrossRef] [PubMed]
14. Xu, C.; Shin, P.; Cao, L.; Gao, D. Preferential Growth of Long ZnO Nanowire Array and Its Application in Dye-Sensitized Solar Cells. *J. Phys. Chem. C* **2009**, *114*, 125–129. [CrossRef]
15. Bai, S.-N.; Wu, S.-C. Synthesis of ZnO nanowires by the hydrothermal method, using sol–gel prepared ZnO seed films. *J. Mater. Sci. Mater. Electron.* **2010**, *22*, 339–344. [CrossRef]
16. Cao, Z.; Wang, Y.; Li, Z.; Yu, N. Hydrothermal Synthesis of ZnO Structures Formed by High-Aspect-Ratio Nanowires for Acetone Detection. *Nanoscale Res. Lett.* **2016**, *11*, 347–360. [CrossRef]
17. Brouri, T.; Leprince-Wang, Y. Schottky junction study for electrodeposited ZnO thin films and nanowires. *Eur. Phys. J. Appl. Phys.* **2014**, *68*, 10401. [CrossRef]
18. Dahiya, A.S.; Morini, F.; Boubenia, S.; Nadaud, K.; Alquier, D.; Poulin-Vittrant, G. Organic/Inorganic Hybrid Stretchable Piezoelectric Nanogenerators for Self-Powered Wearable Electronics. *Adv. Mater. Technol.* **2017**, *3*, 1700249. [CrossRef]
19. Semple, J.; Rossbauer, S.; Anthopoulos, T.D. Analysis of Schottky Contact Formation in Coplanar Au/ZnO/Al Nanogap Radio Frequency Diodes Processed from Solution at Low Temperature. *ACS Appl. Mater. Interfaces* **2016**, *8*, 23167–23174. [CrossRef]
20. Dahiya, A.S.; Morini, F.; Boubenia, S.; Justeau, C.; Nadaud, K.; Rajeev, K.P.; Alquier, D.; Poulin-Vittrant, G. Zinc oxide nanowire-parylene nanocomposite based stretchable piezoelectric nanogenerators for self-powered wearable electronics. *J. Phys. Conf. Ser.* **2018**, *1052*, 012028. [CrossRef]
21. Opoku, C.; Dahiya, A.S.; Oshman, C.; Cayrel, F.; Poulin-Vittrant, G.; Alquier, D.; Camara, N. Fabrication of ZnO Nanowire Based Piezoelectric Generators and Related Structures. *Phys. Procedia* **2015**, *70*, 858–862. [CrossRef]
22. Kathalingam, A.; Rhee, J.-K. ZnO nanowire based piezo and photoelectric effects coupled nanogenerator. In Proceedings of the International Conference on Advanced Nanomaterials and Emerging Engineering Technologies, Chennai, India, 24–26 July 2013; pp. 419–423. [CrossRef]
23. Hansen, B.J.; Liu, Y.; Yang, R.; Wang, Z.L. Hybrid Nanogenerator for Concurrently Harvesting Biomechanical and Biochemical Energy. *ACS Nano* **2010**, *4*, 3647–3652. [CrossRef] [PubMed]
24. Hinchet, R.; Lee, S.; Ardila, G.; Montès, L.; Mouis, M.; Wang, Z.L. Performance Optimization of Vertical Nanowire-based Piezoelectric Nanogenerators. *Adv. Funct. Mater.* **2013**, *24*, 971–977. [CrossRef]

Article

Effects of Waste-Derived ZnO Nanoparticles against Growth of Plant Pathogenic Bacteria and Epidermoid Carcinoma Cells

Titiradsadakorn Jaithon [1], Jittiporn Ruangtong [1], Jiraroj T-Thienprasert [2,3] and Nattanan Panjaworayan T-Thienprasert [1,*]

1. Department of Biochemistry, Faculty of Science, Kasetsart University, Bangkok 10900, Thailand; titiradsadakorn.j@ku.th (T.J.); jittiporn.rt@gmail.com (J.R.)
2. Department of Physics, Faculty of Science, Kasetsart University, Bangkok 10900, Thailand; chorawut.t@ku.th
3. Thailand Center of Excellence in Physics, Ministry of Higher Education, Science, Research and Innovation, Bangkok 10400, Thailand
* Correspondence: fscinnp@ku.ac.th; Tel.: +66-86-789-9433

Abstract: Green synthesis of zinc oxide nanoparticles (ZnO NPs) has recently gained considerable interest because it is simple, environmentally friendly, and cost-effective. This study therefore aimed to synthesize ZnO NPs by utilizing bioactive compounds derived from waste materials, mangosteen peels, and water hyacinth crude extracts and investigated their antibacterial and anticancer activities. As a result, X-ray diffraction analysis confirmed the presence of ZnO NPs without impurities. An ultraviolet–visible absorption spectrum showed a specific absorbance peak around 365 nm with an average electronic band gap of 2.79 eV and 2.88 eV for ZnO NPs from mangosteen peels and a water hyacinth extract, respectively. An SEM analysis displayed both spherical shapes of ZnO NPs from the mangosteen peel extract (dimension of 154.41 × 172.89 nm) and the water hyacinth extract (dimension of 142.16 × 160.30 nm). Fourier transform infrared spectroscopy further validated the occurrence of bioactive molecules on the produced ZnO NPs. By performing an antibacterial activity assay, these green synthesized ZnO NPs significantly inhibited the growth of *Xanthomonas oryzae* pv. *oryzae*, *Xanthomonas axonopodis* pv. *citri*, and *Ralstonia solanacearum*. Moreover, they demonstrated potent anti-skin cancer activity in vitro. Consequently, this study demonstrated the possibility of using green-synthesized ZnO NPs in the development of antibacterial or anticancer agents. Furthermore, this research raised the prospect of increasing the value of agricultural waste.

Keywords: green synthesis; ZnO nanoparticles; antibacterial activity; anticancer activity; fruit peels; water hyacinth

1. Introduction

Zinc oxide nanoparticles (ZnO NPs) are widely used metal oxides due to their dominant properties including photocatalytic properties [1], antimicrobial activity [2–4], anti-inflammatory response [5], anti-cancer activity [6], thermal stability [7], and biocompatibility [8]. Hence, ZnO NPs are applied in a variety of industries, such as agriculture, food processing, cosmetics, medicine, and textiles [9–11].

Chemical and physical approaches are the most common ways to synthesize ZnO NPs [12]. The term "green synthesis" has recently gained popularity due to its simplicity, environmental friendliness, and economic effectiveness. Green synthesis utilizes phytochemicals present in plants, microbes, and fruit peel waste for the bioreduction of metal ions to their corresponding nanoparticles [13,14], which have a variety of particle sizes and shapes [15]. Previous research has shown that using higher concentrations of banana peel extracts resulted in a higher purity of green-produced ZnO NPs with less zinc hydroxide production [16], implying that the phytochemicals present in the extracts act as metal capping agents and/or reducing agents [17]. Subsequently, plant extracts have been proposed to play a role in three different phases of synthesis: (i) reducing agents for the bioreduction of metal ions or metal salts; (ii) the nanoparticle growth phase; and (iii) stabilizing agents for nanoparticles' final shapes [18,19]. Examples of fruit or vegetable peel extracts that have previously been reported to synthesize ZnO NPs are banana peels [16], citrus peels (orange, lemon, and grape) [20], drumstick peels [21], and onion peels [22].

Mangosteen (*Garcinia mangostana* L.) peels contain abundant amounts of many phytochemicals such as xanthones, flavonoids, tannins, and anthocyanins [23] and have been demonstrated in vitro to possess antioxidant, anti-inflammatory, antiallergy, antibacterial, and anticancer activities [24,25]. Notably, phytochemicals such as phenolic compounds, terpenoids, alkaloids, vitamins, amino acids, proteins, and glycosides have been used as reducing, capping, and stabilizing agents in nanoparticle formation in a green synthesis method [26–28]. Furthermore, bioactive compounds such as alkaloids, terpenoids, steroids, glycosides, phenols, and flavonoids have also been discovered in water hyacinth (*Eichhornia crassipes*) [29], which is the world's worst invasive aquatic weed [30].

The goal of this study was to utilize a green synthesis approach to synthesize ZnO NPs from crude extracts of mangosteen peels and water hyacinth. Subsequently, antibacterial activity against plant pathogen diseases including *Xanthomonas oryzae* pv. *oryzae* (rice bacterial blight pathogen), *Xanthomonas axonopodis* pv. *citri* (citrus canker disease), and *Ralstonia solanacearum* (bacterial wilt disease) and anticancer activity against epidermoid cancer cells (A431) of newly synthesized ZnO NPs were investigated.

2. Materials and Methods

2.1. Preparation of Crude Extracts

In May, mangosteen peels were collected from households in Phayao District, Phayao province, Northern Thailand. Freshwater hyacinth was collected from the canal in Bann Na District, Nakhon Nayok, Thailand. All samples were shed dried and ground into powder. Next, 2 g of ground mangosteen peels were extracted in 500 mL of deionized water at room temperature for 30 min. Then, a crude extract was prepared as described in [6]. For water hyacinth extraction, the preparation was performed as described in [31]. After filtration, the crude extracts were refrigerated at 4 °C until next use.

2.2. Green Synthesis of ZnO NPs

For the synthesis of ZnO NPs from mangosteen peels, 50 mL of 2 M zinc acetate solution ($Zn(CH_3COO)_2$, LOBA chemie, Maharashtra, India) was prepared in deionized water for 20 min at room temperature with continual stirring. After that, each burette was filled with 100 mL of the crude extract and 100 mL of the precursor, which were then dropped into a beaker dropwise. The mixed solution was kept stirring for 1 h at room temperature. The mixture was then mixed dropwise with 2 M of NaOH (KEMAUS, Sydney, Australia) until it reached pH 12 and further agitated for 1 h. The mixture was further centrifuged for 30 min at $8000 \times g$ at 4 °C. The precipitates were filtered and dried overnight at 80 °C. For the green synthesis of ZnO NPs using water hyacinth, the procedure was performed as described in [31].

2.3. X-ray Diffraction (XRD) Analysis

The diffraction pattern was analyzed using an X-ray diffractometer (Bruker D8 Advance, MA, USA) with CuKα radiation and wavelength (λ) = 1.541 Å. The diffraction intensity was measured in the 2θ ranging between 20° and 80°. After that, the phases of the ZnO wurtzite were compared using JCPDS number 00036-1451 [32]. Then, Rietveld refinement was used to determine the lattice parameters and crystalline sizes using MAUD software (Trento, Italy) [33–35].

2.4. Ultraviolet-Visible (UV-Vis) Spectroscopy

In the photoluminescence mode of a UV-1800 spectrophotometer (SHIMADZU, Kyoto, Japan), the dispersed ZnO NPs in deionized water were investigated for absorption spectra between 300 and 600 nm. The optical band gap of the generated ZnO NPs was then calculated using the Tauc plot [36].

2.5. Scanning Electron Microscopy (SEM)

The ZnO NPs were prepared using a Mini Sputter Coater (SC7620, Quorum Technologies Ltd., Kent, UK), and their morphology was analyzed using FEI Quanta 450 (OR, USA) as described in [6]. ImageJ program [37] was also used to determine the particle sizes.

2.6. Fourier Transform Infrared (FT-IR) Analysis

To investigate the functional group of the synthesized ZnO NPs, FT-IR analysis was performed using an FT-IR spectrometer (Bruker, MA, USA). The samples were operated using a potassium bromide (KBr) approach with an infrared region of 400 to 4000 cm^{-1} and a 4 cm^{-1} resolution. The covalent bonds between the zinc metal and oxygen atoms (Zn-O) were observed in a range between 400 to 600 cm^{-1}.

2.7. Antibacterial Activity Test

The antibacterial activity of the ZnO NPs was tested against *Xanthomonas oryzae* pv. *oryzae* (rice bacterial blight pathogen), *Xanthomonas axonopodis* pv. *citri* (citrus canker disease), and *Ralstonia solanacearum* (bacterial wilt disease). These bacterial strains were isolated, characterized, and identified by the Plant Protection Research and Development Office, Bangkok, Thailand. Bacteria cultures were prepared by inoculating the strains in Luria broth (LB) with shaking at 37 °C until the OD_{600} reached 0.6. Then, 500 µL of the cultures were treated with various concentrations of ZnO NPs ranging from 0–10 mg/mL and further incubated with shaking at 37 °C for 24 h. Next, the UV-Vis spectroscopic analysis was performed at 600 nm (SHIMADZU, Kyoto, Japan). The experiment was performed in two independent experiments with quadruplicate. The percentage of cell growth was determined as follows: Bacteria Growth (%) = [($A_{ZnO\ NPs} - A_{Blank}$)/($A_{Control} - A_{Blank}$)] × 100], where $A_{ZnO\ NPs}$ is the mean absorbance value for the culture treated with ZnO NPs, A_{Blank} is the mean absorbance value for the ZnO NPs solution, and $A_{Control}$ is the mean absorbance for the culture only (without ZnO NPs treatment). Then, the half-maximal inhibitory concentration (IC_{50}) values of the ZnO NPs against the bacterial growth were calculated using GraphPad QuickCalcs (GraphPad software Prism 9, CA, USA).

2.8. Cell Viability Assay, MTT Assay

Skin cancer cells (A431; ATCC® CRL1555™) and an immortalized human keratinocyte cell line (HaCaT; CLS 300493) were seeded approximately 1.5×10^4 cells per well into 96 well-plates (Falcon® a Corning brand, USA) at 37 °C. After 24 h of incubation, the cells were treated with dispersed ZnO NPs in deionized water in different concentrations ranging from 7.8–1000 µg/mL and further incubated at 37 °C for 96 h. The cells were then incubated for 3 h with an MTT solution. To dissolve the formazan, 50 µL of dimethyl sulfoxide (Fisher Scientific, Hampton, NH, USA) was added. Next, the absorbance was measured at 570 nm using a microplate reader (Sunrise-Basic TECAN, Männedorf, Switzerland). The cell viability (%) was calculated as described in [38].

2.9. Statistical Analysis

The significant differences between the samples were compared by ANOVA Tukey's multiple comparison (GraphPad Software Prism9, San Diego, CA, USA).

3. Results

3.1. Green Synthesis of ZnO NPs Using Mangosteen Peels and Water Hyacinth Crude Extracts

The green synthesis was carried out in four steps: the preparation of a water crude extract, the preparation of a precursor solution, a green synthesis reaction, and particle collection, as shown in Figure 1.

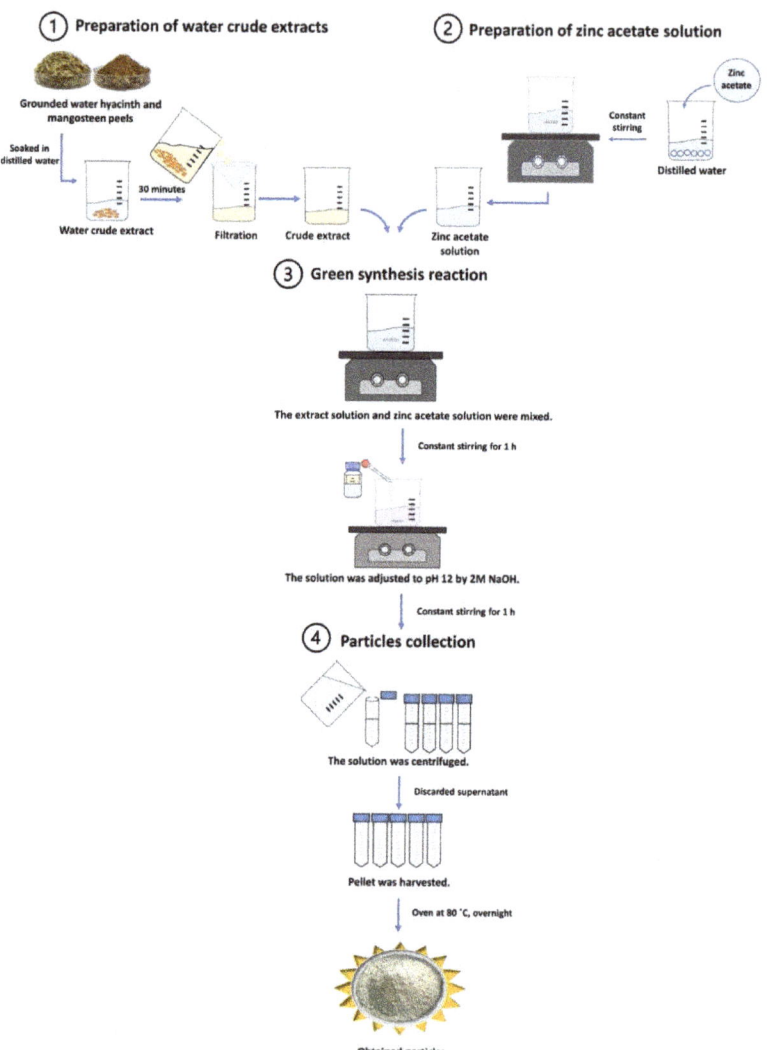

Figure 1. Schematic illustration of green synthesis using mangosteen peels and water hyacinth.

For the synthesis of ZnO NPs using mangosteen peel extract, the color of crude extract appeared in pale yellow. The reaction mixture of zinc acetate and the crude extract was purple-blueish in appearance. After pH adjustment, the precipitant was observed to be white in color, implying the formation of ZnO NPs. Likewise, the zinc acetate and crude extract reaction mixture was white in color after pH correction, despite the pale green color of the water hyacinth crude extract. The precipitant was also white, suggesting the presence of ZnO NPs (data not shown).

Notably, the particles obtained from the synthesis of the mangosteen (*Garcinia mangostana* L.) peel extract were designated as ZnO-Gm, whereas ZnO-Ec was obtained from the green synthesis of the water hyacinth (*Eichhornia crassipes*) extract.

3.1.1. The ZnO Wurtzite Structure Has Been Identified in All Synthesis Samples

Using XRD analysis, the XRD pattern confirmed that the synthesis process successfully synthesized the ZnO wurtzite structure from both the mangosteen peels and water hyacinth crude extracts without the impurity of other particles. The higher peak intensity obtained from ZnO-Ec indicates the better crystallinity of ZnO NPs (Figure 2).

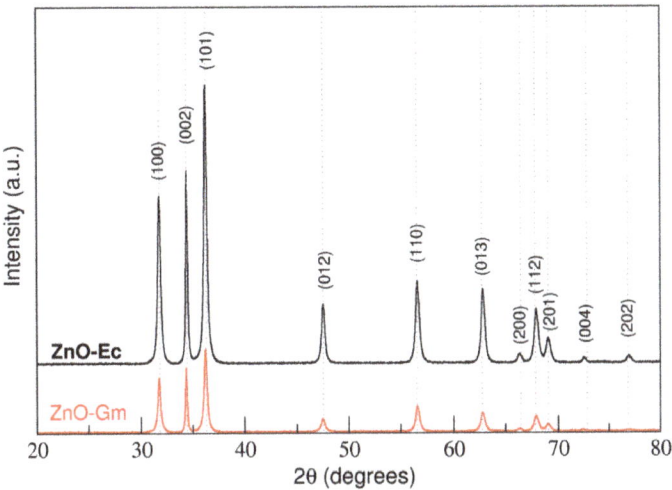

Figure 2. XRD analysis of synthesized ZnO particles prepared by mangosteen (*G. mangostana*) peel extract, ZnO-Gm; and whole water hyacinth (*E. crassipes*), ZnO-Ec.

Next, Rietveld refinement was used to calculate the lattice parameters and crystalline sizes of the synthesized ZnO NPs using MAUD software. ZnO-Gm and ZnO-Ec had lattice parameters of a = 3.2536 Å, c = 5.2155 Å, and a = 3.2545 Å, c = 5.2126 Å, respectively. ZnO-Gm (290.42 Å) showed slightly smaller estimated crystalline sizes than ZnO-Ec (318.99 Å). (Table 1).

Table 1. The lattice parameters and crystalline sizes for synthesized ZnO NPs determined from XRD data after Rietveld refinement.

ZnO NPs	a (Å)	c (Å)	Crystalline Size (Å)
ZnO-Gm	3.2536	5.2155	290.42
ZnO-Ec	3.2545	5.2126	318.99

3.1.2. UV–Visible Absorption Spectra and Optical Band Gap of Newly Synthesized ZnO Particles

Both ZnO-Gm and ZnO-Ec revealed absorption spectra in the UV region at 365 nm. The ZnO-Gm and ZnO-Ec samples have average energy band gaps of 2.79 eV and 2.88 eV, respectively (Figure 3).

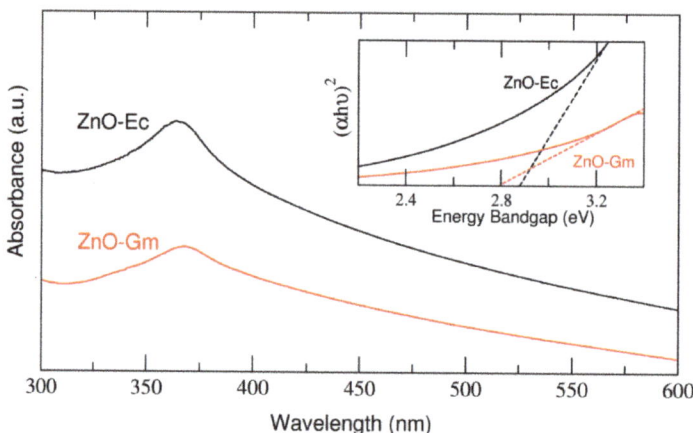

Figure 3. UV-Vis spectra and energy band gap of green synthesized ZnO samples. ZnO-Gm nanoparticles produced from mangosteen peel crude extract whereas ZnO-Ec samples generated from water hyacinth crude extract.

3.1.3. Morphology and Size of Newly Synthesized ZnO NPs

To determine the morphology and sizes of ZnO-Gm and ZnO-Ec, SEM analysis was performed. Despite the use of different types of crude extracts in the synthesis, SEM images demonstrated round, almost spherical shapes of the synthesized ZnO particles, (Figure 4). The ZnO-Gm particles were 154.41 × 172.89 nm in size, while the ZnO-Ec particles were slightly smaller, averaging 142.16 × 160.30 nm (Figure 4).

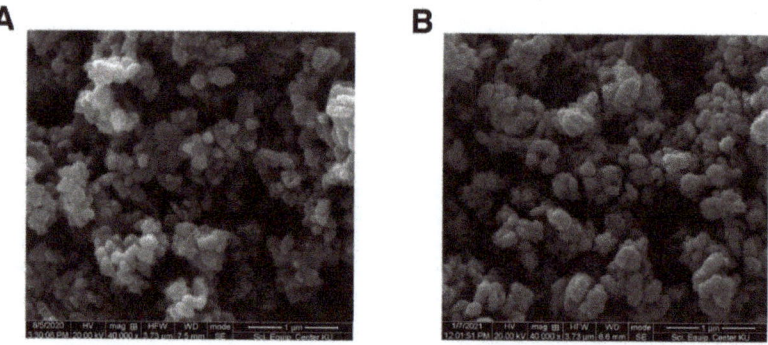

Figure 4. SEM images of synthesized ZnO NPs. (**A**) ZnO-Gm (mangosteen peel extract) and (**B**) ZnO-Ec (water hyacinth).

3.1.4. FTIR Analysis of Synthesized ZnO NPs

Next, FTIR was then used to identify the functional groups involved in the formation of the ZnO NPs. As a result, the spectral peaks between 700 and 500 cm^{-1} indicated the formation of ZnO NPs in both ZnO-Gm and ZnO-Ec. Furthermore, the broad peak at around 3500 cm^{-1} implied the stretching vibration of O—H stretching. Both ZnO-Gm and ZnO-Ec also presented peaks in the region around 1500 cm^{-1}, suggesting carbonyl stretching (C—O) (Figure 5).

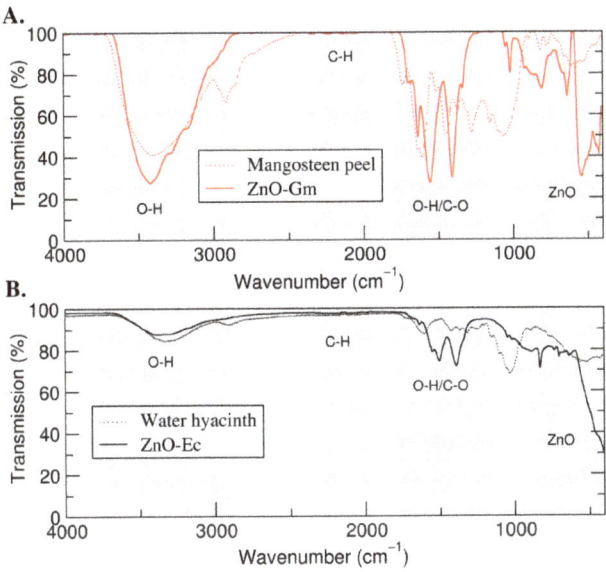

Figure 5. FTIR analysis of green synthesized ZnO NPs. (**A**) ZnO-Gm generated from mangosteen peel extract; (**B**) ZnO-Ec produced from water hyacinth extract.

3.2. Synthesized ZnO NPs Drastically Inhibited Growth of Plant Pathogenic Bacteria

From Figure 6, the results showed that both ZnO-Gm and ZnO-Ec particles significantly inhibited all the tested plant pathogenic bacteria. Furthermore, ZnO-Gm demonstrated almost 2-fold greater antibacterial activity than ZnO-Ec, with IC_{50} values of 1.887 mg/mL, 1.802 mg/mL, and 1.800 mg/mL, against *X. oryzae* pv. *oryzae*, *R. solanacearum*, and *X. axonopodis* pv. *citri*, respectively (Figure 6D). In addition, the IC_{50} values of ZnO-Gm for *X. oryzae* pv. *oryzae*, *R. solanacearum*, and *X. axonopodis* pv. *citri* were 3.970 mg/mL, 3.835 mg/mL, and 3.385 mg/mL, respectively (Figure 6D). The inhibitory effects were caused by the synthesized ZnO NPs, not the mangosteen peel extract or water hyacinth extract, because the extract at the same concentration as the ZnO NPs showed no antibacterial activity (data not shown).

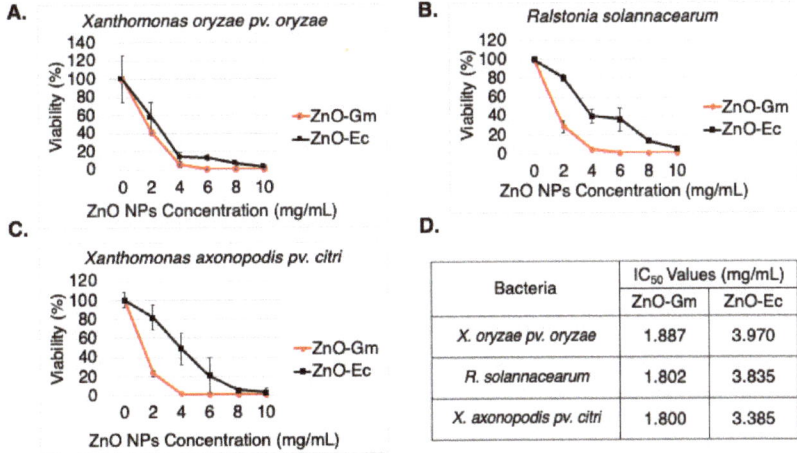

Figure 6. Synthesized ZnO NPs possessed antibacterial activity (**A**) Viability (%) of *Xanthomonas oryzae* pv. *oryzae* (rice bacterial blight pathogen), (**B**) Viability (%) of *Xanthomonas axonopodis* pv. *citri*

(citrus canker disease), (**C**) Viability (%) of *Ralstonia solanacearum* (bacterial wilt disease), (**D**) IC_{50} values of ZnO-Gm and ZnO-Gc against tested bacteria in the study. The data are represented in quadruplicate of mean ± SD from at least three independent experiments.

3.3. Synthesized ZnO NPs Possessed Anticancer Activity against Skin Cancer Cells

To investigate anti-skin cancer activity, a non-melanoma skin cancer cell (A431) [39] and an intermediate cancerous skin carcinoma cell, HaCaT [40], were treated with different concentrations of either ZnO-Gm and ZnO-Ec ranging from 0–1000 µg/mL. The experiments also included a normal cell, Vero, and the effects of crude extracts, mangosteen peel and water hyacinth. As a result, both ZnO-Gm and ZnO-Ec dramatically reduced cell viability in a dose-dependent manner, with greater inhibitory effects against A431 and HaCaT cells than Vero cells (Figures 7 and 8). In contrast, the mangosteen peel extract and water hyacinth extract showed no inhibitory effects on any cells at the concentrations tested (Figures 7 and 8). The IC_{50} values of ZnO-Gm were 28 µg/mL, 39 µg/mL, and 145.6 µg/mL for HaCaT, A431, and Vero cells, respectively (Figure 7).

On the other hand, the IC_{50} values of ZnO-Ec were 79.5 µg/mL, 10.12 µg/mL, and 162 µg/mL (Figure 8).

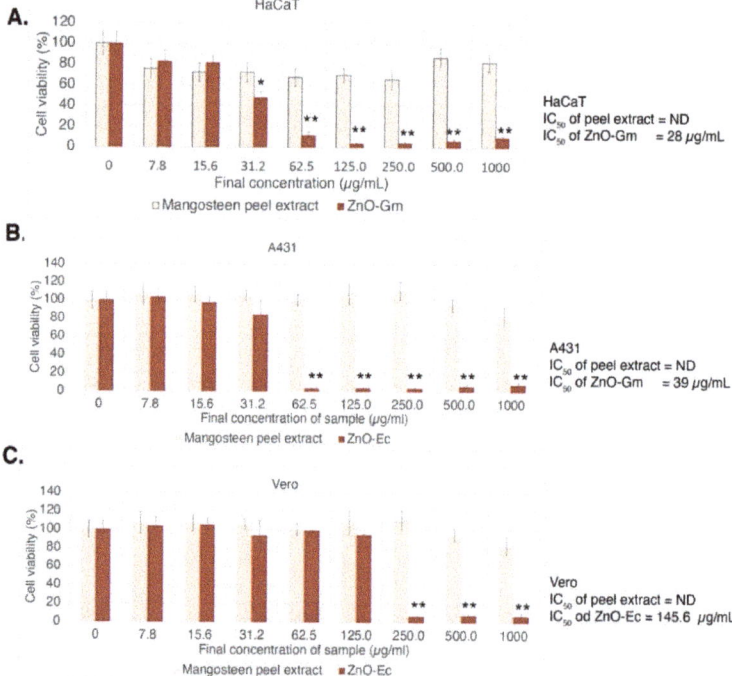

Figure 7. Cytotoxicity effects of ZnO-Gm and mangosteen extract in vitro. (**A**) HaCaT, (**B**) A431, and (**C**) Vero. ND stands for not determined. IC_{50} values were shown as indicated. The data are represented in quadruplicate of mean ± SD from three independent experiments. The significant differences between the samples were shown as * $p < 0.05$, ** $p < 0.01$ (by ANOVA, Tukey's test).

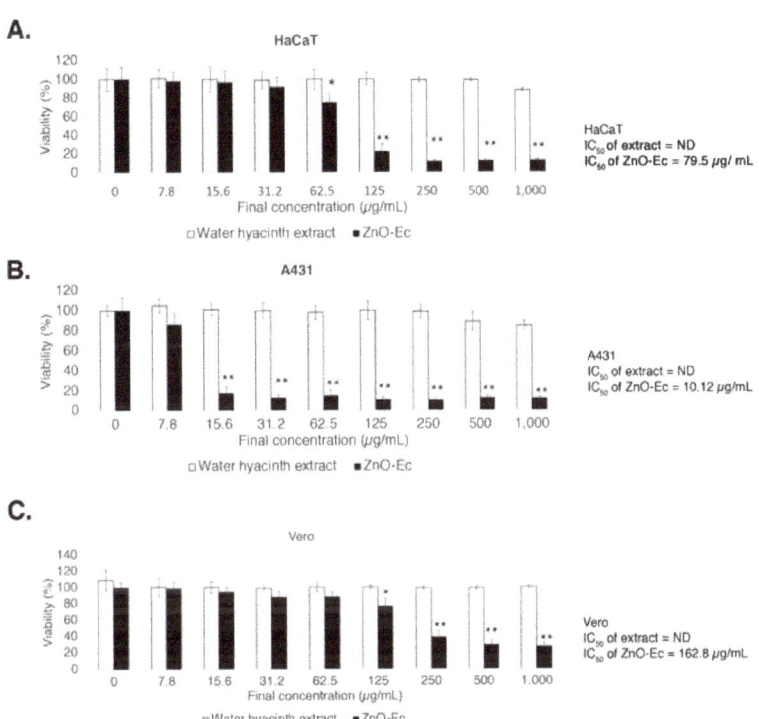

Figure 8. Cytotoxicity effects of ZnO-Ec and water hyacinth crude extract in vitro. (**A**) HaCaT, (**B**) A431, and (**C**) Vero. ND stands for not determined. IC$_{50}$ values were shown as indicated. The results are presented as mean ± SD of quadruplicate data from three independent experiments. * $p < 0.05$, ** $p < 0.01$ (using ANOVA, Tukey's test) were used to indicate significant differences between the samples.

4. Discussion

Using a green synthesis method, this study newly synthesized ZnO NPs from mangosteen peel and water hyacinth crude extracts, which were designated as ZnO-Gm and ZnO-Ec, respectively. The results implied that mangosteen peel and water hyacinth extracts contained bioactive compounds that served as reducing agents and capping agents that react with zinc acetate solution to form ZnO NPs. Phytochemicals found in mangosteen peels include xanthones, flavonoids, tannins, and anthocyanins [23], whereas water hyacinth contains alkaloids, terpenoids, steroids, glycosides, phenols, and flavonoids [29]. Both ZnO-Gm and ZnO-Ec displayed absorption maxima about 365 nm, which is consistent with previous results using *Coccinia abyssinica* [41], *Cratoxylum formosum* [6], and *Coriandrum sativum* [42,43], but differs from ZnO bulk, which occurs at around 373 nm [44,45]. Notably, ZnO NPs typically have a band gap of 3.37 eV [46], and thus synthesized ZnO-Gm and ZnO-Ec are narrow-band-gap ZnO NPs with 2.79 eV and 2.88 eV, respectively. This narrow band gap is likely due to the organic molecules of the extracts attached on the surface of ZnO NPs [47]. According to prior research, the plant species, concentration of extract, precursor concentration, duration of synthesis, pH condition, and calcination temperature are six key characteristics that can influence ZnO NPs morphology (see review [48]). Despite using distinct types of plant crude extracts, this study generated green ZnO NPs in spherical shapes with slightly different sizes from mangosteen peel extract and water hyacinth extract.

According to antibacterial activity assays, this study provided additional knowledge indicating that the green synthesized ZnO NPs from mangosteen peel and water hyacinth extracts effectively inhibited various plant pathogenic bacteria including *Xanthomonas oryzae* pv. *oryzae* (rice bacterial blight pathogen), *Xanthomonas axonopodis* pv. *citri* (citrus canker disease), and *Ralstonia solanacearum* (bacterial wilt disease). In comparison to previous studies, the IC_{50} of the generated ZnO NPs (1.80–3.97 mg/mL) demonstrated more potent suppression against *X. oryzae* pv. *oryzae* than the IC_{50} of methanol extracts of *Piper sarmentosum* fruit and leaves (8.41 mg/mL and 24.69 mg/mL, respectively) [49], but the IC_{50} of the produced ZnO NPs against *X. oryzae* pv. *oryzae* were less strong than IC_{50} of melittin (about 9–10 µM) [50] and IC_{50} of resveratrol (11.67 ± 0.58 µg/mL) [51]. The synthesized ZnO NPs, on the other hand, displayed lower IC_{50} against *X. axonopodis* pv. *citri* and *R. solanacearum* than streptomycin sulfate (6.94 µg/mL and 7.63 µg/mL, respectively) [52]. Our antibacterial activity assay was carried out in the dark inside an incubator, and thus there were no photocatalytic effects, which could lead to poorer bacterial growth suppression. We are currently evaluating the antibacterial properties of synthesized ZnO NPs against plant pathogens on crop fields to prove this hypothesis. Interestingly, despite their similar form and size, ZnO-Gm had more potent antibacterial activity than ZnO-Ec. We hypothesized that it was because ZnO-Gm presents more higher functional groups of phytochemicals on its surface than ZnO-Ec based on the FTIR and energy band gap analyses. Previous research has also proposed that the addition of phytochemicals on the surface of ZnO NPs could improve their anticancer efficacy [6]. Furthermore, Kalachyova et al. (2017) demonstrated that the bonded chemical functional groups were important for the light-induced antibacterial activities of surface-modified gold multibranched nanoparticles [53].

In contrast to chemotherapeutic drugs, ZnO NPs have been found to exhibit low toxicity, biodegradability, and therapeutic effects with a high degree of cancer selectivity [54,55]. This study showed that low doses of ZnO-Gm and ZnO-Ec (<80 µg/mL) significantly reduced cell viability by more than 50% inhibition against epidermoid carcinoma cells (A431) and very early-stage cells in skin tumorigenesis (HaCaT) without causing cytotoxicity in normal cells. Likewise, the green synthesis of ZnO NPs from rhizomes of *Alpinia calcarata* also inhibited the growth of A431 [56]. Furthermore, green ZnO NPs from a *Cratoxylum formosum* leaf extract were previously shown to drastically inhibit A431 by upregulating transcripts involved in the inflammatory response and downregulating transcripts that promote cell proliferation [6]. Even though ZnO NPs have shown significant promise in the treatment of skin cancer, further research and the in-depth understanding of cellular and molecular pathways, as well as clinical studies, will be required in the future for the development of cancer therapies.

5. Conclusions

This study highlighted the green synthesis of ZnO NPs from mangosteen peel extract (ZnO-Gm) and water hyacinth extract (ZnO-Ec). The spherical forms of ZnO-Gm (dimensions of 154.41 × 172.89 nm) and ZnO-Ec (dimensions of 142.16 × 160.30 nm) were obtained without the presence of additional crystalline impurities. The energy band gaps of ZnO-Gm were 2.79 eV, whereas for ZnO-Ec, they were 2.88 eV. Both synthesized ZnO NPs showed a specific absorbance peak around 365 nm, and their surfaces had bioactive functional groups from the extracts. The green-synthesized ZnO NPs significantly inhibited the growth of pathogenic plant bacteria including *Xanthomonas oryzae* pv. *oryzae*, *Xanthomonas axonopodis* pv. *citri*, and *Ralstonia solanacearum*. Moreover, they possessed potent anti-skin-cancer activity in vitro.

Author Contributions: Conceptualization, T.J. and N.P.T.-T.; methodology, N.P.T.-T.; validation, T.J., J.R., J.T.-T. and N.P.T.-T.; formal analysis, T.J., J.T.-T. and N.P.T.-T.; investigation, T.J. and J.R.; data curation, J.T.-T. and N.P.T.-T.; writing—original draft preparation, T.J., J.T.-T. and N.P.T.-T.; writing—review and editing, N.P.T.-T.; visualization, T.J., J.R., J.T.-T. and N.P.T.-T.; supervision, J.T.-T. and N.P.T.-T.; project administration, N.P.T.-T.; funding acquisition, T.J., J.R., J.T.-T. and N.P.T.-T. All authors have read and agreed to the published version of the manuscript.

Funding: This research was funded by Kasetsart University through the Graduate School Fellowship Program. N.P.T.-T. was funded by the National Research Council of Thailand (Aor-Por-Sor 96/2563). J.T.-T. and N.P.T.-T. have been supported by Kasetsart University Research and Development Institute (KURDI), Bangkok, Thailand.

Institutional Review Board Statement: Not applicable for studies not involving humans or animals.

Informed Consent Statement: Not applicable.

Data Availability Statement: Data available on request. The data presented in this study are available on request from the corresponding author.

Acknowledgments: Our most heartfelt thanks to Kamonrat Sukchom for collecting and processing water hyacinth samples, as well as to Wannaporn Kaewduanglek for preliminary data of green-synthesized ZnO NPs using water hyacinth extract.

Conflicts of Interest: The authors declare no conflict of interest.

References

1. Adeel, M.; Saeed, M.; Khan, I.; Muneer, M.; Akram, N. Synthesis and Characterization of Co–ZnO and Evaluation of Its Photocatalytic Activity for Photodegradation of Methyl Orange. *ACS Omega* **2021**, *6*, 1426–1435. [CrossRef] [PubMed]
2. Jiang, S.; Lin, K.; Cai, M. ZnO Nanomaterials: Current Advancements in Antibacterial Mechanisms and Applications. *Front. Chem.* **2020**, *8*, 580. [CrossRef] [PubMed]
3. Siddiqi, K.S.; ur Rahman, A.; Tajuddin Husen, A. Properties of Zinc Oxide Nanoparticles and Their Activity Against Microbes. *Nanoscale Res. Lett.* **2018**, *13*, 141. [CrossRef] [PubMed]
4. Burmistrov, D.E.; Simakin, A.V.; Smirnova, V.V.; Uvarov, O.V.; Ivashkin, P.I.; Kucherov, R.N.; Ivanov, V.E.; Bruskov, V.I.; Sevostyanov, M.A.; Baikin, A.S.; et al. Bacteriostatic and Cytotoxic Properties of Composite Material Based on ZnO Nanoparticles in PLGA Obtained by Low Temperature Method. *Polymers* **2021**, *14*, 49. [CrossRef]
5. Nagajyothi, P.C.; Cha, S.J.; Yang, I.J.; Sreekanth, T.V.M.; Kim, K.J.; Shin, H.M. Antioxidant and anti-inflammatory activities of zinc oxide nanoparticles synthesized using Polygala tenuifolia root extract. *J. Photochem. Photobiol. B Biol.* **2015**, *146*, 10–17. [CrossRef]
6. Jevapatarakul, D.; T-Thienprasert, J.; Payungporn, S.; Chavalit, T.; Khamwut, A.; T-Thienprasert, N.P. Utilization of Cratoxylum formosum crude extract for synthesis of ZnO nanosheets: Characterization, biological activities and effects on gene expression of nonmelanoma skin cancer cell. *Biomed. Pharmacother.* **2020**, *130*, 110552. [CrossRef]
7. Srivastava, N.; Srivastava, M.; Mishra, P.K.; Ramteke, P.W. Application of ZnO Nanoparticles for Improving the Thermal and pH Stability of Crude Cellulase Obtained from Aspergillus fumigatus AA001. *Front. Microbiol.* **2016**, *7*, 514. [CrossRef]
8. Zhou, J.; Xu, N.S.; Wang, Z.L. Dissolving behavior and stability of ZnO wires in biofluids: A study on biodegradability and biocompatibility of ZnO nanostructures. *Adv. Mater.* **2006**, *18*, 2432–2435. [CrossRef]
9. Verbič, A.; Gorjanc, M.; Simončič, B. Zinc Oxide for Functional Textile Coatings: Recent Advances. *Coatings* **2019**, *9*, 550. [CrossRef]
10. Sabir, S.; Arshad, M.; Chaudhari, S.K. Zinc Oxide Nanoparticles for Revolutionizing Agriculture: Synthesis and Applications. *Sci. World J.* **2014**, *2014*, 925494. [CrossRef]
11. Mishra, P.K.; Mishra, H.; Ekielski, A.; Talegaonkar, S.; Vaidya, B. Zinc oxide nanoparticles: A promising nanomaterial for biomedical applications. *Drug Discov. Today* **2017**, *22*, 1825–1834. [CrossRef] [PubMed]
12. Singh, T.A.; Sharma, A.; Tejwan, N.; Ghosh, N.; Das, J.; Sil, P.C. A state of the art review on the synthesis, antibacterial, antioxidant, antidiabetic and tissue regeneration activities of zinc oxide nanoparticles. *Adv. Colloid Interface Sci.* **2021**, *295*, 102495. [CrossRef] [PubMed]
13. Agarwal, H.; Venkat Kumar, S.; Rajeshkumar, S. A review on green synthesis of zinc oxide nanoparticles—An eco-friendly approach. *Resour.-Effic. Technol.* **2017**, *3*, 406–413. [CrossRef]
14. Noah, N.M.; Ndangili, P.M. Green synthesis of nanomaterials from sustainable materials for biosensors and drug delivery. *Sens. Int.* **2022**, *3*, 100166. [CrossRef]
15. Jamdagni, P.; Khatri, P.; Rana, J.S. Green synthesis of zinc oxide nanoparticles using flower extract of Nyctanthes arbor-tristis and their antifungal activity. *J. King Saud Univ.-Sci.* **2018**, *30*, 168–175. [CrossRef]
16. Ruangtong, J.; T-Thienprasert, J.; T-Thienprasert, N.P. Green synthesized ZnO nanosheets from banana peel extract possess anti-bacterial activity and anti-cancer activity. *Mater. Today Commun.* **2020**, *24*, 101224. [CrossRef]
17. Basnet, P.; Inakhunbi Chanu, T.; Samanta, D.; Chatterjee, S. A review on bio-synthesized zinc oxide nanoparticles using plant extracts as reductants and stabilizing agents. *J. Photochem. Photobiol. B Biol.* **2018**, *183*, 201–221. [CrossRef]
18. Singh, J.; Dutta, T.; Kim, K.-H.; Rawat, M.; Samddar, P.; Kumar, P. 'Green' synthesis of metals and their oxide nanoparticles: Applications for environmental remediation. *J. Nanobiotechnol.* **2018**, *16*, 84. [CrossRef]
19. Makarov, V.V.; Love, A.J.; Sinitsyna, O.V.; Makarova, S.S.; Yaminsky, I.V.; Taliansky, M.E.; Kalinina, N.O. "Green" nanotechnologies: Synthesis of metal nanoparticles using plants. *Acta Nat.* **2014**, *6*, 35–44. [CrossRef]
20. Okpara, E.C.; Fayemi, O.E.; Sherif, E.-S.M.; Junaedi, H.; Ebenso, E.E. Green Wastes Mediated Zinc Oxide Nanoparticles: Synthesis, Characterization and Electrochemical Studies. *Materials* **2020**, *13*, 4241. [CrossRef]

21. Surendra, T.V.; Roopan, S.M.; Al-Dhabi, N.A.; Arasu, M.V.; Sarkar, G.; Suthindhiran, K. Vegetable Peel Waste for the Production of ZnO Nanoparticles and its Toxicological Efficiency, Antifungal, Hemolytic, and Antibacterial Activities. *Nanoscale Res. Lett.* **2016**, *11*, 546. [CrossRef] [PubMed]
22. Modi, S.; Yadav, V.K.; Choudhary, N.; Alswieleh, A.M.; Sharma, A.K.; Bhardwaj, A.K.; Khan, S.H.; Yadav, K.K.; Cheon, J.-K.; Jeon, B.-H. Onion Peel Waste Mediated-Green Synthesis of Zinc Oxide Nanoparticles and Their Phytotoxicity on Mung Bean and Wheat Plant Growth. *Materials* **2022**, *15*, 2393. [CrossRef] [PubMed]
23. Aizat, W.M.; Jamil, I.N.; Ahmad-Hashim, F.H.; Noor, N.M. Recent updates on metabolite composition and medicinal benefits of mangosteen plant. *PeerJ* **2019**, *7*, e6324. [CrossRef] [PubMed]
24. Shan, T.; Ma, Q.; Guo, K.; Liu, J.; Li, W.; Wang, F.; Wu, E. Xanthones from mangosteen extracts as natural chemopreventive agents: Potential anticancer drugs. *Curr. Mol. Med.* **2011**, *11*, 666–677. [CrossRef] [PubMed]
25. Suttirak, W.; Manurakchinakorn, S. In vitro antioxidant properties of mangosteen peel extract. *J Food Sci. Technol.* **2014**, *51*, 3546–3558. [CrossRef]
26. Akhtar, M.S.; Panwar, J.; Yun, Y.-S. Biogenic Synthesis of Metallic Nanoparticles by Plant Extracts. *ACS Sustain. Chem. Eng.* **2013**, *1*, 591–602. [CrossRef]
27. Ghosh, P.R.; Fawcett, D.; Sharma, S.B.; Poinern, G.E.J. Production of High-Value Nanoparticles via Biogenic Processes Using Aquacultural and Horticultural Food Waste. *Materials* **2017**, *10*, 852. [CrossRef]
28. Korbekandi, H.; Iravani SFau-Abbasi, S.; Abbasi, S. Production of nanoparticles using organisms. *Crit. Rev. Biotechnol.* **2009**, *29*, 279–306. [CrossRef]
29. Guna, V.; Ilangovan, M.; Anantha Prasad, M.G.; Reddy, N. Water Hyacinth: A Unique Source for Sustainable Materials and Products. *ACS Sustainable Chem. Eng.* **2017**, *5*, 4478–4490. [CrossRef]
30. Kathiresan, R.M. Allelopathic potential of native plants against water hyacinth. *Crop Prot.* **2000**, *19*, 705–708. [CrossRef]
31. T-Thienprasert, N.P.; T-Thienprasert, J.; Ruangtong, J.; Jaithon, T.; Srifah Huehne, P.; Piasai, O. Large Scale Synthesis of Green Synthesized Zinc Oxide Nanoparticles from Banana Peel Extracts and Their Inhibitory Effects against *Colletotrichum* sp., Isolate KUFC 021, Causal Agent of Anthracnose on *Dendrobium* Orchid. *J. Nanomater.* **2021**, *2021*, 5625199. [CrossRef]
32. Jenkins, R.; Fawcett, T.G.; Smith, D.K.; Visser, J.W.; Morris, M.C.; Frevel, L.K. JCPDS—International Centre for Diffraction Data Sample Preparation Methods in X-Ray Powder Diffraction. *Powder Diffr.* **1986**, *1*, 51–63. [CrossRef]
33. Lutterotti, L. Total pattern fitting for the combined size–strain–stress–texture determination in thin film diffraction. *Nucl. Instrum. Methods Phys. Res. Sect. B Beam Interact. Mater. At.* **2010**, *268*, 334–340. [CrossRef]
34. Lutterotti, L.; Bortolotti, M.; Ischia, G.; Lonardelli, I.; Wenk, H.R. Rietveld texture analysis from diffraction images. *Z. Kristallogr. Suppl.* **2007**, *26*, 125–130. [CrossRef]
35. Lutterotti, L.; Chateigner, D.; Ferrari, S.; Ricote, J. Texture, residual stress and structural analysis of thin films using a combined X-ray analysis. *Thin Solid Film.* **2004**, *450*, 34–41. [CrossRef]
36. Tauc, J. Optical properties and electronic structure of amorphous Ge and Si. *Mater. Res. Bull.* **1968**, *3*, 37–46. [CrossRef]
37. Rueden, C.T.; Schindelin, J.; Hiner, M.C.; Dezonia, B.E.; Walter, A.E.; Arena, E.T.; Eliceiri, K.W. ImageJ2: ImageJ for the next generation of scientific image data. *BMC Bioinform.* **2017**, *18*, 529. [CrossRef]
38. Budchart, P.; Khamwut, A.; Sinthuvanich, C.; Ratanapo, S.; Poovorawan, Y.; T-Thienprasert, N.P. Partially Purified Gloriosa superba Peptides Inhibit Colon Cancer Cell Viability by Inducing Apoptosis Through p53 Upregulation. *Am. J. Med. Sci.* **2017**, *354*, 423–429. [CrossRef]
39. Khamwut, A.; Jevapatarakul, D.; Reamtong, O.; T-Thienprasert, N.P. In vitro evaluation of anti-epidermoid cancer activity of Acanthus ebracteatus protein hydrolysate and their effects on apoptosis and cellular proteins. *Oncol. Lett.* **2019**, *18*, 3128–3136. [CrossRef]
40. Leonard, M.K.; Kommagani, R.; Payal, V.; Mayo, L.D.; Shamma, H.N.; Kadakia, M.P. ΔNp63α regulates keratinocyte proliferation by controlling PTEN expression and localization. *Cell Death Differ.* **2011**, *18*, 1924–1933. [CrossRef]
41. Safawo, T.; Sandeep, B.V.; Pola, S.; Tadesse, A. Synthesis and characterization of zinc oxide nanoparticles using tuber extract of anchote (*Coccinia abyssinica* (Lam.) Cong.) for antimicrobial and antioxidant activity assessment. *OpenNano* **2018**, *3*, 56–63. [CrossRef]
42. Hassan, S.S.M.; El Azab, W.I.M.; Ali, H.R.; Mansour, M.S.M. Green synthesis and characterization of ZnO nanoparticles for photocatalytic degradation of anthracene. *Adv. Nat. Sci. Nanosci. Nanotechnol.* **2015**, *6*, 045012. [CrossRef]
43. Singh, J.; Kaur, S.; Kaur, G.; Basu, S.; Rawat, M. Biogenic ZnO nanoparticles: A study of blueshift of optical band gap and photocatalytic degradation of reactive yellow 186 dye under direct sunlight. *Green Processing Synth.* **2019**, *8*, 272–280. [CrossRef]
44. Li, X.-H.; Xu, J.-Y.; Jin, M.; Shen, H.; Li, X.-M. Electrical and Optical Properties of Bulk ZnO Single Crystal Grown by Flux Bridgman Method. *Chin. Phys. Lett.* **2006**, *23*, 3356–3358. [CrossRef]
45. Debanath, M.K.; Karmakar, S. Study of blueshift of optical band gap in zinc oxide (ZnO) nanoparticles prepared by low-temperature wet chemical method. *Mater. Lett.* **2013**, *111*, 116–119. [CrossRef]
46. Huang, M.H.; Mao, S.; Feick, H.; Yan, H.; Wu, Y.; Kind, H.; Weber, E.; Russo, R.; Yang, P. Room-Temperature Ultraviolet Nanowire Nanolasers. *Science* **2001**, *292*, 1897–1899. [CrossRef]
47. Khan, M.M.; Saadah, N.H.; Khan, M.E.; Harunsani, M.H.; Tan, A.L.; Cho, M.H. Phytogenic Synthesis of Band Gap-Narrowed ZnO Nanoparticles Using the Bulb Extract of Costus woodsonii. *BioNanoScience* **2019**, *9*, 334–344. [CrossRef]

48. Xu, J.; Huang, Y.; Zhu, S.; Abbes, N.; Jing, X.; Zhang, L. A review of the green synthesis of ZnO nanoparticles using plant extracts and their prospects for application in antibacterial textiles. *J. Eng. Fibers Fabr.* **2021**, *16*, 15589250211046242. [CrossRef]
49. Syed Ab Rahman, S.F.; Sijam, K.; Omar, D. Chemical composition of Piper sarmentosum extracts and antibacterial activity against the plant pathogenic bacteria Pseudomonas fuscovaginae and Xanthomonas oryzae pv. oryzae. *J. Plant Dis. Prot.* **2014**, *121*, 237–242. [CrossRef]
50. Shi, W.; Li, C.; Li, M.; Zong, X.; Han, D.; Chen, Y. Antimicrobial peptide melittin against Xanthomonas oryzae pv. oryzae, the bacterial leaf blight pathogen in rice. *Appl. Microbiol. Biotechnol.* **2016**, *100*, 5059–5067. [CrossRef]
51. Luo, H.-Z.; Guan, Y.; Yang, R.; Qian, G.-L.; Yang, X.-H.; Wang, J.-S.; Jia, A.-Q. Growth inhibition and metabolomic analysis of Xanthomonas oryzae pv. oryzae treated with resveratrol. *BMC Microbiol.* **2020**, *20*, 117. [CrossRef] [PubMed]
52. Huang, R.-H.; Lin, W.; Zhang, P.; Liu, J.-Y.; Wang, D.; Li, Y.-Q.; Wang, X.-Q.; Zhang, C.-S.; Li, W.; Zhao, D.-L. Anti-phytopathogenic Bacterial Metabolites From the Seaweed-Derived Fungus *Aspergillus* sp. D40. *Front. Mar. Sci.* **2020**, *7*, 313. [CrossRef]
53. Kalachyova, Y.; Olshtrem, A.; Guselnikova, O.A.; Postnikov, P.S.; Elashnikov, R.; Ulbrich, P.; Rimpelova, S.; Švorčík, V.; Lyutakov, O. Synthesis, Characterization, and Antimicrobial Activity of Near-IR Photoactive Functionalized Gold Multibranched Nanoparticles. *ChemistryOpen* **2017**, *6*, 254–260. [CrossRef] [PubMed]
54. Hanley, C.; Layne, J.; Punnoose, A.; Reddy, K.M.; Coombs, I.; Coombs, A.; Feris, K.; Wingett, D. Preferential killing of cancer cells and activated human T cells using ZnO nanoparticles. *Nanotechnology* **2008**, *19*, 295103. [CrossRef]
55. Aljabali, A.A.A.; Obeid, M.A.; Bakshi, H.A.; Alshaer, W.; Ennab, R.M.; Al-Trad, B.; Al Khateeb, W.; Al-Batayneh, K.M.; Al-Kadash, A.; Alsotari, S.; et al. Synthesis, Characterization, and Assessment of Anti-Cancer Potential of ZnO Nanoparticles in an In Vitro Model of Breast Cancer. *Molecules* **2022**, *27*, 1827. [CrossRef]
56. Chelladurai, M.; Sahadevan, R.; Margavelu, G.; Vijayakumar, S.; González-Sánchez, Z.I.; Vijayan, K.; KC, D.B. Anti-skin cancer activity of Alpinia calcarata ZnO nanoparticles: Characterization and potential antimicrobial effects. *J. Drug Deliv. Sci. Technol.* **2021**, *61*, 102180. [CrossRef]

Article

Hydrothermally Grown ZnO Nanostructures for Water Purification via Photocatalysis

Marie Le Pivert [1,2], Nathan Martin [1,3] and Yamin Leprince-Wang [1,*]

1. ESYCOM Lab, CNRS, University Gustave Eiffel, 77454 Marne-la-Vallee, France; marie.le-pivert@univ-eiffel.fr (M.L.P.); nathan.martin@univ-eiffel.fr (N.M.)
2. COSYS-LISIS Lab, University Gustave Eiffel, 77454 Marne-la-Vallee, France
3. Eden Tech, 75012 Paris, France
* Correspondence: yamin.leprince@univ-eiffel.fr

Abstract: Semiconductor-based photocatalysis is a well-known and efficient process for achieving water depollution with very limited rejects in the environment. Zinc oxide (ZnO), as a wide-bandgap metallic oxide, is an excellent photocatalyst, able to mineralize a large scale of organic pollutants in water, under UV irradiation, that can be enlarged to visible range by doping nontoxic elements such as Ag and Fe. With high surface/volume ratio, the ZnO nanostructures have been shown to be prominent photocatalyst candidates with enhanced photocatalytic efficiency, owing to their being low-cost, non-toxic, and able to be produced with easy and controllable synthesis. Thus, ZnO nanostructures-based photocatalysis can be considered as an eco-friendly and sustainable process. This paper presents the photocatalytic activity of ZnO nanostructures (NSs) grown on different substrates. The photocatalysis has been carried out both under classic mode and microfluidic mode. All tests show the notable photocatalytic efficiency of ZnO NSs with remarkable results obtained from a ZnO-NSs-integrated microfluidic reactor, which exhibited an important enhancement of photocatalytic activity by drastically reducing the photodegradation time. UV-visible spectrometry and high-performance liquid chromatography, coupled with mass spectrometry (HPLC-MS), are simultaneously used to follow real-time information, revealing both the photodegradation efficiency and the degradation mechanism of the organic dye methylene blue.

Keywords: ZnO nanostructures; hydrothermal synthesis; ZnO doping; photocatalysis; water purification; degradation mechanism study

Citation: Le Pivert, M.; Martin, N.; Leprince-Wang, Y. Hydrothermally Grown ZnO Nanostructures for Water Purification via Photocatalysis. *Crystals* **2022**, *12*, 308. https://doi.org/10.3390/cryst12030308

Academic Editor: Claudia Graiff

Received: 4 February 2022
Accepted: 19 February 2022
Published: 22 February 2022

Publisher's Note: MDPI stays neutral with regard to jurisdictional claims in published maps and institutional affiliations.

Copyright: © 2022 by the authors. Licensee MDPI, Basel, Switzerland. This article is an open access article distributed under the terms and conditions of the Creative Commons Attribution (CC BY) license (https:// creativecommons.org/licenses/by/ 4.0/).

1. Introduction

The management of water resources is a continually growing issue, impacting human health, climate change and the global economy [1]. Unfortunately, water pollution problems, notably caused by various dyes originating from the textile, pharmaceutical and food-processing industries, increase at the same time. For this reason, water quality and treatment were set at the top of societal priority action list [2] to find efficient, low-cost, and environmentally friendly purification processes, leading to a huge increase in the research on this topic. Among all the solutions, photocatalytic oxidative processes appeared as promising air and water purification methods due to their ability to degrade and mineralize toxic organic pollutants into harmless compounds such as H_2O, CO_2, NO_3^- ... by using only a light source and a photocatalyst [3,4].

To produce highly efficient materials for water remediation, a variety of photocatalysts and synthesis methods have been developed [5–11]. Nevertheless, even among them, some solutions are using complex, expensive, and energy-consuming production processes. Furthermore, some of the synthesis processes, although efficient, lead to non-eco-friendly photocatalysts. In order to develop new photocatalytic materials, it is therefore needed to use eco-friendly photocatalysts, synthesized with a low-cost method, by using the least chemical products possible and employing the shortest possible fabrication period. Moreover, photocatalyst synthesis must be universal to a large kind of support in order to be developed at different scales, and to be usable by different solutions.

Among the different existing options of photocatalysts, zinc oxide nanostructures (ZnO NSs) have been selected for their ease of growth using soft chemistry conditions. Moreover, ZnO NSs have already shown great potential as eco-friendly photocatalysts for environmental pollution remediation under UV or solar light [12–14]. Besides, ZnO is a multifunctional material with for instance piezo-electric properties, which could improve photocatalysis efficiency [15]. ZnO NSs also present the advantages of being low-cost photocatalysts, whose raw materials are abundant in nature [4], and which could be produced by soft chemical methods, at low temperature, and with a reduced need for dangerous chemical compounds. The easy-to-perform method of the hydrothermal synthesis only requires a low processing temperature ($\leq 100\ ^\circ C$) and a short duration (≤ 4 h), and could be easily scaled-up and adapted to different substrates [12,15,16].

This paper gives an overview of our previous works on the development of different ZnO-NSs-based materials by hydrothermal synthesis and their photocatalytic activity for water purification. The universality and feasibility of this production will be presented on various substrates, from wafer silicon (Si) substrates to civil engineering materials, and from the classic synthesis in an autoclave reactor to an in situ synthesis in microfluidic cells. Different strategies to reduce the ZnO band gap and improve its photocatalytic efficiency will also be introduced. Then, this paper will display photocatalytic results under classic mode and microfluidic mode, both under UV light and solar light. Furthermore, this paper will investigate the photodegradation mechanisms of two organic dyes, methylene blue (MB) and Acid Red 14 (AR14), by UV-visible spectrometry (UV-vis) and high-performance liquid chromatography, coupled with mass spectrometry (HPLC-MS).

2. ZnO Nanostructures Grown by Hydrothermal Synthesis and Strategies to Improve Their Photocatalytic Activity

Usually, hydrothermal synthesis allows the ZnO NSs growth onto different substrates by two simple operating steps: (1) a ZnO seed layer deposition for creating nucleation sites on the substrate to act as nucleation centers to promote the homogeneous growth of the ZnO NS. This step also allows a better control on the morphology and density of the ZnO NSs in the second step; (2) a hydrothermal growth in presence of zinc salt and hexamethylenetetramine (HMTA) to obtain nanowires (NWs) or nanorods (NRs) following the growth mechanism presented in the Equations (1)–(5) [9,17–19]. During step 2, the ZnO NWs and NRs growth follows the (002) plan to minimize the surface energy [20]. Indeed, ZnO polar faces have higher energy than the non-polar ones.

$$(CH_2)_6N_4 + 6\ H_2O \rightarrow 6\ HCHO + 4\ NH_3 \qquad (1)$$

$$NH_3 + H_2O \rightarrow OH^- + NH_4^+ \qquad (2)$$

$$Zn^{2+} + 4\ OH^- \rightarrow Zn(OH)_4^{2-} \rightarrow ZnO + 2\ OH^- + H_2O \qquad (3)$$

$$Zn(NH_3)_4^{3+} + 2\ OH^- \rightarrow ZnO + 4\ NH_3 + H_2O \qquad (4)$$

$$Zn^{2+} + 2\ OH^- \rightarrow Zn(OH)_2 \xrightarrow{\Delta} ZnO + H_2O \qquad (5)$$

This section presents the adaptation of this process to different substrates and the main proofs of the good synthesis of ZnO NSs. The development of the synthesis at different scales with different modes and the strategies to improve photocatalytic activity are also introduced.

2.1. Silicon Functionalization

Si functionalization has been a well-designed synthesis for years now. The seed layer deposition commonly consists of a spin-coating (1 min, 3000 rpm) of the Si substrate with a buffer layer of polyvinyl alcohol (PVA, 10 g) and zinc acetate dihydrate (Zn(Ac)$_2$·2H$_2$O, 1 g) in water (500 mL). Then, the deposited thin film is calcined at 500 °C for 3 h in ambient atmosphere to remove the PVA and form ZnO nanocrystallites as seeds. The growing process of ZnO NWs is finally achieved at 90 °C during 4 h into a sealed Teflon reactor containing an aqueous solution of zinc nitrate hexahydrate (Zn(NO$_3$)$_2$·6H$_2$O) and HMTA. By playing with the concentration, the morphology and the defect concentration in the ZnO band gap could be tuned (Figure 1) [20,21]. It is noteworthy to mention that our results presented in Figure 1 are in line with literature [20,21].

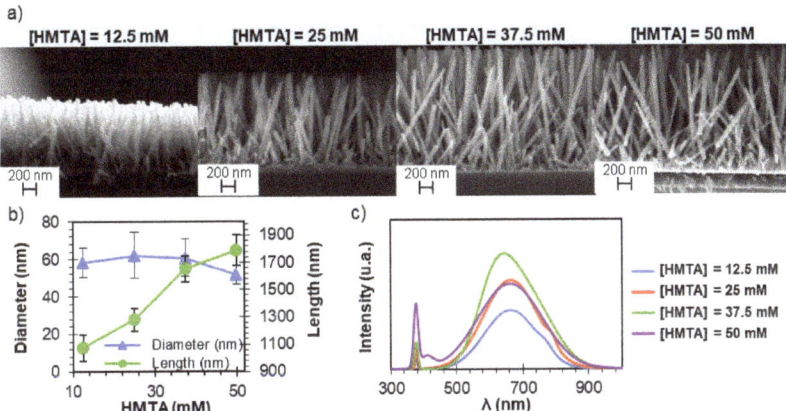

Figure 1. (a) SEM images, (b) diameter and length, and (c) photoluminescence characterization of ZnO NWs grown at fixed Zn(NO$_3$)$_2$ concentration of 0.05 M and variable HMTA concentration and post-annealed at 500 °C.

In our previous work, two concentration conditions were selected depending on the NWs wanted morphology (ratio diameter vs. length) and defect concentration: (1) 75 mM of Zn(NO$_3$)$_2$ and 37.5 mM of HMTA, corresponding to a growth solution with 37.5 mM of Zn(NO$_3$)$_2$ and 18.75 mM of HMTA (C1); (2) 50 mM of Zn(NO$_3$)$_2$ and 50 mM of HMTA corresponding to an equimolar growth solution at 25 mM (C2) [22–24]. At the end of the growing process, the Si substrates covered by ZnO NWs were washed with DI water, dried under hot airflow (~30 s at ~53 °C), and post-annealed in an oven at 350 °C for 30 min in ambient atmosphere to improve the ZnO crystallinity. Scanning electron microscope images (SEM, Zeiss FE-SEM NEON 40, Iéna, Germany) demonstrated that well organized and homogeneous ZnO NWs grown onto Si (1.55 cm^2) were obtained, with a measured height of 1.10 ± 0.05 µm and a measured mean diameter of 85 ± 5 nm in concentration conditions (C1) and a measured height of 1.80 ± 0.1 µm and a measured mean diameter of 51 ± 5 nm in concentration conditions (C2) [24,25]. Previous characterization works also proved the good crystallinity of the as-obtained ZnO NWs by ultraviolet–visible spectrophotometry (Maya2000 Pro from Ocean Optics, Dunedin, FL, USA), with a mean measured gap value around 3.21 ± 0.03 eV, and by X-ray diffraction (XRD, CuKα, λ = 1.5418 Å, Rigaku Smartlab, Neu-Isenburg, Germany), with ZnO Wurtzite peaks obtention [22–25].

2.2. Engineering Materials Functionalization

The synthesis method presented in Section 2.1 proved its efficiency not only on silicon (Figure 2a), but also on other substrates such as quartz glass [26,27]. However, depending on the used substrate, this synthesis must be adapted—notably, the seed layer deposition step. Indeed, the annealing at 500 °C degree could damage several kinds of samples, such as building construction materials. To overcome this problem, the seed layer buffer solution of PVA deposition by spin-coating was replaced by a horizontal impregnation with a Zn(Ac)$_2$ ethanolic solution (0.01 M). This new deposition method led to a shorter and softer annealing at 350 °C during 30 min only (Figure 2b) [28]. In short, a ZnO seed layer was deposited on engineering materials, such as tiling, concrete, and rocks aggregate, by a "horizontal impregnation" method with the zinc acetate ethanolic solution, followed by an annealing at 350 °C for 30 min. Then, a classical hydrothermal growth using equimolar aqueous solutions of HMTA and Zn(NO$_3$)$_2$ or Zn(Ac)$_2$ at 0.025 M was carried out in an autoclave at 90 °C. According to our previous work about the optimal growth conditions depending on the used substrate [29], 2 h was selected as an optimal growth duration for tiling or rock aggregate substrates, and 1 h 30 min for concrete substrates. Finally, the as-synthesized samples were annealed for 30 min at 350 °C to remove all the potential residues from the synthesis process and to improve the ZnO NSs crystallinity.

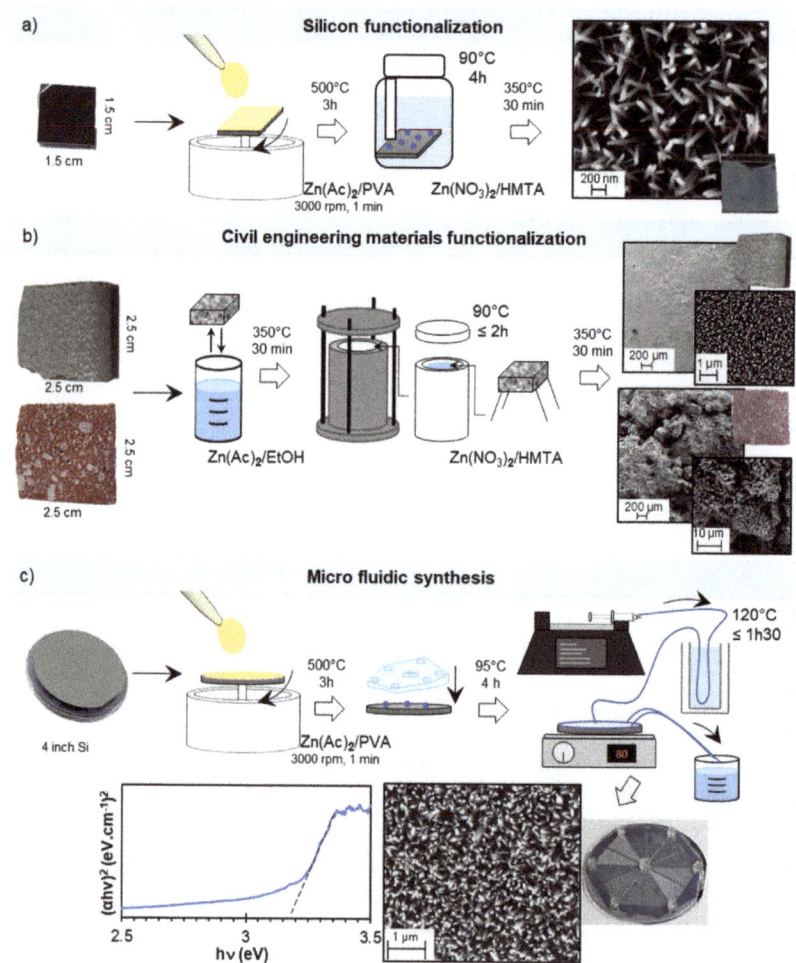

Figure 2. Experimental set-up schema of different developed ZnO NSs synthesis from silicon functionalization (**a**) and civil engineering materials functionalization (**b**) to micro fluidic reactor production (**c**) and obtained SEM images.

Results demonstrated that this hydrothermal synthesis method allowed us to grow ZnO NSs directly on non-conventional substrates such as tiling and concrete (6.25 cm^2).

Tiling samples showed NSs similar to ZnO NWs grown onto silicon in terms of morphology (~55 ± 17 nm), gap value (~3.20 ± 0.03 eV), and XRD pattern [28,29]. Conversely, a huge discrepancy was recorded on the morphologies of the ZnO NSs synthesized on concrete substrates. Depending on the growth time (1 h 30 min to 4 h), NSs vary from complex ZnO nanosheet structures with a lower gap value (~3.11 ± 0.04 eV) to NWs and NRs with a gap value of (~3.17 ± 0.04 eV) [29]. This might be due to a possible influence of the basic surface pH value and the complex chemical composition of concrete. NSs growth could be also influenced by the textural properties (porosity, roughness, etc.) of concrete and local micro-turbulences [28–30]. The band gap variation could be caused by the morphological modifications of ZnO NSs, and could also be associated with their crystal quality, dislocations density, impurities, size, and thickness [30–32]. Indeed, nanosheets are supposed to contain more oxygen defects, which could reduce the band gap by acting as indirect donor energy levels below the conduction band [33–35]. Finally, the feasibility of the upscaling of this process was proved by going from a reactor containing 250 mL of growth solution and a single sample production (6.25 cm^2) to a reactor with 8 L of growth solution with 3 samples per synthesis, without any modification of the ZnO NSs properties (210.25 cm^2) [12].

2.3. Microfluidic Reactor Production

In the same way as the ease of the scale up, the ease of transposing this hydrothermal synthesis method to the microfluidic mode was proved [30,36]. In order to change from laboratory scale to industrial scale, it is natural to consider going from batch experiments to continuous degradation. Integrating our NSs in a microreactor allows performing this change while avoiding scale-up issues in the synthesis, such as increasing the synthesis reactor size or consuming more growth solution. Furthermore, performing the photocatalytic degradation at microfluidic scale presents several advantages compared to the meso- or macro-fluidic scales: the reduced size of the channels increases the contact between the photocatalyst and the organic pollutants, while diminishing the quantity of light absorbed by the liquid. This is especially important when trying to degrade compounds that absorb the same wavelength as the photocatalyst, or compounds that tend to block the light, as industrial dyes. The continuous stream of liquid to be degraded, combined with the size of the channels, also leads to a situation where the mass transfer of the pollutants is no longer limiting the reaction kinetics, improving the reaction rate [37–40]. All these advantages lead to a generally faster reaction, meaning a higher photocatalytic efficiency than in bigger reactors.

However, one of the major drawbacks of microfluidic reactors is the small flowrate they can deliver, often in the range of 1 mL·min^{-1}, and the large pressure losses in the microchannels, leading to reactors needing either high pressure at their inlets to ensure the circulation of the liquid, or the consumption of energy to make the liquid flow. To circumvent these problems, the start-up Eden Tech developed a microreactor with a web-like design, allowing higher flowrates with lower pressure losses. This microreactor design is thermally pressed into a Flexdym™ (a thermoset compound created by Eden Tech [41]) chip at 165 °C during 760 s to create the superior part of the microreactor. The other part of the reactor consists of a circular 4 inches Si wafer, onto which a seed-layer solution has been deposited following the spin-coating process shown in part 2.1 (Figure 2c). The two parts of the microreactor are then bonded together in an oven at 120 °C during 4 h. After these two steps, the synthesis *per se* takes place: the growth solution, a volume balance mix of a solution of Zn(NO_3)$_2$ at 75 mM and a solution of HMTA at 37.5 mM, is continuously injected into a double-layer beaker where it is heated to the desired temperature before entering the microreactor, filling the channels, before exiting the microreactor. The functionalized microreactor obtained at the end of the production process is pictured in Figure 3. Playing with the solution flowrate (from 100 µL·min^{-1} to 400 µL·min^{-1}), the solution temperature (from 80 °C to 90 °C) and the growth time (from 30 min to 1 h 30 min) leads to changes in the ZnO NWs morphology and density, resulting in huge differences in their photocatalytic efficiency, as shown in our previous work [30].

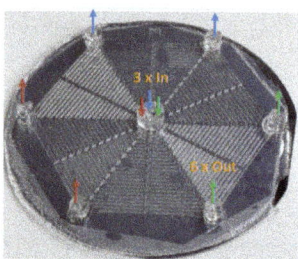

Figure 3. Picture of the final microreactor produced according to our method. The inlets (in the center) and outlets (at the edge) of the reactor are highlighted in color.

The best results, in terms of their photocatalytic efficiency, were obtained with the following parameters: a flowrate of 200 µL·min^{-1}, a solution temperature of 80 °C and a growth time of 1 h. Those parameters allow the growth of ZnO NWs with a mean diameter of 58 ± 5 nm and a mean density of 64 ± 5 NWs/µm^2. Increasing the temperature or the flowrate leads to the creation of aggregates and zones devoid from NSs in the microchannels, whereas decreasing the flowrate means smaller NWs diameter and density. Changing the growth time still leads to the NWs obtaining, but has an effect on their density and mean diameter, and, thus, on their photocatalytic efficiency [30].

The effect of the flowrate on the NSs is linked to the availability of the precursors in the solution: when the flowrate is too low (i.e., under 100 µL·min^{-1}), the precursors are not regenerated quickly enough compared to the reaction kinetics, leading to a limited total reaction rate, and as such, less and smaller NWs in the same reaction time. On the other hand, when the flowrate increases too much (above 200 µL·min^{-1}), the size of the depletion layer, defined as the layer in the flow above the bottom of the samples, where there is no precursor, as they are consumed by the reaction, decreases. This means that the precursors are more available for the reaction, promoting the NWs growth and their agglomeration into big clusters [30].

When the temperature increases above 80 °C, the solubility of the gases in the solution diminishes. As the ZnO growth reaction produces NH$_3$ (see Equation (4)), this leads to the apparition of gas bubbles in the microchannels, creating zones where the liquid does not flow, and where the reaction cannot take place. Combined with the augmentation of the reaction rate with the temperature, this causes the apparition of zones devoid from any NSs, and zones where the NSs agglomerate and form clusters [30].

Finally, the effect of the growth time is simply linked to the reaction kinetics: a smaller growth time leads to smaller and less dense NWs, and a higher growth time leads to bigger and more dense NWs [30]. The effect of the different growth conditions on the NSs was measured thanks to SEM observations. The obtained images for each of the different growth conditions are summarized in Figure 4.

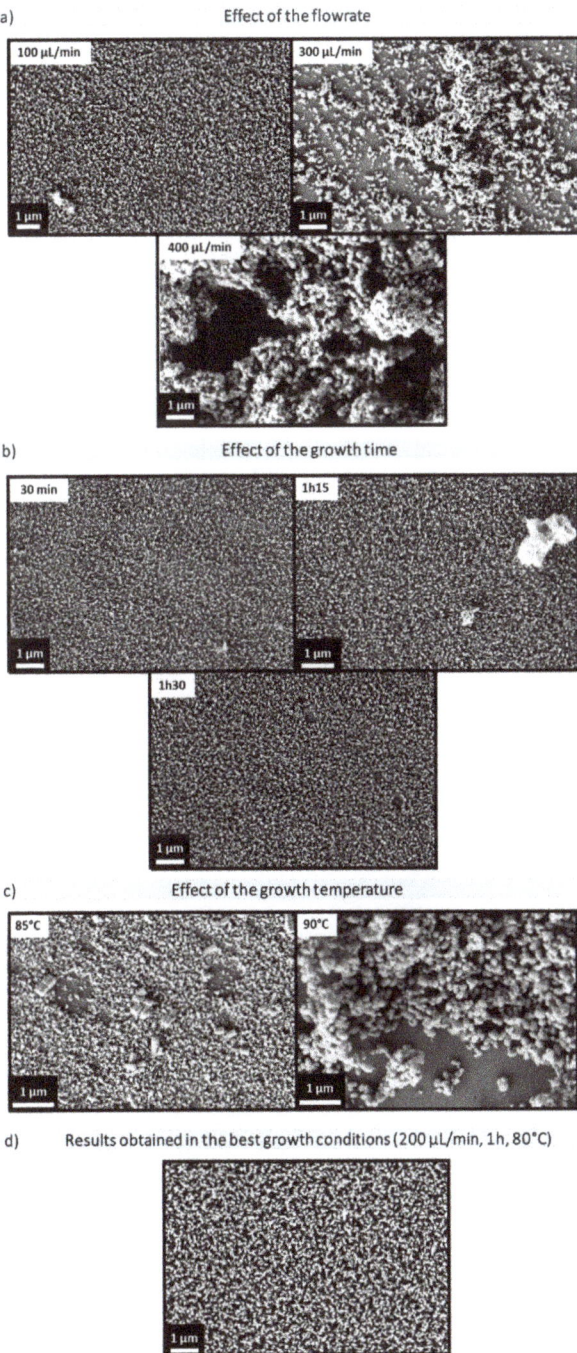

Figure 4. SEM images of the ZnO NWs obtained in the microreactor depending on the different growth conditions: effect of flowrate (**a**), effect of growth time (**b**), effect of the growth temperature (**c**) and the best growth conditions: 200 µL/min, 1 h, 80 °C (**d**) (reproduced from [30] with permission of the Royal Society of Chemistry).

2.4. ZnO Photocatalytic Activity Improvement Strategies

As a large band gap semiconductor, ZnO is particularly interesting as a stable photocatalyst with appropriate band-level potential energy to oxidize water and reduce oxygen, to provide radicals for the photocatalytic degradation of organic pollutants. Nevertheless, in view of an application under natural sunlight, extending the ZnO light absorption will be necessary. Indeed, ZnO is only able to absorb ~5% of natural sunlight, which corresponds to UV light. Increasing the quantity of light that the samples are able to absorb would help accelerate the photocatalytic reaction, and so, increase their photocatalytic efficiency. Another objective is to maximize the lifetime of the photo-generated electron-hole pairs, as this will increase their chances of producing the radicals needed by the reaction. Thus, until now, many strategies were developed to improve the ZnO light absorption, such as playing on the synthesis parameters and oxygen defects in the structure [42,43], by introducing non-metallic ions such as carbon (C), nitrogen (N), fluorine (F) [4] or metallic ions, such as iron (Fe), lead (Pb) or chromium (Cr) in the hydrothermal growth solution, and thus in the ZnO band gap [44,45], or the synthesis of ZnO composites with other semi-conductors [5].

In our work, we studied the modification of ZnO with carbon. According to the literature, the addition of carbon can be done by several routes, such as nanocomposites synthesis [46–48], surface modification by adsorption and calcination of carbonaceous species on the surface of the as-grown ZnO NSs [49,50], and carbon introduction in the ZnO lattice by adding carbonaceous species during the synthesis of the NSs [51,52]. To introduce carbon in the ZnO lattice, the $Zn(NO_3)_2$ salt was replaced by $Zn(Ac)_2$ in the hydrothermal growth solution (Figure 2a). SEM observations showed the good distribution of ZnO NWs grown with nitrate zinc salt (ZnO–N) with a diameter of 50 ± 7 nm, a length of 1.79 ± 0.10 µm and a density of $\sim 3 \times 10^9$ NWs/cm². A good distribution of NWs grown with acetate zinc salt (ZnO–A) was also observed with a diameter of 65 ± 8 nm, a length of 2.655 ± 0.1 µm and a density of $\sim 2 \times 10^9$ NWs/cm² (Figure 5a). As morphology investigation revealed, ZnO–A NWs are larger and longer than ZnO–N NWs. In accordance with the literature, an impact of the counter ion used is observed [53]. The use of zinc acetate salt for NWs growth enhances the capping effect, inhibits the rates dissolution and aids the aging and growth of the ZnO NWs [54]. Indeed, acetate ions will be selectively adsorbed during the growth and the intercalation will extend the NWs along the c-axis [48,55]. Gap measurements suggested the acetate ions adsorbed during the growth and post-annealing seem to contribute to the ZnO band gap reduction (3.18 eV against 3.21 eV) [56].

As already mentioned, another studied strategy was to dope ZnO with transition metals. This strategy is based on the addition of transition metal ions in the growth solution. In short, the samples are synthesized by a simple hydrothermal method, already described in Figure 2a, with a growth solution containing 1.125 mM of $Zn(NO_3)_2$, 0.5625 mM of HMTA and the appropriate quantity of the doping solution ($FeCl_3$, $AgNO_3$ or $Co(NO_3)_2$) to obtain a molar percentage of 1%, 2%, or 3% in dopant product [44,45]. SEM study showed quite homogeneous ZnO NWs arrays for all Fe and Co doped samples with aspect ratios and morphologies similar to those of the undoped samples (Figure 5b). The Ag-doped samples exhibited a dependence on the dopant concentration, leading to larger and smaller NWs with an increase in Ag% (Figure 5c). Concerning the gap values, the band gaps decrease with the increasing dopant quantity (3.10 eV against 3.21 eV).

Unfortunately, this method, for which the efficiency was already demonstrated [44,45], could not be applied on construction materials due to possible interactions between the non-conventional substrate and the dopant ions, which could have an adverse effect on the photocatalytic efficiency under natural sunlight [22]. Thus, for this application, a post-grown ZnO co-catalyst synthesis was developed to reduce the contact between the substrate and the doping solution, and thus avoid any interaction between them. The Fe(III) ion was selected to improve the ZnO NWs light-absorption and photocatalytic activity, owing to its low-cost, relatively eco-friendly properties and previous results [45]. The Fe(III) solution was dropped on ZnO NWs grown onto Si (ZnO–A), as described in Figure 2a, then dried and calcinated to fix the iron oxide on the surface of ZnO NWs. A volume of

69.8 µL of Fe(NO$_3$)$_3$, whose iron concentration range is included between 8.9 ×10^{-9} mol and 4.5 × 10^{-7} mol, was applied on the sample surface before a drying at 50 °C for 30 min in an oven. Finally, the sample was annealed at 350 °C for 1 h in ambient atmosphere to convert and fix the iron oxide on the ZnO surface (ZnO/Fe$_x$O$_y$). This co-catalyst synthesis has shown its effectiveness to extend the range of absorbed light thanks to the deposition of iron oxide on ZnO NWs as it leads to a decrease of the gap value from 3.27 eV to 3.14 eV and higher visible light absorption [22]. XRD works did not permit to determine the iron oxide phase, but by drawing parallel with literature [57,58], strong assumptions on the FeOOH and Fe$_2$O$_3$ combined presence were made. SEM observations demonstrated a non-homogeneous deposition of iron oxide and, thus, the need of improving the deposition process (Figure 5d) [22].

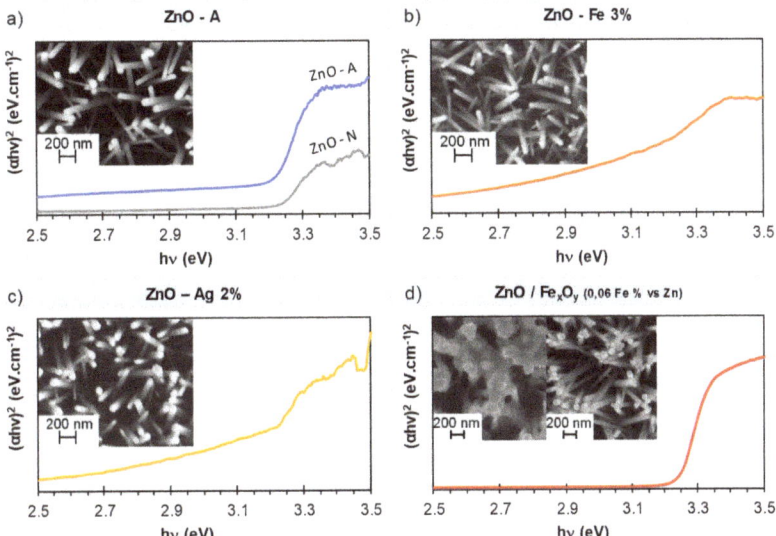

Figure 5. UV–visible spectral plots with Tauc-Lorentz model and SEM images of different modified ZnO NWs: ZnO-A (a), ZnO–Fe 3% (b), ZnO–Ag 2% (c), and ZnO/Fe$_x$O$_y$ (0.06 Fe % vs Zn), (d) respectively.

3. Photocatalysis for Water Purification

The photocatalytic performances of these developed materials were proved for the photodegradation of model organic molecules commonly used in the pharmaceutical, food and textile industries. The selected molecules were three organic dyes: acid red 14 (AR14), methylene blue (MB), and methyl orange (MO). The different dye solutions had an initial concentration of 10 µM. Samples of ZnO-decorated civil engineering materials were immersed into 30 mL of dye-contaminated aqueous solution and irradiated with a UV lamp (Hamamatsu LC8, 4500 mW/cm^2, λ = 365 nm) under magnetic stirring. The photocatalytic process was monitored by UV-visible spectrophotometry every 15 min for 3 h and the degradation efficiency X(%) was estimated thanks to Equation (6):

$$X(\%) = \frac{A_0 - A}{A_0} \times 100 \quad (6)$$

where A_0 and A, respectively, stand for the initial and actual absorption peak values at the wavelength of the maximum absorption for the studied dye (λ$_{max}$ = 665 nm for MB, 515 nm for AR14, and 464 nm for MO). All Si-functionalized samples and functionalized civil engineering materials demonstrated their ability to degrade the three selected organic dyes in less than 3 h [22–24,28]. MO, the most difficult to degrade of our model pollutants due to its stability, was degrade more slowly (~105 min) than MB and AR14 (~90 min). The recyclability of the samples was also proved with no losses of efficiency after 5 cycles of uses [28].

Depending on the growth conditions, either C1 or C2 (Section 2.1), ZnO–N grown on silicon demonstrated different efficiencies (Figure 6a). These differences can be assigned to the bigger ZnO–C2 surface area. The same should explain the higher photocatalytic activity of functionalized tiling and concrete samples after 4 h of hydrothermal growth. Indeed, the silicon substrate area is around 2.25 cm^2 and the civil engineering material substrate areas are around 6.25 cm^2. Larger photocatalytic differences could be expected in regard to the surface areas difference. The weak gap in efficiency could be explained by the non-optimal agitation and reactor for civil engineering materials. Then, the photocatalytic efficiency of functionalized tiling and concrete samples were improved for AR14 degradation by reducing its growth time synthesis [29].

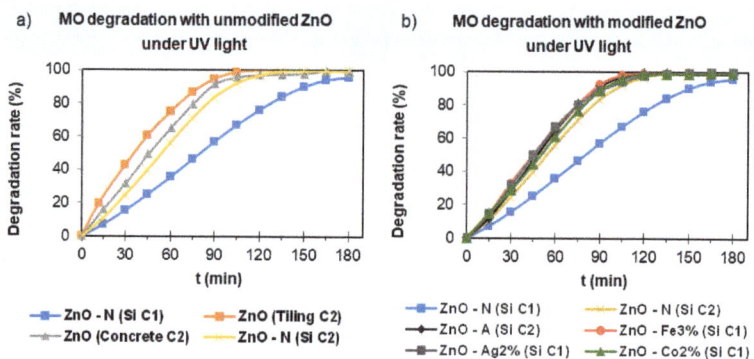

Figure 6. MO degradation with unmodified (**a**) and modified (**b**) ZnO grown onto different substrates under UV light (~365 nm, $I_{received\ by\ the\ sample}$ = 35 mW/cm^2).

Previous results on ZnO grown in the conditions C1 demonstrated that all doped samples showed better efficiency than undoped samples, the most efficient for each dopant being the ZnO doped with 3% of Fe (ZnO-Fe3%), and 2% of Ag or Co (ZnO–Ag2% and ZnO–Co2%) [44]. The photocatalytic efficiency enhancement (Figure 6b) can be attributed to two effects: the reduction of the band-gap energy, allowing more photons to be absorbed; and the reduced electron–hole recombination rate, thanks to the increase in oxygen vacancies and to the dopant ions on the NWs surface. Similar results were obtained with ZnO–A (C2) samples, which efficiency is better than the one of the ZnO–N (C1 and C2) samples, demonstrating the efficiency of the modification of ZnO strategies to increase the photocatalytic efficiency of the samples.

Then, the photocatalytic performances of these optimal functionalized samples were proved for dye photodegradation under natural sunlight. Under sunny weather, MB and AR14 were fully degraded in 4 h over the optimal concrete sample and were removed from water with degradation rates of 98% and 60% for the tiling optimal sample. MO, which is the most difficult dye to degrade, was degraded at 82% and 58% over concrete and tiling samples, respectively, in the same conditions [29]. Cloudy weather leads to a slight decrease in the efficiency with the lowered UV intensity received by the samples, but still offers excellent photocatalytic activity. In a previous study [22], the efficiency of the ZnO/iron oxide samples was also evaluated and proved for the MO photodegradation under natural sunlight. Modifying the ZnO with an optimal ratio of 0.06% of iron (Fe vs. Zn M) improved the photocatalytic efficiency under natural sunlight with an increase of 11% of the MO degradation, compared to the reference MO degradation with ZnO, after 5 h of irradiation with no iron leaching [22].

In parallel to these experiments, the photocatalytic efficiency of the ZnO-NSs-based microfluidic reactors—for which production was described in Section 2.3—was proved by degrading continuous streams of water polluted with AR14, under UV irradiation (Figure 7). The obtained results lead to the determination of the optimal synthesis parameters for the best photocatalytic efficiency (200 µL·min^{-1}, 80 °C, 1 h). Thanks to these NSs-based microfluidic reactor, a 98% photodegradation rate was reached after 4 passes in the reactor, which is roughly equivalent to 40 s of UV light irradiation [30]. In this same work, the photocatalysis flow rate influence was studied, and while the optimum was observed at 200 µL·min^{-1}, very good results were obtained when the flowrate was increased to 500 µL·min^{-1} (more than 80% of degradation obtained in 5 passes, thus, 50 s of irradiation). These results are highly encouraging in the aim of designing high flowrate microreactors.

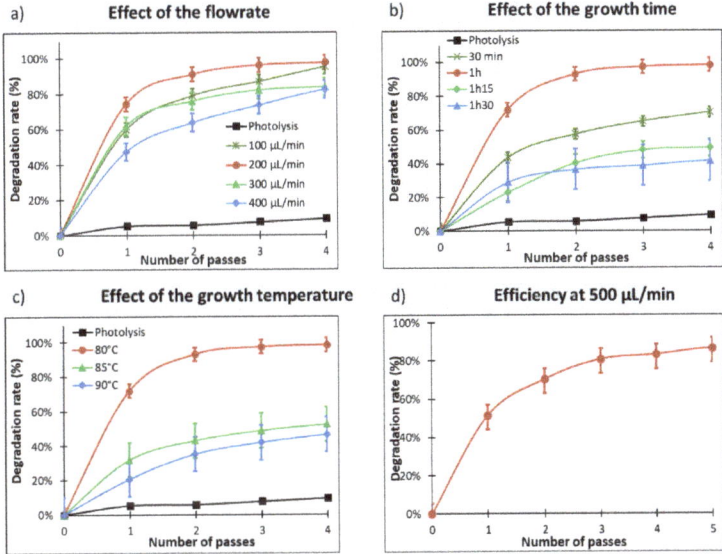

Figure 7. Effect of the different growth conditions (a) flowrate, (b) growth time and (c) growth temperature and effect of the augmentation of the photocatalysis flowrate (d) on the photocatalytic efficiency of the samples for AR14 degradation under UV light (~365 nm, $I_{received\ by\ the\ sample}$ = 35 mW/cm^2, reproduced from [30] with permission of the Royal Society of Chemistry).

Thus, these results imply that ZnO NSs could be used on different substrates and under different modes for environmental remediation by photocatalysis under UV light and natural sunlight, with no particular need to dope or modify the ZnO NSs. Nevertheless, modified ZnO led to an improved photocatalytic efficiency.

4. HPLC-MS Analysis and Comparison with UV-Visible Results

The photocatalytic results presented in the previous parts were obtained by UV-visible spectroscopy. This kind of analysis is very efficient to quickly estimate the degradation rate of the pollutants but does not give any information on the degradation mechanism or on the remaining compounds in the effluent solution. This could lead to considering effluents in which dangerous compounds remain as safe. Thus, a more sensitive analytic method is required to ensure the harmlessness of the effluents.

High-Performance Liquid Chromatography coupled with Mass Spectrometry (HPLC-MS) is a very powerful method [59], widely used in the detection and quantification of several different compounds. This method is, for example, used to detect active compounds in medicinal herbs [60], or to identify medicaments that are new [61,62] or counterfeit [63].

However, it requires much more preparation and more expensive equipment than UV-visible spectrometry.

We already followed the degradation of BM by ZnO NWs samples, synthesized under the conditions C1 (see Section 2.1), by UV-visible spectroscopy and by HPLC-MS analysis in our previous work [24]. We found that, even if MB absorbance reaches zero, HPLC-MS is still able to detect MB in the effluents, and that the concentrations calculated from UV-visible spectroscopy are lower than the ones obtained from the HPLC-MS results. However, according to the two different methods, even if there is still MB in the effluents at the end of the photodegradation, the final effluents are safe for humans, in regard to their MB concentration (14 µg·L^{-1}), and proving that the photocatalytic degradation by ZnO NWs is an efficient method to purify water from organic pollutants.

In light of the results obtained with MB, we performed the same experimental protocol on AR14, still following the degradation by UV spectroscopy and HPLC-MS. The UV-visible results are presented in Figure 8a, and showed a degradation rate of 95% after 90 min, and a degradation rate higher than 99% after 135 min. The mobile phase composition of the HPLC-MS analysis was adapted from literature [64], and the HPLC-MS parameters are summarized in Table 1.

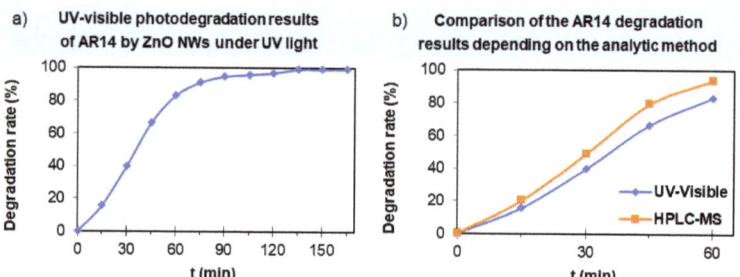

Figure 8. (a) UV-visible photodegradation results of AR14 by ZnO NWs (grown conditions (1), 4 h, 90 °C) under UV light and (b) comparison of the AR14 degradation results depending on the analytic method (UV light: ~365 nm, $I_{received\ by\ the\ sample}$ = 35 mW/cm^2).

Table 1. HPLC-MS parameters of the analysis of AR14.

HPLC	
Mobile phase	Acetonitrile 20% (volumic)
	Aqueous ammonium acetate (0.05 M) 80% (volumic)
Flowrate	0.2 mL·min^{-1}
Injection volume	20 µL
Mass spectrometer	
Polarity	Negative
Frequency	Extended Dynamic Range 2 GHz
Mass range	50–1700 m/z

Examining the AR14 degradation by HPLC-MS showed the disappearance of both the AR14 chromatography and MS peaks after 90 min under UV light. The HPLC-MS results could only be used for the calculation of the remaining AR14 concentration for the first 60 min, as the AR14 detection decreased below the quantification limit of the device after that time stamp. The obtained results are presented in Table 2.

Table 2. HPLC-MS photodegradation results of AR 14 by ZnO NWs under UV light.

Time (min)	m/z 288 Peak Area	Calculated Concentration (µM)	Calculated Degradation Rate
0	2,963,405	8.59	0%
15	2,370,033	6.87	20.02%
30	151,015	4.38	49.01%
45	605,947	1.76	79.55%
60	174,369	0.51	94.12%

Comparing the results obtained from HPLC-MS and from the UV-vis demonstrates that, conversely to what was observed for MB, HPLC-MS seems to overestimate the degradation results. This could be due to the parasitic molecules we detected in the reaction mix, which could be increasing the absorbance of the reaction mix, thus decreasing the degradation rate calculated from the UV-visible results. The comparison curves of the two results are shown in Figure 8b.

Unfortunately, no degradation product was identified in the reaction mix, and we could not provide a degradation mechanism for the degradation of AR14. This might be due to a variety of factors, such as the short lifespan of the degradation products, making them disappear before the analysis, or their quantity being too low to be detected even by the HPLC. However, we could calculate the concentration of AR14 in the reaction mix after 60 min, and found it to be of 0.51 µM, which corresponds to 254 µg·L^{-1}. Considering the degradation continues for 105 more minutes, we can make the hypothesis that the actual final concentration will be even lower than 254 µg·L^{-1}.

It is still possible to compare the concentration we calculated after 60 min to the safety limits recommended by literature and governments. In the EU, the quantity of AR14 in food and drinks must stay in the 50–500 mg·L^{-1} interval, depending on the considered food and drink types [65]. According to the AR14 Safety Data Sheet, its toxicity threshold for algae is 34.82 mg·L^{-1} for 3 days. Finally, chronic toxicity is reached with a concentration of 1% AR14 in water [66]. Thus, with their calculated concentration of 254 µg·L^{-1}, we can state that our effluents are safe, both for humans and for the environment.

5. Conclusions

This work is a review of our previous works on the development of ZnO NSs-based photocatalyst materials by hydrothermal synthesis and their photocatalytic activity for organic dyes removal. Firstly, the universality and the feasibility of the synthesis method were presented from the initial Si substrates to non-conventional substrates such as civil engineering materials, allowing for the obtaining of ZnO NSs with gap values from 3.18 eV to 3.23 eV and ZnO Wurtzite XRD peaks. In this section, a short summary on microfluidic cells development and their main results are also presented. Strategies developed to reduce the ZnO band gap and improve its photocatalytic efficiency are also introduced by demonstrating how modifying ZnO with carbon, doping ZnO with transition metal ions, and synthesizing ZnO/iron oxide co-catalyst via simple processes are easily adaptable to the industry, improving the photodegradation efficiency. Then, this paper displayed the photocatalytic results of the different kinds of samples under classic and microfluidic modes, and under UV and solar light, for the degradation of MO, AR14 and MB. Excellent results were obtained with a total degradation of the three dyes after 3 h under UV light, as well as very promising results under natural sunlight, even on cloudy days. Furthermore, the effect of the used substrates and the growth conditions on the kinetics and efficiency of the photodegradation were discussed. Lastly, this paper investigated the photodegradation results obtained with HPLC-MS for MB and AR 14, by comparing them with the UV-visible results, and used the previous method to demonstrate that the final effluents after the degradation are safe for humans and the environment.

Author Contributions: M.L.P.: Conceptualization, Methodology, Investigation, Formal analysis, Writing-original draft, Validation; N.M.: Methodology, Investigation, Formal analysis, Writing-original draft; Y.L.-W.: Conceptualization, Supervision, Validation, Project administration. All authors have read and agreed to the published version of the manuscript.

Funding: This research received no external funding.

Institutional Review Board Statement: Not applicable.

Informed Consent Statement: Not applicable.

Data Availability Statement: The study did not report any data.

Conflicts of Interest: The authors declare no conflict of interest.

References

1. Corcoran, E.; Nellemann, C.; Baker, E.; Bos, R.; Osborn, D.; Savelli, H. *Sick Water? The Central Role of Wastewater Management in Sustained Development*; UN-HABITAT, GRID-Arendal; United Nations Environment Programme: Nairobi, Kenya, 2010; pp. 1–88.
2. World Economic Forum. *The Global Risks Report*, 11th ed.; WEF: Cologny, Switzerland, 2016; pp. 1–103.
3. Singh, V.P.; Sandeep, K.; Kushwaha, H.S.; Powar, S.; Vaish, R. Photocatalytic, Hydrophobic and antimicrobial characteristics of ZnO nano needle embedded cement composites. *Construct. Build. Mater.* **2018**, *185*, 285–294. [CrossRef]
4. Lee, K.M.; Lai, C.W.; Ngai, K.S.; Ruan, J.C. Recent developments of zinc oxide based photocatalyst in water treatment technology: A review. *Water Res.* **2016**, *88*, 428–448. [CrossRef] [PubMed]
5. Liu, Y.; Yu, L.; Hu, Y.; Guo, C.; Zhang, F.; Lou, X.W. A magnetically separable photocatalyst based on nest-like γ-Fe_2O_3/ZnO double-shelled hollow structures with enhanced photocatalytic activity. *Nanoscale* **2012**, *4*, 183–187. [CrossRef] [PubMed]
6. Sun, C.; Xu, Q.; Xie, Y.; Ling, Y.; Hou, Y. Designed synthesis of anatase–TiO_2 (B) biphase nanowire/ZnO nanoparticle heterojunction for enhanced photocatalysis. *J. Mater. Chem.* **2018**, *6*, 8289–8298. [CrossRef]
7. Gosh, S.; Amoin Kouame, N.; Remita, L.; Ramos, L.; Goubard, F.; Aubert, P.H.; Dazzi, A.; Deniset-Besseau, A.; Remita, H. Visible-light active conducting polymer nanostructures with superior photocatalytic activity. *Sci. Rep.* **2015**, *5*, 18002. [CrossRef]
8. Chen, F.; Liu, H.; Bagwasi, S.; Shen, X.; Zhang, J. Photocatalytic study of BiOCl for degradation of organic pollutants under UV irradiation. *J. Photochem. Photobiol. A Chem.* **2010**, *215*, 76–80. [CrossRef]
9. Manthina, V.; Agrios, A.G. Single-pot ZnO nanostructure synthesis by chemical bath deposition and their applications. *Nano-Struct. Nano-Objects* **2016**, *7*, 1–11. [CrossRef]
10. Wojnarowicz, J.; Chudoba, T.; Lojkowski, W. A Review of Microwave Synthesis of Zinc Oxide Nanomaterials: Reactants, Process Parameters and Morphologies. *Nanomaterials* **2020**, *10*, 1086. [CrossRef]
11. Noman, M.T.; Amor, N.; Petru, M. Synthesis and applications of ZnO nanostructures (ZONSs): A review. *Crit. Rev. Solid State Mater. Sci.* **2021**, 1–43. [CrossRef]
12. Le Pivert, M.; Kerivel, O.; Zerreli, B.; Leprince-Wang, Y. ZnO nanostructures based innovative photocatalytic road for air purification. *J. Clean. Prod.* **2021**, *318*, 128447. [CrossRef]
13. Qiu, Y.; Wang, L.; Xu, L.; Shen, Y.; Wang, L.; Liu, Y. Shaped-controlled growth of sphere-like ZnO on modified polyester fabric in water bath. *Mater. Lett.* **2021**, *288*, 129342. [CrossRef]
14. Choudhary, S.; Sahu, K.; Bisht, A.; Satpati, B.; Mohapatra, S. Rapid synthesis of ZnO nanowires and nanoplates with highly enhanced photocatalytic performance. *Appl. Surf. Sci.* **2021**, *541*, 148484. [CrossRef]
15. Zhang, Y.; Huang, X.; Yeom, J. A flotable piezo-photocatalytic platform based on semi-embedded ZnO nanowire array for high-performance water decontamination. *Nano-Micro Lett.* **2019**, *11*, 11. [CrossRef]
16. Hossain, M.F.; Naka, S.; Okada, H. Fabrication of perovskite solal cells with ZnO nanostructures prepared on seedless ITO substrate. *J. Mater. Sci. Mater. Elec.* **2018**, *29*, 13864–13871. [CrossRef]
17. Vayssieres, L. Growth of Arrayed Nanorods and Nanowires of ZnO from Aqueous Solutions. *Adv. Mater.* **2003**, *15*, 464–466. [CrossRef]
18. Vayssieres, L.; Keis, K.; Lindquist, S.E.; Hagfeldt, A. Purpose-Built Anisotropic Metal Oxide Material: 3D Highly Oriented Microrod Array of ZnO. *J. Phys. Chem. B* **2001**, *105*, 3350–3352. [CrossRef]
19. Bao, Y.; Wang, C.; Ma, J.Z. Morphology control of ZnO microstructures by varying hexamethylenetetramine and trisodium citrate concentration and their photocatalytic activity. *Mater. Des.* **2016**, *101*, 7–15. [CrossRef]
20. McPeak, K.M.; Le, T.P.; Britton, N.G.; Nickolov, Z.S.; Elabd, Y.A.; Baxter, J.B. Chemical Bath Deposition of ZnO Nanowires at Near—Neutral pH Conditions without Hexamethylenetetramine (HMTA): Understanding the Role of HMTA in ZnO Nanowire Growth. *Langmuir* **2011**, *27*, 3672–3677. [CrossRef]
21. Greene, L.E.; Yuhas, B.D.; Law, M.; Zitoun, D.; Yang, P. Solution-grown Zinc oxide nanowires. *Inorg. Chem.* **2006**, *45*, 735–7543. [CrossRef]
22. Le Pivert, M.; Suo, H.; Tang, G.; Qiao, H.; Zhao, Z.; Martin, N.; Liu, C.; Leprince-Wang, Y. Improving natural sunlight photocatalytic efficiency of ZnO nanowires decorated by iron oxide cocatalyst via a simple drop method. *Mater. Chem. Phys.* **2022**, *275*, 125304. [CrossRef]
23. Habba, Y.G.; Capochichi-Gnambodoe, M.; Serairi, L.; Leprince-Wang, Y. Enhanced photocatalytic activity of ZnO nanostructure for water purification. *Phys. Status Solidi B* **2016**, *253*, 1480–1484. [CrossRef]

24. Martin, N.; Leprince-Wang, Y. HPLC-MS and UV–Visible Coupled Analysis of Methylene Blue Photodegradation by Hydrothermally Grown ZnO Nanowires. *Phys. Status Solidi A* **2021**, *218*, 2100532. [CrossRef]
25. Fathy, A.; Le Pivert, M.; Kim, Y.J.; Ba, M.O.; Erfan, M.; Sabry, Y.M.; Khalil, D.; Leprince-Wang, Y.; Bourouina, T.; Gnambodoe-Capochichi, M. Continuous Monitoring of Air Purification: A Study on Volatile Organic Compounds in a Gas Cell. *Sensors* **2020**, *20*, 934. [CrossRef] [PubMed]
26. Tian, J.-H.; Hu, H.; Li, S.-S.; Zhang, F.; Liu, J.; Shi, J.; Li, X.; Tian, Z.-Q.; Chen, Y. Improved seedless hydrothermal synthesis of dense and ultralong ZnO nanowires. *Nanotechnol.* **2011**, *22*, 245601. [CrossRef]
27. Gao, L. A Dual Approach for Water Purification Based on Solar Energy. Ph.D. Thesis, Degree-Granting Gustave Eiffel University, Paris, France, 2022.
28. Le Pivert, M.; Poupart, R.; Capochichi-Gnambodoe, M.; Martin, N.; Leprince-Wang, Y. Direct growth of ZnO nanowires on civil engineering materials: Smart materials for supported photodegradation. *Microsyst. Nanoengin.* **2019**, *5*, 57. [CrossRef]
29. Le Pivert, M.; Zerelli, B.; Martin, N.; Capochichi-Gnambodoe, M.; Leprince-Wang, Y. Smart ZnO decorated optimized engineering materials for water purification under natural sunlight. *Construct. Build. Mater.* **2019**, *257*, 119592. [CrossRef]
30. Martin, N.; Lacour, V.; Perrault, C.M.T.; Roy, E.; Lerince-Wang, Y. High flow rate microreactors integrating in situ grown ZnO nanowires for photocatalytic degradation. *React. Chem. Eng.* **2022**, 1–8. [CrossRef]
31. Pastor, A.; Balbuena, J.; Cruz-Yusta, M.; Pavlovic, I.; Schànchez, L. ZnO on rice husk: A sustainable photocatalyst for urban air purification. *Chem. Eng. J.* **2019**, *368*, 659–667. [CrossRef]
32. Abdulrahman, A.F.; Ahmed, S.M.; Hamad, S.M.; Almessiere, M.A.; Ahmed, N.M.; Sadaji, S.M. Effect of different pH values on growth solutions for the ZnO nanostructures. *J. Phys.* **2021**, *71*, 175–189. [CrossRef]
33. Khoa, N.T.; Kim, S.W.; Thuan, D.V.; Yoo, D.H.; Kim, E.J.; Hahn, S.H. Hydrothermally controlled ZnO nanosheet self-assembled hollow spheres/hierarchical aggregates and their photocatalytic activities. *CrystEngComm* **2013**, *16*, 1344–1350. [CrossRef]
34. Samadi, M.; Shivaee, H.A.; Pourjavadi, A.; Moshfegh, A.Z. Synergism of oxygen vacancy and carbonaceous species on enhanced photocatalytic activity of electrospun ZnO-carbon nanofibers: Charge carrier scavengers mechanism. *Appl. Catal. A Gen.* **2013**, *466*, 153–160. [CrossRef]
35. Yu, Z.; Moussa, H.; Liu, M.; Schneider, R.; Moliere, M.; Liao, H. Solution precursor plasma spray process as an alternative rapid one-step route for the development of hierarchical ZnO films for improved photocatalytic degradation. *Ceram. Int.* **2018**, *44*, 2085–2092. [CrossRef]
36. Erfan, M.; Gnambodoe-Capochichi, M.; Sabry, Y.M.; Khalil, D.; Leprince-Wang, Y.; Bourouina, T. Spatiotemporal dynamics of nanowire growth in a microfluidic reactor. *Microsyst. Nanoeng.* **2021**, *7*, 77. [CrossRef]
37. McPeak, K.M.; Baxter, J.B. ZnO Nanowires Grown by Chemical Bath Deposition in a Continuous Flow Microreactor. *Cryst. Growth. Des.* **2009**, *9*, 4538–4545. [CrossRef]
38. He, Z.; Li, Y.; Zhang, Q.; Wang, H. Capillary microchannel-based microreactors with highly durable ZnO/TiO$_2$ nanorod arrays for rapid, high efficiency and continuous-flow photocatalysis. *Appl. Catal. B* **2010**, *93*, 376–382. [CrossRef]
39. McPeak, K.M.; Baxter, J.B. Microreactor for High-Yield Chemical Bath Deposition of Semiconductor Nanowires: ZnO Nanowire Case Study. *Ind. Eng. Chem. Res.* **2009**, *48*, 5954–5961. [CrossRef]
40. Wang, N.; Zhang, X.; Wang, Y.; Yu, W.; Chan, H.L.W. Microfluidic reactors for photocatalytic water purification. *Lab Chip* **2014**, *14*, 1074–1082. [CrossRef]
41. Cargou, S. Dispositif Fluidique D'assainissement d'un Fluide et Procédé D'assainissement Associé. Patent number WO2019/053219, 21 March 2019.
42. Dash, P.; Manna, A.; Mishra, N.C.; Varma, S. Synthesis and characterization of aligned ZnO nanorods for visible light photocatalysis. *Physica E* **2019**, *107*, 38–46. [CrossRef]
43. Bora, T.; Sathe, P.; Laxman, K.; Dobrestov, S.; Dutta, J. Defect engineered visible light active ZnO nanorods for photocatalytic treatment of water. *Catal. Today* **2017**, *284*, 11–18. [CrossRef]
44. Martin, N.; Capochichi-Gnambodoe, M.; Le Pivert, M.; Leprince-Wang, Y. A Comparative Study on the Photocatalytic Efficiency of ZnO Nanowires Doped by Different Transition Metals. *Acta Phys. Pol.* **2019**, *135*, 471–474. [CrossRef]
45. Habba, Y.G.; Capochichi-Gnambodoe, M.; Leprince-Wang, Y. Enhanced Photocatalytic Activity of Iron-Doped ZnO Nanowires for Water Purification. *Appl. Sci.* **2017**, *7*, 1185. [CrossRef]
46. Chen, Y.C.; Katsumata, K.I.; Chiu, Y.H.; Okada, K.; Matsushita, N.; Hsu, Y.J. ZnO-graphene composites as practical photocatalysts for gaseous acetaldehyde degradation and electrolytic water oxidation. *Appl. Catal. A Gen.* **2015**, *90*, 1–9. [CrossRef]
47. Bechambi, O.; Sayadi, S.; Najjar, W. Photocatalytic degradation of bisphenol A in the presence of C-Doped ZnO: Effect of operational parameters and photodegradation mechanism. *J. Ind. Eng. Chem.* **2015**, *32*, 201–210. [CrossRef]
48. Liu, X.; Du, H.; Sun, X.W.; Liu, B.; Zhao, D.; Sun, H. Visible-light photoresponse in a hollow microtube-nanowire structure made of carbon-doped ZnO. *CrystEngComm* **2012**, *8*, 2886–2890. [CrossRef]
49. Akir, S.; Hamdi, A.; Addad, A.; Coffinier, Y.; Boukherroub, R.; Omrani, A.D. Facile synthesis of carbon-ZnO nanocomposite with enhanced visible light photocatalytic performance. *Appl. Surf. Sci.* **2017**, *400*, 461–470. [CrossRef]
50. Chakraborty, P.; Majumder, T.; Dhar, S.; Mondal, S.P. Nonenzymetic glucose sensing using carbon functionalized carbon doped ZnO nanorod arrays. *AIP Conf. Proc.* **2018**, *1942*, 050074. [CrossRef]
51. Liu, S.; Li, C.; Yu, J.; Xiang, Q. Improved visible-light photocatalytic activity of porous carbon self-doped ZnO nanosheet-assembled flowers. *CrystEngComm* **2011**, *13*, 2533–2541. [CrossRef]

52. Chen, L.C.; Tu, Y.J.; Wang, Y.S.; Kan, R.S.; Huang, C.M. Characterization and photoreactivity of N-, S-, and C- doped ZnO under UV and visible light illumination. *J. Photochem. Photobio. A Chem.* **2008**, *199*, 170–178. [CrossRef]
53. Yogamalar, N.R.; Bose, A.C. Tuning the aspect ratio of hydrothermally grown ZnO by choice of precursor. *Solidi State Chem.* **2011**, *184*, 12–20. [CrossRef]
54. Govender, K.; Boyle, D.S.; Kenway, P.B.; O'Bien, P. Understanding the factors that govern the deposition and morphology of thin films of ZnO from aqueous solution. *J. Mater. Chem.* **2004**, *14*, 2575–2591. [CrossRef]
55. Akgun, M.C.; Kalay, Y.E.; Unalan, H.E. Hydrothermal zinc oxide nanowire growth using zinc acetate dihydrate salt. *J. Mater. Res.* **2012**, *27*, 1445–1451. [CrossRef]
56. Le Pivert, M. Nanostructures de ZnO Pour le Développement de Route Dépolluantes Photocatalytiques. Ph.D. Thesis, Degree-Granting University Paris-Est, Paris, France, 2020.
57. Shi, W.; Gao, T.; Zhang, L.; Ma, Y.; Liu, Z.; Zhang, B. Tailoring the surface structures of iron oxide nanorods to support Au nanoparticles for CO oxidation. *Chin. J. Catal.* **2020**, *40*, 1884–1894. [CrossRef]
58. Achouri, F.; Corbel, S.; Aboulaich, A.; Balan, L.; Ghrabi, A.; Ben Said, M.; Schneider, R. Aqueous synthesis and enhanced photocatalytic activity of ZnO/Fe_2O_3 heterostructures. *J. Phys. Chem. Solid.* **2014**, *75*, 1081–1087. [CrossRef]
59. Fernandez, C.; Soledad Larrechi, M. An analytical overview of processes for removing organic dyes from wastewater effluents. *Trend. Anal. Chem.* **2010**, *29*, 1202–1211. [CrossRef]
60. Steinmann, D.; Ganzera, M. Recent advances on HPLC/MS in medicinal plant analysis. *J. Pharm. Biomed. Anal.* **2011**, *55*, 744–757. [CrossRef] [PubMed]
61. Ackermann, B.L.; Berna, M.J.; Murphy, A.T. Recent Advances in use of LC/MS/MS for Quantitative High-Throughput Bioanalytical Support of Drug Discovery. *Curr. Top. Med. Chem.* **2002**, *2*, 53–66. [CrossRef]
62. Korfmacher, W.A. Foundation review: Principles and applications of LC-MS in new drug discovery. *Drug Discov. Today* **2005**, *20*, 1357–1367. [CrossRef]
63. Martino, R.; Malet-Martino, M.; Gilard, V.; Balayssac, S. Counterfeit drugs: Analytical techniques for their identification. *Anal. Bioanal. Chem.* **2010**, *398*, 77–92. [CrossRef]
64. Wang, A.; Qu, J.; Liu, H.; Ge, J. Degradation of azo dye Acid Red 14 in aqueous solution by electrokinetic and electrooxidation process. *Chemosphere* **2004**, *55*, 1189–1196. [CrossRef]
65. Peksa, V.; Jahn, M.; Stolcova, L.; Schulz, V.; Proska, J.; Prochazka, M.; Weber, K.; Cialla-May, D.; Popp, J. Quantitative SERS Analysis of Azorubine (E 122) in Sweet Drinks. *Anal. Chem.* **2015**, *87*, 2840–2844. [CrossRef]
66. European Food Safety Authority. Scientific opinion on the re-evaluation of azorubine/carmoisine (E 122) as a food additive. *EFSA J.* **2009**, *7*, 1332. [CrossRef]

Review

Recent Advances in ZnO-Based Nanostructures for the Photocatalytic Degradation of Hazardous, Non-Biodegradable Medicines

K. M. Mohamed [1], J. John Benitto [1], J. Judith Vijaya [1,*] and M. Bououdina [2]

1 Catalysis and Nanomaterials Research Laboratory, Department of Chemistry, Loyola College, Chennai 600 034, India
2 Department of Mathematics and Sciences, Faculty of Humanities and Sciences, Prince Sultan University, Riyadh 12435, Saudi Arabia
* Correspondence: judithvijaya@loyolacollege.edu

Abstract: Antibiotics are pervasive contaminants in aqueous systems that pose an environmental threat to aquatic life and humans. Typically, antibiotics are developed to counteract bacterial infections; however, their prolonged and excessive use has provoked unintended consequences. The presence of excessive amounts of antibiotics and anti-inflammatory, anti-depressive, and contraceptive drugs in hospital and industrial wastewater poses a significant threat to the ecosystem, with groundwater containing drug concentrations of <1 mg/L to hundreds of µg/L. According to the literature, 33,000 people die directly from drug-resistant bacterial infections in Europe annually, which costs EUR 1.5 billion in health care and productivity loss. Consequently, the continuous spread of antibiotics in the ecosystem has led to greater interest in developing a sustainable method for effective antibiotic removal from wastewater. This critical review aims to present and discuss recent advances in the photocatalytic degradation of widely used drugs by ZnO-based nanostructures, namely (i) antibiotics; (ii) antidepressants; (iii) contraceptives; and (iv) anti-inflammatories. This study endows a comprehensive understanding of the degradation of antibiotics using ZnO-based nanomaterials (bare, doped, and composites) for effective treatment of wastewater containing antibiotics. In addition, the operational conditions and mechanisms involved during the photocatalytic degradation process are systematically discussed. Finally, particular emphasis is devoted to future challenges and the corresponding outlook with respect to toxic effects following the utilization of ZnO-based nanomaterials.

Keywords: antibiotics; photodegradation; ZnO-based nanoparticles; mechanisms

Citation: Mohamed, K.M.; Benitto, J.J.; Vijaya, J.J.; Bououdina, M. Recent Advances in ZnO-Based Nanostructures for the Photocatalytic Degradation of Hazardous, Non-Biodegradable Medicines. *Crystals* **2023**, *13*, 329. https://doi.org/10.3390/cryst13020329

Academic Editors: Yamin Leprince-Wang, Guangyin Jing and Basma El Zein

Received: 31 December 2022
Revised: 5 February 2023
Accepted: 8 February 2023
Published: 15 February 2023

Copyright: © 2023 by the authors. Licensee MDPI, Basel, Switzerland. This article is an open access article distributed under the terms and conditions of the Creative Commons Attribution (CC BY) license (https://creativecommons.org/licenses/by/4.0/).

1. Introduction

Antibiotic use among human beings and animals has massively increased with the swift evolution of the pharma and medical industries. Antibiotics can prevent the onset of many illnesses by curing infections quickly [1]. The widespread use of antibiotics averts bacterial infections in humans and animals, hence saving numerous lives. Apart from the extreme water pollution, the persistence and difficult-to-degrade characteristics of antibiotics leads to important environmental issues, such as the development of drug-resistant bacteria [2]. Pharmaceutical drugs are released further and further into the environment, which is a serious threat to the environment. The presence of antibiotics in the water for a prolonged time span will make it easier for bacteria to develop antibiotic resistance, posing a more serious danger to human health and the efficacy of antibiotic medications [3]. The current review aims to portray the response to the critical environmental challenges caused by antibiotics, antidepressants, contraceptives, and anti-inflammatory drugs. In this context, diverse approaches, including adsorption, photocatalysis, biodegradation, electrochemical treatment, and others, have been effectively employed to address the troubles brought on by antibiotic contamination [4]. As a unique and appealing catalytic technique with several advantages, such as being green, eco-friendly, and economically viable, photocatalysis was developed to address issues with earlier catalytic technologies. Owing to the desire to exploit abundant solar energy as a sustainable energy source, photocatalysis has emerged as a hot topic in recent years [5–12]. Photocatalysis is a field of chemistry that studies chemical reactions triggered by light and a photocatalyst (a semiconductor that enhances reaction kinetics). Photocatalysts change the reaction rate by absorbing light and acting as a catalyst in chemical reactions [13].

The semiconductor ZnO has drawn great attention from scientists mainly because of its energy band gap, which allows light to be absorbed for photocatalytic reactions to take place [14]. In addition to photocatalysis, ZnO is a potential candidate in transparent thin-film transistors, transducers, transparent ohmic contacts, light absorption amplification structures for GaN-based light-emitting diodes (LEDs), and other possible optical devices [15]. Numerous review and research papers addressing the synthesis and validation of ZnO nanomaterial synthesis have previously been published. Therefore, the applicability of ZnO nanomaterials to the degradation of antibiotics, antidepressants, contraceptives, and anti-inflammatory drugs is significantly emphasized. The following review primarily addresses the structural characteristics of ZnO and the photocatalytic degradation performance of pristine, doped, and composite ZnO towards the aforementioned drugs under diverse conditions, including temperature, pH, and irradiation contact time.

2. Principal Approaches Driving Photocatalysis

Photocatalysts are essentially semiconductors that catalyze the reaction upon exposure to light. In recent years, photocatalysts evolved as benchmark green catalysts owing to their hazardless nature, unlike other energy sources. Upon exposure to light, an electron-hole pair is generated within the semiconductor material. The energy band gap is a major determinant of the physical characteristics of semiconductors. The energy band gap (E_g) represents the difference in energy between the valence band (HOMO) and conduction band (LUMO) [16]. Figure 1 illustrates the energy band gaps of different materials.

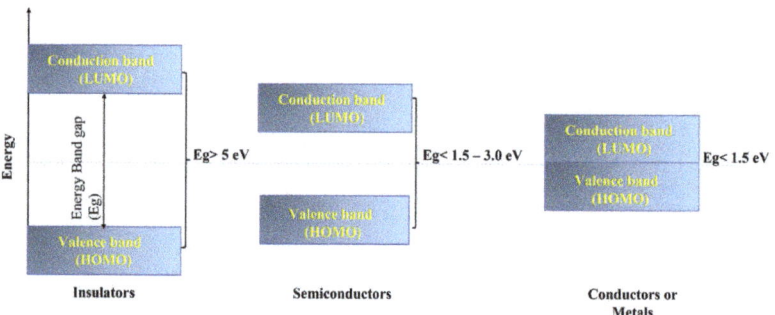

Figure 1. Material-based energy band gap (E_g).

Since it aids in resolving the issue associated with quick charge recombination, semiconductor-mediated photocatalysis has attracted much interest and metal oxides possess the required characteristics, such as the necessary electronic structure, light-absorbing capabilities, and charge transport characteristics. Figure 2 depicts a schematic illustration of photocatalytic degradation by ZnO.

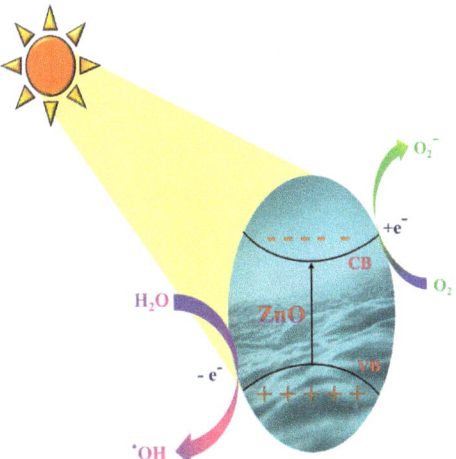

Figure 2. Schematic representation of photocatalytic degradation by ZnO.

2.1. Degradation Mechanism of ZnO

Hermann et al. [17] demonstrated a detailed mechanism for the photocatalytic oxidation steps involved in the use of ZnO: (i) initially, the pollutants disseminate from the liquid phase to the outer surface of ZnO nanoparticles so adsorption takes place on ZnO. (ii) During the adsorption process, redox reactions take place and desorption of the products occurs. (iii) Finally, the polluted products are removed from the interface region. Typically, when a ZnO photocatalyst is photo-induced by UV/solar light with a photonic energy (hv) higher than the excitation energy (E_g), the electrons present in the filled valence band (VB) are transferred to the empty conduction band (CB). During the photo-induced process, electron-hole (e^-/h^+) pairs are produced. These electron-hole pairs travel to the ZnO surface to undergo redox reactions where H^+ combines with water and hydroxide ions to generate hydroxyl radicals, while electrons combine with oxygen to generate superoxide radical anions and then produce hydrogen peroxide. The generated hydrogen peroxide reacts with superoxide radicals to produce hydroxyl radicals. Subsequently, the powerful hydroxyl radicals, as oxidizing agents, attack the adsorbed contaminants present at the surface of ZnO to rapidly generate intermediate compounds [18,19]. These compounds are converted to produce H_2O, CO_2, and mineral acids. Figure 3 illustrates the detailed mechanism of ZnO photocatalysis [20].

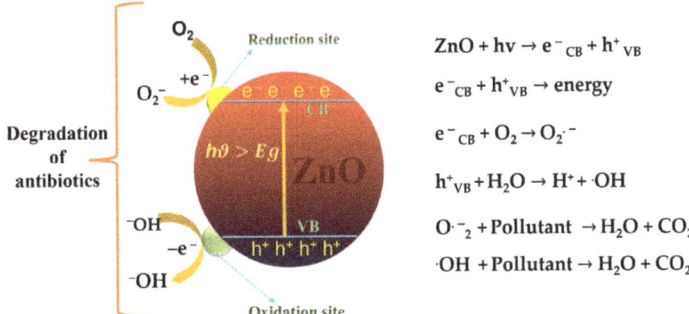

Figure 3. Photocatalysis mechanism of ZnO.

Therefore, the ZnO photocatalyst is controlled by its ability to generate photogenerated electron-hole pairs. However, the major limitation of ZnO photocatalysts is the high rate of photogenerated electron-hole recombination, which complicates the photodegradation mechanism. The wide energy band gap of ZnO only adsorbs UV light, which limits photocatalytic activity to the UV region. Hence, improving the efficiency of ZnO photocatalysts is a top research topic in recent decades.

2.2. Enhancing Semiconductor Performance: An Overview of Techniques and Approaches

There are four main approaches to improve a semiconductor's performance, including: employing a semiconductor with a low E_g; creating a localized state just above the valence band or creating a localized state just beneath the conduction band; forming a color center in the band gap; and surface modification. Consequently, the following techniques are used to achieve the required modifications: (i) metal and non-metal doping, (ii) co-doping, (iii) composites, (iv) substitution, (v) sensitization, and (vi) various other methods.

2.3. Metal and Non-Metal Doping

Metal and non-metal doping is regarded as a component of energy band gap engineering, which entails introducing an electron or hole into the semiconductor used in photocatalysis. By introducing new energy levels (also referred to as the impurity state) between the valence and conduction bands, metal and non-metal doping boosts the photocatalyst's photoresponsiveness to the visible region. Metal dopants (impurities) such as Cu, Zn [21], Mn [22], Co, Cr [23], Fe, Ni, Mo [24], etc., give rise to a new band below the conduction band, whereas non-metals such as N [25], P [26], F, Si, S, Cl, Se, Br, I, etc., [27] give rise to a new band above the valence band. The subsequent explanations highlight why the addition of dopants to photocatalysts improves their performance. Doping prevents electron-hole recombination, promotes surface area, increases particle size in porous structure, enhances crystallinity, and increases sensitivity across a diverse range [28]. Studies have shown that non-metals, such as S, N, F, and C, can alter the energy band gap of ZnO via doping, resulting in increased oxygen vacancy defects and enhanced photocatalytic activity under visible light [29]. S-doping specifically modifies electrical, optical, and photocatalytic properties due to the difference in electronegativity and size between S and O. Doping with sulfur is an effective method to narrow ZnO's energy band gap and shift its threshold wavelength towards visible light [30].

2.4. Co-doping

Co-doping raises the valence band edge while dropping the conduction band edge to minimize the band gap. This approach also enables the resolution of challenges such as the host material's lack of responsiveness to visible light, the carrier recombination solubility limit, and poor carrier mobility. In contrast to pure TiO_2, the integration of ZnO and Fe_2O_3 into TiO_2 improved the photodegradation of phenol. This TiO_2 co-doping led to a coordinated rise in activity, which was attributed to the synergistic interaction between the co-dopants and the energy bands of TiO_2 [31].

2.5. Composites

The fabrication of composites is an alternate approach to boost the photoresponsiveness of semiconductors in the visible range of the whole spectrum. The semiconductors used to develop new composites must have varying band gaps. A semiconductor with a small energy band gap and greater negative conduction band level is typically coupled with a semiconductor with a large energy band gap. As a result, the conduction band electrons are transferred from the semiconductor with a small energy band gap to the semiconductor with a large band gap. The ZnO/γ-Fe_2O_3 nanocomposite catalyst was found to be easily removed from water after photocatalytic treatment due to the presence of paramagnetic γ-Fe_2O_3 nanoparticles. This made it simple to recover the catalyst and reuse it in future degradation cycles with the application of a weak external magnetic field [32].

2.6. Substitution

An additional approach to modify the activity of a photocatalyst is to replace one metal with another, called substitution. For example, when W is replaced in WO_3 by another metal of the same valency, such as Cr or Mo, the E_g decreases due to the movement of the conduction band edge to the lower state. In contrast, the introduction of lower valency metals, such as Ti, Zr, and Hf, causes an elevation in E_g owing to the upward movement of the conduction band edge [33]. The results of a photocatalytic degradation experiment showed that 5% Pb-substituted ZnO nanoparticles were an effective photocatalyst for degradation [34]. When exposed to visible light irradiation, the photocatalytic degradation of organic pollutants was significantly enhanced by substituting Al^{3+} with Zn^{2+} at Zn^{2+} sites within the ZnO host lattice [35]

2.7. Sensitization

Organic and inorganic compounds that chemisorb or physisorb on the surface of semiconductors are known as sensitizers. Sensitization is another intriguing concept for photocatalyst surface modification. Because of their redox capabilities and sensitivity to visible light, dyes and complexes can be employed in photocatalytic devices, including solar cells, to enhance the photocatalytic activity. Some chemical compounds containing chromophores, such as dyes or natural pigments, might be employed to improve the photosensitivity of semiconductors. When photosensitizers are exposed to visible light, absorbed light injects electrons into the semiconductor's conduction band, triggering a catalytic process. Even when exposed to indoor vis-LED lighting, C-sensitized and N-doped TiO_2 had the ability to purify and mineralize water by effectively removing any harmful chemicals [36].

2.8. Other Methods

Undoped TiO_2 nanoparticles possess an energy band gap of 3.1 eV, which can be narrowed significantly up to 2.2 eV by the formation of a midgap state-induced energy gap during synthesis. Thus, undoped TiO_2 synthesized using the mixed phase solution method eventually possessed greater surface area. The as-synthesized material possessed enhanced photocatalytic activity [37]. The utilization of synthetic techniques for surface functionalization is an efficient approach to address challenges such as chemical reactivity in solution and inherent flaws that impede the incorporation of catalysts into practical applications. These techniques encompass everything from complete coverings to the utilization of low-dimensional elements such as nanoparticles, photosensitive dyes, quantum dots, and organic compounds. This straightforward and reliable method may be used to modify the surface charge of different kinds of nano photocatalysts and enhance their photocatalytic activities. The photodegradation rate and adsorption effectiveness of a dye were altered by employing TiO_2 nanoparticles with various surface charges [38].

3. Structural and Electronic Aspects of ZnO

ZnO is a discrete semiconductor with a suitable energy band gap ($\approx$3.30 eV) and an unusually large exciton binding energy of 60 meV [39–42]. The greatest ionization energy of any element in the sixth group of the periodic table is oxygen, which leads to the strongest bonding between Zn (3d) and O (2p) orbitals [43]. Substantial electromechanical incorporating effects in piezoelectric and pyroelectric properties can be applied in piezoelectric sensors and mechanical actuators owing to the deficiency of a center of symmetry in ZnO's wurtzite structure [44–48]. Due to its unique characteristics, ZnO has attracted much attention, particularly because of its hexagonal wurtzite-type structure. Cubic rocksalt and blende forms are two further structural variations ZnO, but the wurtzite form is thermodynamically stable at moderate temperature and pressure [49]. High-temperature and high-power operation, reduced noise production, larger breakdown voltages, and the capacity to withstand strong electric fields are positive characteristics related to the large bandgap of ZnO. For both low and high electric fields, electron transport in semiconductors can be taken into account (i) in a suitably weak electric field that has no impact on the energy distribution of electrons and (ii) in a suitably high electric field where electron distribution function differs considerably from its equilibrium [15]. SEM at a magnification of 10,000× was utilized to identify the surface morphological characteristics of the materials prior to and after calcination (Figure 4) [50]. The uncalcined material's morphology (Figure 4a) showed the aggregation of particles. However, the morphology of the calcined materials showed nanoprisms and nanorods at various temperatures. The TEM images (Figure 4b) confirmed that the uncalcined material was agglomerated and great homogeneity and crystallinity were observed when the temperature rose.

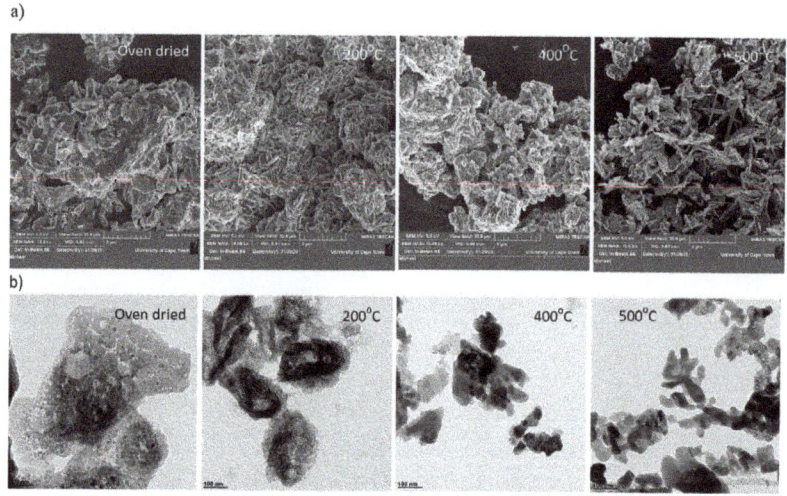

Figure 4. SEM (**a**) and TEM (**b**) images of ZnO at different annealing temperatures.

The Raman spectra of spherical and commercial ZnO particles in the range of 250–680 cm^{-1} are shown in Figure 5 [51]. No apparent Raman peaks of flower-like ZnO were observed. Better optical quality was achieved in as-prepared ZnO crystals than in commercial ZnO, which had a hexagonal wurtzite structure with four atoms per unit cell and corresponded to the space group C46m (P63mc).

Nanoparticles are favored due to their increased surface area to volume ratio. This property allows nanoparticles to absorb more energy and, as an outcome, generate more hydroxyl groups, which aids in the oxidation of organic contaminants [52]. The fundamental drawback of ZnO, like all other semiconductor materials, is that the recombination of h$^+$/VB–e$^-$/CB has an adverse influence on photocatalytic performance. Using simple control techniques, ZnO/Ag/CdO, Fe$_x$Zn$_{1-x}$O, ZnO/CeO$_2$, and certain nanocomposites were synthesized. The photocatalytic efficiency undoubtedly increased since the surface of ZnO was altered. ZnO has been synthesized through a variety of routes as a matrix material with varied structure and forms. The catalytic performance of nanoparticles is significantly influenced by their shape. ZnO's enhanced photocatalytic performance is made possible by its crystallinity and spherical form [53]. The photodegradation of tetracycline under visible light irradiation showed maximum efficiency, which was related to the synergistic action of tetracycline desorption because of its high surface area. Employing ZnO nanoparticles will greatly enhance the economic future of photocatalysis by massively decreasing the required dosage due to its greater surface area and more active sites. For example, 50% of ciprofloxacin was eliminated with just 20 mg/L of nano-ZnO. Moreover, utilizing 20 mg/L of nano-ZnO achieved higher efficiency, i.e., 90% and 59% elimination of tetracycline and ibuprofen, respectively [54].

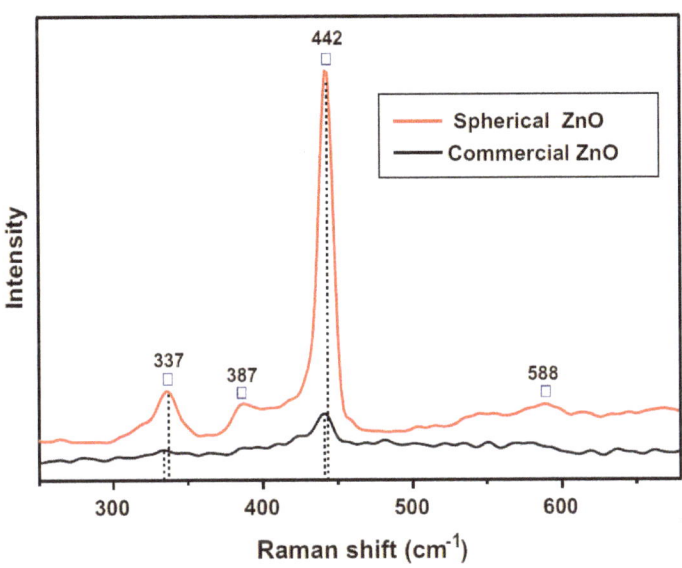

Figure 5. Raman spectra of spherical and commercial ZnO (□—Prominent peaks).

3.1. The Influence of pH and Zeta Potential on the Photocatalytic Degradation of Antibiotics in Aqueous Solutions

The pH of a solution plays a crucial role in photodegradation processes as it affects the size and charge of the photocatalyst [55]. For example, ZnO nanowires showed improved photodegradation of cephalexin under alkaline conditions (pH 7.2–9.2). However, the ZnO nanowires' surface charge changed from positive to negative at pH 6.8, thereby affecting the species involved in the oxidation process. The stability of colloidal solutions depends on the zeta potential value, which is in turn influenced by the surface charge. The calcination temperature can also affect the activity of photocatalysts, with the highest efficiency (82%) achieved at 4000 °C. However, higher calcination temperatures can reduce photocatalytic activity due to a decrease in surface area and increase in crystal size [56].

3.2. Structural Stability and Reusability of ZnO as Photocatalysts

ZnO has a broad UV absorption spectrum and excellent photostability, biocompatibility, and degradability. Recyclability is a crucial factor to consider when examining the structural stability and reuse of a photocatalyst. The ZnO photocatalyst synthesized using the solvothermal method to eliminate the antibiotic ofloxacin exhibited good cycling stability and retained 95% efficiency even after three cycles. This showed the ZnO photocatalyst's benefits of outstanding structural stability and enticing reusability [57]. Maximum photocatalytic activity was shown by ZnO photocatalysts prepared using the hydrothermal method in the removal of the antibiotic ofloxacin even after undergoing a third cycle. The ZnO photocatalyst continued to perform at a high level, demonstrating its superior cycling capacity [58]. N,S-doped carbon quantum dot-embedded ZnO nanoflowers had a high degree of stability and could be utilized repeatedly [59]. In five successive cycles, the Cr_2O_3@ZnO photocatalyst tested for the degradation of ciprofloxacin showed 100% destruction and high reusability [60].

3.3. Corrosion Effects of ZnO

ZnO's efficiency and stability as a photocatalyst are reduced when exposed to UV radiation due to its susceptibility to corrosion. Thus, ZnO is being combined with carbon-based compounds, noble metals, or other metal oxides in an effort to increase its performance [61]. For instance, the durability and photocatalytic efficacy of ZnO nanoparticles were enhanced by coating them with reduced graphene oxide. ZnO can have improved visible light responsiveness and greater photo corrosion resistance when doped with Ag, Au, or Pt, as demonstrated by the Ag-ZnO photocatalyst's degradation of the antibiotic ofloxacin [62].

4. Pristine ZnO Photocatalysts

Zinc oxide is a well-known n-type semiconductor that has garnered widespread use as a photocatalyst. This is due to its high absorption efficiency and large exciton binding energy, which result in its exceptional mechanical, electrical, and optical properties. The use of ZnO-based materials is widespread across a variety of applications, including biological and catalytic processes. Among these materials, pristine ZnO nanoparticles are particularly favored for their exceptional photocatalytic activity [63]. Furthermore, the production cost of ZnO nanoparticles is much lower than that of other materials, as reported by Liang et al. [64]. However, the incorporation of dopants in ZnO nanoparticles and the fabrication of composite-based materials can achieve improved efficiency.

The combination of ZnO with other materials to produce a hybrid photocatalyst system subsequently improves the separation of photogenerated electron-hole pairs [65,66]. Above all, nanostructured ZnO is divided into zero-, one-, two-, and three-dimensions and each of these are subdivided into planar, dots, and quantum arrays. One-dimensional ZnO arrays include nanofibers, nanotubes, nanorods, nanoneedles, and nanowires, while two- and three-dimensional ZnO arrays include nanoflowers. Luo et al. [67] reported that ZnO has a higher specific surface area, which is advantageous in the photocatalytic degradation process as more pollutants can be easily adsorbed. Additionally, the lower crystallinity of ZnO promotes the trapping of photoinduced electron-hole pairs, which enhances easy separation.

There are several approaches for the synthesis of ZnO nanostructures (Figure 6), but vapor phase and solution-based approaches are primarily employed [68]. The solution-based approach is the easiest and least energy-consuming technique to improve the morphology and control the size of the nanostructure [69]. The techniques used to produce ZnO in a solution-based approach include hydrothermal, precipitation, solvothermal, sol-gel, microwave, electrospinning, and wet chemical methods. Among them, sol-gel is the most commonly used and effective technique due to its advantages [70]. Ciciliati et al. used Fe-based ZnO nanoparticles for photocatalysis, which were synthesized using the sol-gel method [71]. The vapor phase approach includes physical vapor deposition, plasma enhanced chemical vapor deposition, and metal-organic chemical vapor deposition methods [72].

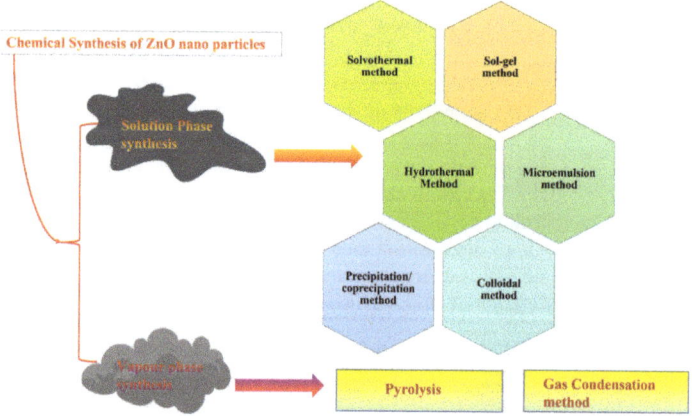

Figure 6. Methods used for the synthesis of ZnO nanostructures.

5. Metal/Non-Metal-Doped ZnO Photocatalysts

The improvement of ZnO as a photocatalyst can be achieved via metal doping and fabrication of composites. Generally, ZnO contains a tetrahedral bonding configuration with a largely ionic nature that is classified between a covalent and ionic semiconductor. The ZnO photocatalyst forms three crystalline structures, including hexagonal wurtzite (P6_3mmc), cubic rock-salt (Fm3m), and cubic zinc-blende (F43m). ZnO hexagonal wurtzite is the most stable at ambient conditions [73]. Metal doping has been adopted to modify the physicochemical properties of ZnO such as to alter the valence band energy upward and minimize the energy band gap to the ultraviolet-visible region. On the other hand, non-metal doping is also used to shift the bandgap of ZnO. Specifically, fluorine, oxygen, carbon, and nitrogen interfere with lattice interstices and bind to atoms via the oxidation process. Among all, carbon is an effective candidate as a non-metal dopant due to its explicit properties, such as good mechanical strength and chemical resistance, and electronic properties. Moreover, the incorporation of metal/non-metals produces greater OH· radicals, thereby leading to enhanced photocatalytic degradation of pollutants [74]. This is owing to the fact that the metal/non-metal dopants act as scavengers, control the recombination of electron-hole pairs, and produce H+ for the formation of OH· radicals. Transition metal incorporation into the ZnO crystal lattice is highly regarded due to its tunability of the energy band gap of ZnO to produce an effective visible light photocatalyst. The transition metal can significantly modify the particle size and shape of ZnO and controls ZnO growth, which leads to the generation of ultrafine nanostructures with higher surface area. Transition metal oxides such as Fe and Cu have oxidation states that improve the photocatalytic activity of the doped ZnO catalyst [75]. The Ag-doped ZnO catalyst is a commonly used photocatalyst. Li et al. reported that the properties of Ag helped to improve the doped ZnO photocatalyst in which Ag acted as an electron acceptor [76]. Mn is another efficient material that enhances the photocatalytic activity of the doped ZnO photocatalyst due to changes in the optical properties. The charge transfer between the valence band and Mn-induced levels and d-d transition in the crystal field induces visible-light irradiation. The charge carrier defect eliminates the photo-induced electron/hole pair, thereby increasing the lifetime of the photocatalyst [77]. Copper is a cost-effective element that can be easily incorporated in substitutional sites of ZnO, which helps to modify the emission and absorption spectra into the visible-light region [78]. Polat et al. synthesized a Cu-doped ZnO photocatalyst for effective dye degradation [79]. Fu et al. utilized a Cu-doped ZnO photocatalyst to examine degradation under UV light. However, the addition of high concentrations of copper acted as a recombination center, which decreased the photocatalytic activity [80]. Likewise, Mohan et al. demonstrated that Cu-doped ZnO produced more holes, which were a surface defect that enhanced the photocatalytic activity. The cobalt-doped ZnO catalyst has received enormous attention due to its great response as photocatalyst [81]. The incorporation of Co in the ZnO lattice shifts the optical absorption edge toward the visible-light region and narrows the energy band gap. The addition of Co hinders the growth of ZnO nanoparticles, thus producing a smaller particle size and favoring photocatalytic activity. Fe doping is similarly used for ZnO modification [82]. Fe doping of ZnO causes a narrow energy band gap and increases the electron excitation rate. The introduction of Fe reduces the crystallite size and increases the oxygen vacancies, which enhances the electron/hole separation efficiency [83]. Yu et al. demonstrated that Al-doped ZnO adsorbed large amounts of organic pollutants compared to pure ZnO [84]. The incorporation of Al in ZnO generates more oxygen defects, which increase the charge density of ZnO and therefore enhance its performance. Alam et al. reported that magnesium-doped ZnO produced a higher response than pure ZnO due to the unpaired electrons [85]. Mrindha et al. showed that Al-doped ZnO exhibited a better photocatalytic response than the pristine ZnO [86]. Further, Dai et al. produced Cu-doped ZnO nanoparticles with improved photocatalytic activity over five consecutive cycles [87]. However, crystal distortion and high levels of metal concentrations reduced the photocatalytic activity.

6. ZnO Composite Photocatalysts

The commercialization of ZnO as a photocatalyst is constrained by the effects of photocorrosion, which can be prevented by preparing composites of ZnO with other metal oxides [88]. The composites have the tendency to modify the energy band gap and efficiently allow visible light. Wang et al. synthesized the MnO-ZnO nanocomposite with a reduced energy band gap that moved the absorption from UV to visible light and enhanced the photocatalytic activity [89]. Sigh et al. demonstrated the preparation of the CuO-ZnO nanocomposite and produced lower energy band gap [90]. Carbon-based nanocomposites are attractive materials owing to their improved activities in many applications. Wang et al. developed the ZnO-rGO nanocomposite as a visible light photocatalyst [91]. The photogenerated electrons in the conduction band of ZnO migrated to rGO, thereby suppressing their recombination with the hole in ZnO. Similarly, Tran et al. synthesized the ZnO-rGO nanocomposite with a reduced energy band gap for the effective photocatalytic degradation of dye under visible light. Spinel ferrite-based ZnO photocatalysts are commonly used as effective photocatalysts [92]. Azar et al. developed the $ZnAl_2O_4$–ZnO nanocomposite with improved photocatalytic activity [93]. Similarly, Dlugosz et al. used the combination of magnetic and catalytic activities to improve the degradation activity of dye [94]. Zhang et al. synthesized the SnO_2/ZnO composite with a reduction of the energy band gap for improved photocatalytic activity of dye [95]. Trakulmututa et al. developed the CuO/ZnO composite with modified optical properties for the degradation of antibiotics [96]. A study by Hemnil et al. produced a novel Fe_2O_3/ZnO nanocomposite with high absorbance in the ultraviolet and visible regions and proposed the synthesized photocatalyst with a suitable energy band gap for effective photodegradation [97]. Krishnan et al. delivered a suitable MOS_2/ZnO nanocomposite for biological photocatalytic degradation under visible light [98]. Zhang et al. proposed a tertiary $ZnO/ZnFe_2O_4/TiO_2$ nanocomposite with enhanced photocatalytic activity under visible light degradation [99]. The ZnO/SiO_2 nanocomposite developed by Fatimah et al. was found to be an effective reusable photocatalyst with a reduced energy band gap for improved photocatalytic degradation of dye pollutants [100]. Sabri et al. fabricated a novel $ZnO/CuBi_2O_4$ nanocomposite for the photocatalytic degradation of dye contaminants under visible light. The proposed photocatalyst had improved photocatalytic activity due to its larger surface area, formation of a heterojunction at ZnO and $CuBi_2O_4$ and better absorption of visible light. The small energy band gap significantly absorbed visible light, and the nanocomposite was effectively utilized as an efficient composite-based photocatalyst [101].

7. Impact of Antibiotic and Pharmaceutical Pollution on the Environment and Human Health

Antibiotics are pharmaceutically active compounds (PACs) that alleviate bacterial ailments and are necessary to protect overall survival [102]. Antibiotics are spilling into the ecosystem to a greater extent and are becoming a prominent contributor to water contamination. The menace that antibiotic waste poses to the purity of aquatic systems and human health is one of the greatest ecological concerns of the twenty-first century [103]. Tetracyclines (TC) are a category of expansive antibiotics prevalently prescribed to combat numerous infectious diseases in both people and animals [104]. More than 70% of TC deployed for pharmacologic therapy is eliminated through renal excretion in metropolitan wastewater and animal liquid effluents due to inefficient adhesion and metabolic activity by animals and people. Tetracyclines have a plethora of repercussions on soil microbes and photosynthetic marine creatures, which correlate with the surge in substantial human health complications and antibiotic-resistant pathogens [105–107]. The World Health Organization (WHO) cited antibiotic resistance as a global health concern in 2019 [108] and estimated that veterinary utilization accounted for over 80% of all antibiotic consumption. The United Nations General Assembly has highlighted livestock consumption of antibiotics as one of the major factors contributing to the escalation in antibiotic resistance, with an anticipated total usage of 200,000 tons annually [109,110]. Antibiotic-resistant genes can be acquired by bacterium when antibiotic dosages are very low, barring them from affecting the bacterium. In turn, other bacterial spores can accumulate these genes. If the prevailing trend continues, 10,000,000 people may perish annually by 2050 from contagious diseases spurred by bacterium with diverse impedances [111–114]. Macrolides, primarily spiramycin (SPY), serve as the most ubiquitous antibiotics in industrial effluents. As a result of their putative assimilation by crops fertilized with product from sewage treatment plants, antibiotics may impede recycling of waste in agriculture [115,116].

A prevalent antibiotic, tetracycline hydrochloride (TC-HCl), is produced in enormous quantities and holds second place in usage worldwide. Through the metabolic pathways of individuals and livestock, and pharmaceutical corporation waste, TC-HCl is transferred into the ambient aquatic system and then reaches the food system of humans. Drug-resistant infectious diseases have evolved as a consequence of the deposition of TC-HCl, as well as other pharmaceuticals, resulting in a serious influence on both public welfare and marine life [117,118]. The World Health Organization (WHO) stated in 2014 that the inclusion of gemifloxacin (GMF), a quinolone drug, in numerous ecosystems could result in significant health and ecological risks, such as cytotoxic effects and bacterium development of resistance [119]. Investigations conducted in Spain, Germany, Australia, Italy, Brazil, Canada, Greece, and the United States have revealed that more than 80 distinct types of pharma drugs and metabolites produced as a consequence of consuming multiple medications have contaminated marine ecosystems. Substantial concentrations of pharma drugs per liter were observed in materials collected from surface fluids downstream of urban sewage treatment facilities, inlet effluent, and contaminated water [120,121].

Naproxen (NAP) and ibuprofen (IBU) are non-steroidal, anti-inflammatory pharmaceuticals implicated in health risks to people, such as renal dysfunction and intestinal problems, as per toxicity testing [122]. One of the most pervasive hazardous pollutants is diclofenac (DCF), a non-steroidal, anti-inflammatory medicine frequently employed to relieve arthritis or rheumatism. Despite having a low toxic effect, diclofenac can incentivize the emergence of drug-resistant pathogens that pose a risk to human wellbeing and aquatic creatures and are thus destructive to the ecosystem and mankind [123].

The tricyclic antidepressant that receives the most usage is amitriptyline hydrochloride (AMI). Moreover, it is also employed in the treatment of sleeplessness, acute headaches, and migraines [124,125]. In France, amitriptyline hydrochloride has been found in drinking water at a concentration of 1.4 ng/L, whereas ground waterways in the UK have been found to contain amitriptyline hydrochloride at a concentration of 0.5 to 21 ng/L. Furthermore, AMI was detected in solid sewage treatment effluent in Canada at a concentration of 448 ng/g [126–128].

Humans, aquatic life, and other species in the environment are gravely endangered by water pollution by endocrine-disrupting pollutants such as estrone, estradiol, ethinylestradiol, and estriol steroid hormones [129]. Endocrine-disrupting compounds (EDCs) are agents that interfere with the synthesis, secretion, transport, binding, action, or elimination of natural hormones in the body that are responsible for the maintenance of homeostasis, reproduction, development, and behavior [130,131]. Waterways in the Czech Republic have been reported to possess steroid hormones at 34–41 ng/L [132]. Most frequently encountered in oral contraceptive tablets, artificial hormones such as 17α-ethinylestradiol (EE2) are also utilized to cure prostatic disease and irregular menstrual cycles. They remain active as an outcome of interactions between hormone residues in the environment and the ensuing deconjugation of EE2 during incomplete degradation at wastewater plants, and their leftovers are also a substantial pollutant in the environment [27,133]. The impacts of EE2 encompass testicular weight reduction in Japanese quail, rapid development in female fish, and reproductive issues [134]. Comparable effects, including lowering of sperm counts and elevated risks of ovarian and breast cancer, are reported in humans [135].

8. Photocatalytic Degradation of Antibiotics

Various methods, including carbon filtration, ozonation [136], catalytic membranes [137], Fenton-like catalysis [138], sorption, and biodegradation [139] have been employed to eliminate such enduring antibiotic substances. As a result, wastewater accumulates effluents from corporations, clinics, and farmland. Sewage treatment is typically regarded as the preferred method for treating these antibiotics. However, investigations have demonstrated that the established methods do not eliminate these pollutants, which are primarily water-soluble but neither volatile nor compostable [140,141].

Photocatalysis has gained a lot of interest as an effective method for eliminating antibiotic contaminants since it is inexpensive, effective, and environmentally benign as it eliminates antibiotics in sunlight and under ambient conditions [142,143]. The reactions endured by a semiconductor and the potential to absorb photons with energies higher than its energy band gap is called "photocatalytic degradation" [144]. The degradation of antibiotics by photocatalysis can be categorized into five major steps: (1) the passage of antibiotics from the fluid to the surface, (2) antibiotic sorption, (3) change during the adsorption phase, (4) desorption of the product, (5) and product separation from the interface area [145,146]. Due to their superior light absorption under UV, visible light, or both, together with their biocompatibility, safety, and stability when subjected to various circumstances, metal oxide-based photocatalysts have recently attracted much attention [147].

8.1. Antibiotic Degradation Utilizing Pristine ZnO Photocatalyst

ZnO has multiple benefits because of its antigen-free, antimicrobial, antitumor, and wound-healing characteristics [148]. ZnO is an affordable material with minimal technical aspects, rendering it the more beneficial choice compared to TiO_2 as a photocatalyst. ZnO has a significant exciton binding energy of 60 meV and good capacity to absorb UV irradiation. ZnO exhibited better photocatalytic performance than TiO_2 due to its photogenerated electron-hole pairs, mobility, and isolation [149].

Effective photodegradation of ciprofloxacin (CIP) was exhibited by ZnO nanoparticles synthesized using an ethanolic root extract of Japanese knotweed with specific operational conditions, as shown in Table 1. A UV-vis spectrophotometer at 271 nm was used to analyze the level of CIP while the degradation process was being carried out under ultraviolet light (λ_{max} = 365 nm). It is necessary to highlight that the λ_{max} of CIP changed from 272 to 265 nm from the start to completion of degradation, emphasizing the disturbance of the chromophoric conjugation network. The perfect nanoparticle size of 14 nm had a perceived rate constant (k_{app}) of 0.038 min^{-1} and completely degraded CIP during exposure for 100 min [150]. The degradation of CIP in water solutions with varied pH values under exposure to UV light was examined with ZnO nanoparticles synthesized using a chemical precipitation method. After 60 min of exposure, the maximal recorded CI antibiotic degradation efficacy was around 18% at pH 4, 42% at pH 7, and 50% at pH 10. Pseudo-first order kinetics govern the photocatalytic degradation of ciprofloxacin. The free hydroxyl ions interact with holes (h^+) and generate hydroxyl radicals (OH·) that possess strong oxidizing potential, thereby enhancing the photocatalytic degradation of ciprofloxacin. At pH 10, ZnO nanoparticles notably demonstrated enhanced degradation of ciprofloxacin (Table 1).

Table 1. Photocatalytic degradation of antibiotics by pristine ZnO as photocatalyst.

Catalyst	Antibiotic	Temperature (T) °C and pH	Light Source	Degradation Efficiency (%)	Duration	Ref.
ZnO (thin film)	Paracetamol 15 ppm	T = 25 °C pH = 5.58–6.69	UV lamp (18W)	14%	4 h	[150]
ZnO powder 20 mg	Ciprofloxacin 10 mg/L	T/pH = n.a.	UV light (15 W)	100%	100 min	[150]

Table 1. Cont.

Catalyst	Antibiotic	Temperature (T) °C and pH	Light Source	Degradation Efficiency (%)	Duration	Ref.
ZnO 1 g/L	Spiramycin 10 mg/L	T = 20 °C pH = 5.5	Xenon arc lamp (450 W)	95–99%	80 min	[151]
ZnO 25 to 75 mg	Ofloxacin (OFL) 5–20 ppm	T = n.a. pH = 10	UV light irradiation (Mercury lamp, 125 W)	95%	120 min	[151]
Nano ZnO 0.02 g/L	Ciprofloxacin 5 mg/L	T = n.a. pH = 10	Xenon lamp	50%	60 min	[152]
ZnO (thin film)	Chloramphenicol 8 ppm	T = 25 °C pH = 5.31–6.49	UV lamp (18W)	40%	4 h	[153]
Nano ZnO 0.02 g/L	Metronidazole 10 mg/L	pH = n.a. T = 30 °C	Ultrasound irradiation	100%	27 min	[154]
ZnO nanorod array	Tetracycline hydrochloride 10 mg/mL	T/pH = n.a.	Xenon light (500 W)	69.80%	140 min	[155]

n.a., not available.

From the above findings, it can be inferred that the pH might modify the photocatalyst's surface charge features and presumably the chemical morphology of the molecule; as a corollary, the photocatalysis process is pH-dependent. It has been revealed that thin ZnO films possessed photocatalytic performance in the photocatalytic decomposition of the drugs chloramphenicol (levomycetin) and paracetamol with conditions (Table 1). When the catalyst is irradiated by ultraviolet light with either an energy equal to or higher than that of the energy band gap, an electron–hole pair is formed. Water adsorption over the ZnO layer or hydroxyl groups occurs in connection with photo-induced holes at the valence band. The outcome of such activity is the powerful OH· radical. Conduction band electrons react with electrophiles, such as O_2, deposited on the surface or dispersed in the water, leading to the production of super oxide radicals. By reacting with paracetamol and chloramphenicol, the ensuing strongly reactive radicals produced intermediates that degraded to give ionic species. The degradation efficiency for paracetamol and chloramphenicol under UV light was 14% and 40% with conditions (Table 1) [153]. Table 1 also summarizes numerous ZnO photocatalysts that effectively photodegraded various drug compounds, such as spiramycin (95–99%), ofloxacin (95%), metronidazole (100%), and tetracycline hydrochloride (69.8%), respectively.

8.2. Antibiotic Degradation Utilizing Metal-Integrated ZnO Photocatalysts

ZnO nanoparticles have several beneficial usages in piezoelectric devices, semiconductors, photovoltaic cells, polymers, skincare, and pharmaceuticals. However, they have specified drawbacks, such as: (i) potential to degrade in water solutions as a consequence of the photocorrosive effect; (ii) scattering of light, which can be minimized by modifying the catalyst dosage rate; (iii) a high energy band gap facilitates performance in ultraviolet light, which may not be realistic for massive sewage treatment; (iv) due to the recombination pathway being more rapid than the surface redox reaction, impulsive e^-/h^+ recombination in the photocatalyst; (v) minimal reuse. In an attempt to tackle these issues, ZnO nanomaterials have been doped with metal nanoparticles. By adding defective modes and tunning the active energy band gap, doping helps to improve the surface region of ZnO nanoparticles and also boosts their ability to absorb photons. Platinum (Pt), gold (Au), and silver (Ag) are the noble metals utilized as dopants. Noble metals easily absorb in the visible part of the electromagnetic spectrum through the surface plasmon resonance (SPR) mechanism. ZnO becomes a visible light-sensitive photocatalyst by incorporating noble metals into its crystal lattice or on its interface [156]. Transition metals, alkali metals, alkaline earth metals, as well as heavy metals such as lanthanides have also been utilized to dope ZnO photocatalysts in addition to noble metals. For instance, the minor misalignment between magnesium (Mg) and ZnO crystal structures leads to the dopant raising the energy band gap of ZnO. The basic pathway of antibiotic photodegradation by metal-doped ZnO is as follows: (i) the antibiotic drug adheres to the photocatalyst's surface; (ii) production of electron-hole pairs; and (iii) reactive oxygen species (ROS) are caused by redox processes to decompose the antibiotic [157].

Comparing the photocatalytic activity of the ZnO photocatalyst under UV light irradiation (duration: 120 min) and the Ag-ZnO photocatalyst under solar light irradiation (duration: 80 min) against the antibiotic ofloxacin revealed that 95% degradation efficiency was achieved by the ZnO photocatalyst and 100% degradation efficiency was achieved by Ag-ZnO photocatalyst (5 wt% silver) under similar operating conditions. Due to its higher electron-hole separating efficacy than pure ZnO, the Ag-doped ZnO photocatalyst exhibited complete degradation of ofloxacin. The Schottky barrier arises at the Ag/ZnO interface as metal silver is coated onto the ZnO photocatalyst, which leads to better photocatalytic activity and greater quantum efficacy [150,158]. Under ideal conditions, the Ag-ZnO photocatalyst exhibited significant degradation rates for medicinal residues, i.e., 70.2% for atenolol (ATL) and 90.8% for acetaminophen (ACT) (Table 2).

Table 2. Photocatalytic degradation of antibiotics by metal-integrated ZnO as photocatalyst.

Catalysts	Antibiotic	Temperature (T) °C and pH	Light Source	Degradation Efficiency (%)	Duration	Ref.
Ag-doped ZnO 25 mg to 75 mg	Ofloxacin 5–20 ppm	T = n.a. pH = 5–9	Solar light irradiation	100%	80 min	[150]
Mg/ZnO 1 g/L	Alprazolam 0.03 mM	T/pH = n.a.	High-pressure Hg lamp (125 W)	87%	10 min	[154]
Ag/ZnO 0.25 g/L	Tetracycline 15 mg/L	T = n.a. pH = 5–9	Visible light tungsten lamp (60 W)	100%	30 min	[155]
Ce-ZnO crystals 1 g/L	Nizatidine 5 mg/L	T = 25 °C pH = 6.7	UV-B mercury lamp (8 W)	96%	4 h	[159]
Ce-ZnO crystals 1 g/L	Levofloxacin 5 mg/L	T = 25 °C pH = 5.6	UV-B mercury lamp (8 W)	96%	4 h	[159]
Ce-ZnO crystals 1 g/L	Acetaminophen 5 mg/L	T = 25 °C pH = 6.8	UV-B mercury lamp (8 W)	65%	4h	[159]
Fe^{3+}-doped ZnO 1 g/L	2-Chlorophenol 50 mg/L	T = 28–38 °C pH = n.a	Solar radiation (23 W/m^2)	85%	90 min	[160]
F-ZnO 1.48 g/L	Sulfamethoxazole (SMX)	T = 21 ± 1 °C pH = 4.7	UVC lamp (10 W)	97%	30 min	[161]
La-doped ZnO	Paracetamol 100 mg/L	T/pH = n.a	Visible light irradiation (20 W) compact fluorescent lamp	99%	3 h	[162]
Ag-ZnO NPs 1 g/L	Atenolol (ATL) 5 mg/L	T = n.a. pH = 5–9	Visible light Tungsten halogen lamp (300 W)	70.20%	120 min	[163]
Ag-ZnO NPs 1 g/L	Acetaminophen (ACT) 5 mg/L	T = n.a. pH = 5–9	Visible light Tungsten halogen lamp (300 W)	90.80%	120 min	[163]

n.a., not available.

ATL and ACT degradation is governed by pseudo-first-order kinetics. Increased surface area, improved charge transport with both ZnO and Ag, and their synergistic influence were the main factors that contributed to the increase in efficiency. An additional finding about the photocatalyst mechanism revealed that the main method for removing ATL and ACT was the OH· pathway [151]. The Ag/ZnO synthesized using the rapid, one-pot, surfactant-free, microwave-assisted, aqueous solution method showed complete degradation of the antibiotic tetracycline (TC) in 30 min under visible light [154]. Silver nanoparticles act as an electron sink in Ag-doped ZnO nanoparticles, improving the charge separation. This enhances the generation of hydroxyl radicals in the reaction media and consequently improves the photocatalytic performance of ZnO nanoparticles [146]. Employing a conventional solid-state process, Mg-doped ZnO nanocrystallites were synthesized and exhibited an 87% alprazolam degradation rate. Mg-doped ZnO annealed at 700 °C showed excellent photocatalytic properties, reaching 100% degradation within 20 min. Meanwhile, under similar conditions, bare ZnO showed 78% removal of alprazolam. Due to excellent electron-hole separation and good textural characteristics, doping ZnO nanoparticles with Mg enhanced the photocatalytic performance triggered by sunlight and the results indicated that doping ZnO with Mg promoted alprazolam elimination by 10% [153]. Under UV light, a hydrothermally prepared Ce-doped ZnO photocatalyst demonstrated 96% degradation rates for nizatidine and levofloxacin and a 65% degradation rate for acetaminophen [159]. Research revealed that the metal doping of ZnO was necessary for controlling the ZnO nanoparticle surface and optical features. It must be highlighted that the photodegradation of metal-doped ZnO depends on the light source. This is due to ZnO's ability to absorb UV light whereas metal-doped ZnO absorbs light primarily in the visible spectrum. Therefore, due to the fact that sunlight carries UV along with visible light, the excitonic formation under solar irradiation may arise by either ZnO or the doped metal [146].

8.3. Antibiotic Degradation Utilizing ZnO Composites Photocatalysts

The major concern with ZnO in photocatalysis is mainly its high rate of electron-hole recombination. With the aim of enhancing the performance of photocatalysis, heterojunction semiconductors have been developed and used to prevent electron-hole recombination [147], consequently enhancing the degradation efficiency with better regeneration and recycling.

The photodegradation of tetracycline by the ZnO/γ-Fe$_2$O$_3$ composite showed 88.52% degradation efficiency under UV-visible light over 150 min. Both the catalyst's surface and pore volume were increased by the inclusion of iron oxide in its structure, thereby promoting the analyte's adsorption on the nanostructure's surface, which is a crucial step in improving photocatalytic degradation. Additionally, it was observed that ZnO served a vital function in the photocatalytic degradation supported by γ-Fe$_2$O$_3$, raising the rate of TC degradation to 20% [163]. The photodegradation activity of the ZnO globular-gC$_3$N$_4$ nanocomposite showed that 78.4% of tetracycline degraded in 50 min and 63.5% of oxytetracycline degraded in 50 min. Reactive species were regarded as the primary cause in the photodegradation mechanism [160]. As Ag loading rose during the photocatalytic degradation of the antibiotic tetracycline hydrochloride, the catalytic performance of the Ag@ZnO/BiOCl composite first rose and then fell. ZnO has minimal photocatalytic performance and achieved a degradation rate of 38.5% over 80 min. The degradation rate achieved by ZnO/BiOCl was moderately greater, attaining 42.7%. It is noteworthy that the degradation rates of Ag-loaded samples were better than that of ZnO/BiOCl, and the Ag nanocomposite achieved the maximum degradation rate with 80.4% removal of tetracycline hydrochloride (Table 3).

Table 3. Photocatalytic degradation of antibiotics by ZnO nanocomposites as photocatalysts.

Catalyst	Antibiotic	Temperature (T) °C and pH	Light Source	Degradation Efficiency (%)	Duration (min)	Ref.
ZnO/g-C$_3$N$_4$ 20 mg	Tetracycline (TC) 20 mg/L	T/pH = n.a.	Visible light (300 W)	78.40%	50	[162]
ZnO/g-C$_3$N$_4$ 20 mg	Oxytetracycline 20 mg/L	T/pH = n.a.	Visible light	63.50%	50	[162]
ZnO/γ-Fe$_2$O$_3$ composite 0.5 mg/mL	Tetracycline (TC) 30 mg/L	T = RT pH = 6.7	UV-visible light (100 mW cm^{-2})	88.52%	150	[163]
Ag Loaded ZnO/BiOCl 32 mg	Tetracycline hydrochloride 20 mg/L	T = 25 °C pH = 8	Simulated solar light	80.40%	80	[164]
SnO$_2$/ZnO nanocomposite 50 mg	Ciprofloxacin 20 mg/L	T/pH = n.a.	Mercury lamp (300 W)	91.23%	60	[165]
SnO$_2$/ZnO nanocomposite 50 mg	Ofloxacin 20 mg/L	T/pH = n.a.	Mercury lamp (300 W)	91.26%	60	[166]
SnO$_2$/ZnO nanocomposite 50 mg	Norfloxacin 20 mg/L	T/pH = n.a.	Mercury lamp (300 W)	88.39%	60	[167]
ZnO/GO/DES nanocomposite 0.532 g/L	Cefxime trihydrate 20.13 mg/L	T = 40.0 ± 1 °C pH = 4.03	UV-A irradiation (15 W)	86%	60	[168]
GO@Fe$_3$O$_4$/ZnO/SnO$_2$ composite 1 g/L	Azithromycin 30 mg/L	T = n.a. pH = 3	UV-C irradiation (6 W)	90.06%	120	[169]
Ag–ZnO/GP composite 0.5 g/L	Metronidazole (MNZ) 30 mg/L	T = n.a. pH = 9	UV-lamp (100 W)	88.50%	60	[170]
Ag–ZnO/GP composite 0.5 g/L	Metronidazole (MNZ) 30 mg/L	T = n.a. pH = 9	Solar irradiation (500 W)	97.30%	180	[170]

n.a., not available.

This was achieved as a result of enhanced absorption of visible light, better charge separation as a function of surface plasmon resonance, the potential of Ag to capture electrons, and enhanced surface catalysis. Hence, adding noble metals to a semiconductor with a wide energy band gap is a practical approach to improving its photocatalytic performance [164]. A simple hydrothermal method was utilized to synthesize the SnO$_2$/ZnO nanocomposite investigated for the photocatalytic degradation of quinolone antibiotics (ciprofloxacin, ofloxacin, and norfloxacin). Over 60 min, ciprofloxacin, ofloxacin, and norfloxacin were all degraded at rates of 91.23%, 91.26%, and 88.39%, respectively. Strong oxidation potential is present in SnO$_2$, whereas high reduction ability is found in ZnO. This can substantially increase the separation efficacy of photo-induced e$^-$/h$^+$. Therefore, a simple method for improving the photocatalytic properties is to combine SnO$_2$ with ZnO to produce a composite photocatalyst [165]. The photodegradation of the most used antibiotics by several ZnO nanocomposites is summarized in Table 3. Ag–ZnO/GP composite, SnO$_2$/ZnO nanocomposite, GO@Fe$_3$O$_4$/ZnO/SnO$_2$ composite, and ZnO/γ-Fe$_2$O$_3$ composite are recognized for exhibiting the maximum degradation for the frequently prescribed antibiotics metronidazole, ciprofloxacin and ofloxacin, azithromycin, and tetracycline, respectively. Therefore, ZnO nanocomposites are potential and effective preferences for the photocatalytic degradation of antibiotics.

9. Photocatalytic Degradation of Anti-Depressives, Anti-Inflammatories, and Contraceptives

Pharmaceuticals and endocrine-disrupting compounds (EDCs) constitute a substantial category of emerging contaminants (ECs) that are ingested in enormous quantities around the world. These pollutants are pervasive in the environment, primarily as a result of their limited degradability and high resistance. Therefore, it is critically necessary to find effective strategies for removing or lowering the concentrations of such developing pollutants. Because of its nontoxicity, accessibility, lack of mass transfer restriction, chemical stability, and potential functioning at room temperature, photocatalytic degradation has been proven to be a viable approach.

ZnO is probably the most common catalyst for the photocatalytic removal of organic contaminants [171]. The advantages and mechanisms of ZnO photocatalysts have already been mentioned above. The most popular tricyclic antidepressant is amitriptyline hydrochloride (AMI), which is used to treat migraines, sleepiness, and severe headaches. The ZnO photocatalyst in the degradation of AMI showed 94.3% degradation efficiency under solar irradiation with variable operating parameters (Table 4).

Table 4. Photocatalytic degradation of antidepressants, anti-inflammatories, and contraceptives by ZnO/ZnO based nanomaterials as photocatalysts.

Catalyst	Antibiotic	Temperature (T) °C and pH	Light Source	Degradation Efficiency (%)	Duration	Ref.
ZnO 100 mg	Ibuprofen (IBU) 20 mg/L	T = n.a. pH = neutral	Ultraviolet (UV) light	94.50%	120 min	[172]
ZnO 100 mg	Naproxen (NAP) 10 mg/L	T = n.a. pH = neutral	Ultraviolet (UV) light	98.70%	120 min	[172]
ZnO NPs 5 mg	Estrogenic hormones (Estrone, Estradiol, Ethinylestradiol, and Estriol) 0.8 mg/L	T/pH = n.a.	UV light	84–93%	60 min	[173]
ZnO 1.0 mg/mL	Amitriptyline Hydrochloride (AMI) 0.0300 mmol/L	T = 25 °C pH = 6.7	Solar irradiation	94.30%	60 min	[174]
ZnO 1 g/L	Venlafaxine 0.58 µg/L	T = RT pH = 7.5	Hg lamp (6 × 8 W)	95%	40 min	[175]
ZnO 0.05 g/L	Carbamazepine 1 mg/L	T = 20 °C pH = 7	Low pressure Hg lamp (150 W)	60%	60 min	[176]
Ag/ZnO 2 g/L	4-Nitrophenol 10 ppm	T/pH = n.a.	UV-light	100%	3 h	[177]
ZnO nanorod array	Estrone 1.85×10^{-3} mmol/L	T/pH = n.a.	UVA irradiation	90%	6 h	[178]
ZnO nanorod array	17β-Estradiol 1.85×10^{-3} mmol/L	T/pH = n.a.	UVA irradiation	80%	6 h	[178]
ZnO nanorod array	17α-Ethinylestradiol 1.85×10^{-3} mmol/L	T/pH = n.a.	UVA irradiation	80%	6 h	[178]
ZnO nanorod array	Bisphenol A 1.85×10^{-3} mmol/L	T/pH = n.a.	UVA irradiation	75%	6 h	[178]

n.a., not available.

The increased mobility, production, and separation of e⁻/h⁺ pairs might be the cause of ZnO's enhanced photocatalytic activity [169]. Under ultraviolet (UV) light irradiation, the ZnO photocatalyst degraded ibuprofen (IBU) and naproxen (NAP) with high rates of 94.5% and 98.7% after 120 min, respectively [167]. Aquatic life, mankind, and other species in the ecosystem are greatly threatened by water contamination with endocrine-disrupting compounds such as estrone (E1), estradiol (E2), 17-ethinylestradiol (EE2), and estriol (E3), which are steroid hormones. ZnO nanoparticles are extremely potent at eliminating estrogenic hormones through photodegradation. The degradation rate of endocrine-disrupting compounds by ZnO nanoparticles was 84–93% altogether, indicating a propensity for fast photocatalytic degradation [168]. Endocrine-disrupting compounds (EDCs), namely estrone (E1), 17-estradiol (E2), 17-ethinylestradiol (EE2), and bisphenol A (BPA) were photodegraded using ZnO nanorod arrays. The simplest EDC to eliminate over six hours was estrone (E1), followed by 17-ethinylestradiol (EE2) and 17-estradiol (E2) with a similar trend. Bisphenol A (BPA) degraded the least quickly of all EDC substances. Electrons will shift from the valence band to the conduction band as ZnO nanorods are exposed to UV light with a light energy greater than or equivalent to their energy band gap, thus creating electron-hole pairs. In subsequent reaction with water, the electrons and holes will produce hydroxyl radicals (OH·). The reactive oxidative species, called OH· radicals, will target EDCs and break them down into CO_2 and water. Additionally, the EDCs may be effectively oxidized by the holes.

10. Conclusions and Future Prospects

This review highlighted the repercussions of hazardous, non-biodegradable medications (antibiotics, antidepressants, anti-inflammatories, and contraceptives). A comprehensive overview of the photocatalytic degradation of such widely used drugs using bare ZnO, metal- and non-metal-doped ZnO, as well as ZnO-based nanocomposites is provided. ZnO demonstrated superior photocatalytic performance because of its photogenerated electron-hole pairs, mobility, and isolation. In practice, heteroatom doping is utilized to improve the catalyst's photocatalytic performance, especially when metal atoms are used as dopants. Moreover, it must be mentioned that metal dopants may act as recombination centers at higher concentrations, which can decrease the effectiveness of a photocatalyst. Consideration should be given to operating parameters such as pH, temperature, catalyst dosage, and pollutant concentration on the complete photodegradation rate. There are limited studies on the photocatalytic activity of ZnO-based catalysts in the degradation of developing pollutants, such as non-biodegradable medicines. Extensive research on multicomponent photocatalytic degradation is required to assess probable competition among a variety of pollutants for the photocatalyst, light absorption, and interactions with oxidizing agents. Decreased rapid recombination and photocorrosion should be addressed because they influence the photocatalyst's potential to effectively degrade pollutants, regenerate, and its lifecycle.

Economic Evaluation of Degradation Techniques

The average retail price of electricity in the United States is approximately USD 0.13 per kilowatt-hour (kWh) and it may vary based on the provider and local regions [179]. A lamp that operates at 500 watts will consume 0.5 kilowatt-hours (kWh) of electricity per hour, and this usage will cost USD 0.07. The cost of a UV lamp can vary greatly depending on the type of lamp, its size, the manufacturer, and the intended application. On average, a simple handheld UV lamp can cost between USD 20 to 100. On average, ZnO nanoparticles cost USD 4.6 per gram, which may vary based on the purity of the chemicals and supplier. The cost of photocatalytic degradation studies can vary depending on several factors, such as the location, size of the study, type of equipment used, and duration of the study.

Author Contributions: K.M.M.: validation, investigation, resources, writing—original draft, writing—review & editing, visualization. J.J.B.: investigation, resources, writing—original draft, writing—review & editing, visualization. J.J.V.: conception, methodology, resources, writing—original draft, writing—review & editing, visualization, supervision.; M.B.: validation, writing—original draft, writing—review & editing, visualization, co-supervision. All authors have read and agreed to the published version of the manuscript.

Funding: No external funding.

Institutional Review Board Statement: Not applicable.

Informed Consent Statement: Not applicable.

Data Availability Statement: Data and material are available from the corresponding author upon reasonable request.

Conflicts of Interest: All authors declare that they have no conflict of interest.

References

1. Zhang, Q.-Q.; Ying, G.-G.; Pan, C.-G.; Liu, Y.-S.; Zhao, J.-L. Comprehensive Evaluation of Antibiotics Emission and Fate in the River Basins of China: Source Analysis, Multimedia Modeling, and Linkage to Bacterial Resistance. *Environ. Sci. Technol.* **2015**, *49*, 6772–6782. [CrossRef] [PubMed]
2. Sharma, V.K.; Johnson, N.; Cizmas, L.; McDonald, T.J.; Kim, H. A Review of the Influence of Treatment Strategies on Antibiotic Resistant Bacteria and Antibiotic Resistance Genes. *Chemosphere* **2016**, *150*, 702–714. [CrossRef] [PubMed]
3. Jiang, Z.; Sun, H.; Wang, T.; Wang, B.; Wei, W.; Li, H.; Yuan, S.; An, T.; Zhao, H.; Yu, J.; et al. Nature-Based Catalyst for Visible-Light-Driven Photocatalytic CO_2 Reduction. *Energy Environ. Sci.* **2018**, *11*, 2382–2389. [CrossRef]
4. Ahmed, M.B.; Zhou, J.L.; Ngo, H.H.; Guo, W.; Thomaidis, N.S.; Xu, J. Progress in the Biological and Chemical Treatment Technologies for Emerging Contaminant Removal from Wastewater: A Critical Review. *J. Hazard. Mater.* **2017**, *323*, 274–298. [CrossRef]
5. Zhou, P.; Yu, J.; Jaroniec, M. All-Solid-State Z-Scheme Photocatalytic Systems. *Adv. Mater.* **2014**, *26*, 4920–4935. [CrossRef]
6. Liu, J.; Fang, W.; Wei, Z.; Qin, Z.; Jiang, Z.; Shangguan, W. Efficient Photocatalytic Hydrogen Evolution on N-Deficient g-C_3N_4 Achieved by a Molten Salt Post-Treatment Approach. *Appl. Catal. B* **2018**, *238*, 465–470. [CrossRef]
7. Fang, W.; Qin, Z.; Liu, J.; Wei, Z.; Jiang, Z.; Shangguan, W. Photo-Switchable Pure Water Splitting under Visible Light over Nano-Pt@P25 by Recycling Scattered Photons. *Appl. Catal. B* **2018**, *236*, 140–146. [CrossRef]
8. Liu, J.; Fang, W.; Wei, Z.; Qin, Z.; Jiang, Z.; Shangguan, W. Metallic 1T-Li x MoS 2 Co-Catalyst Enhanced Photocatalytic Hydrogen Evolution over ZnIn2S4 Floriated Microspheres under Visible Light Irradiation. *Catal. Sci. Technol.* **2018**, *8*, 1375–1382. [CrossRef]
9. Qin, Z.; Fang, W.; Liu, J.; Wei, Z.; Jiang, Z.; Shangguan, W. Zinc-Doped g-C_3N_4/$BiVO_4$ as a Z-Scheme Photocatalyst System for Water Splitting under Visible Light. *Chin. J. Catal.* **2018**, *39*, 472–478. [CrossRef]
10. Fang, W.; Liu, J.; Yu, L.; Jiang, Z.; Shangguan, W. Novel (Na, O) Co-Doped g-C_3N_4 with Simultaneously Enhanced Absorption and Narrowed Bandgap for Highly Efficient Hydrogen Evolution. *Appl. Catal. B* **2017**, *209*, 631–636. [CrossRef]
11. Wei, Z.; Li, R.; Wang, R. Enhanced Visible Light Photocatalytic Activity of BiOBr by in SituReactable Ionic Liquid Modification for Pollutant Degradation. *RSC Adv.* **2018**, *8*, 7956–7962. [CrossRef]
12. Wei, Z.-D.; Wang, R. Hierarchical BiOBr Microspheres with Oxygen Vacancies Synthesized via Reactable Ionic Liquids for Dyes Removal. *Chin. Chem. Lett.* **2016**, *27*, 769–772. [CrossRef]
13. Ameta, R.; Solanki, M.S.; Benjamin, S.; Ameta, S.C. Photocatalysis. In *Advanced Oxidation Processes for Waste Water Treatment*; Academic Press: London, UK, 2018; pp. 135–175. [CrossRef]
14. Davis, K.; Yarbrough, R.; Froeschle, M.; White, J.; Rathnayake, H. band gap Engineered Zinc Oxide Nanostructures via a Sol–Gel Synthesis of Solvent Driven Shape-Controlled Crystal Growth. *RSC Adv.* **2019**, *9*, 14638–14648. [CrossRef]
15. Morkoç, H.; Özgür, Ü. *Zinc Oxide*; Wiley: Hoboken, NJ, USA, 2009. [CrossRef]
16. Ameta, R.; Ameta, S.C. *Photocatalysis*; CRC Press: Boca Raton, FL, USA, 2016. [CrossRef]
17. Wang, H.; Zhang, L.; Chen, Z.; Hu, J.; Li, S.; Wang, Z.; Liu, J.; Wang, X. Semiconductor Heterojunction Photocatalysts: Design, Construction, and Photocatalytic Performances. *Chem. Soc. Rev.* **2014**, *43*, 5234. [CrossRef]
18. Herrmann, J.-M. Heterogeneous Photocatalysis: Fundamentals and Applications to the Removal of Various Types of Aqueous Pollutants. *Catal. Today* **1999**, *53*, 115–129. [CrossRef]
19. Rajamanickam, D.; Shanthi, M. Photocatalytic Degradation of an Organic Pollutant by Zinc Oxide—Solar Process. *Arab. J. Chem.* **2016**, *9*, S1858–S1868. [CrossRef]
20. Rauf, M.A.; Ashraf, S.S. Fundamental Principles and Application of Heterogeneous Photocatalytic Degradation of Dyes in Solution. *Chem. Eng. J.* **2009**, *151*, 10–18. [CrossRef]
21. Li, L.; Wang, M. Advanced Nanomatericals for Solar Photocatalysis. In *Advanced Catalytic Materials—Photocatalysis and Other Current Trends*; InTechOpen: London, UK, 2016. [CrossRef]

22. Khairy, M.; Zakaria, W. Effect of Metal-Doping of TiO$_2$ Nanoparticles on Their Photocatalytic Activities toward Removal of Organic Dyes. *Egypt. J. Pet.* **2014**, *23*, 419–426. [CrossRef]
23. Li, N.; Teng, H.; Zhang, L.; Zhou, J.; Liu, M. Synthesis of Mo-Doped WO 3 Nanosheets with Enhanced Visible-Light-Driven Photocatalytic Properties. *RSC Adv.* **2015**, *5*, 95394–95400. [CrossRef]
24. Bae, S.W.; Borse, P.H.; Hong, S.J.; Jang, J.S.; Lee, J.S.; Jeong, E.D.; Hong, T.E.; Yoon, J.H.; Jin, J.S.; Kim, H.G. Photophysical Properties of Nanosized Metal-Doped TiO$_2$ Photocatalyst Working under Visible Light. *J. Korean Phys. Soc.* **2007**, *51*, 22. [CrossRef]
25. Feng, Y.; Ji, W.-X.; Huang, B.-J.; Chen, X.; Li, F.; Li, P.; Zhang, C.; Wang, P.-J. The Magnetic and Optical Properties of 3d Transition Metal Doped SnO$_2$ Nanosheets. *RSC Adv.* **2015**, *5*, 24306–24312. [CrossRef]
26. Zhang, D.; Gong, J.; Ma, J.; Han, G.; Tong, Z. A Facile Method for Synthesis of N-Doped ZnO Mesoporous Nanospheres and Enhanced Photocatalytic Activity. *Dalton Trans.* **2013**, *42*, 16556. [CrossRef] [PubMed]
27. Kuo, C.-Y.; Wu, C.-H.; Wu, J.-T.; Chen, Y.-R. Synthesis and Characterization of a Phosphorus-Doped TiO$_2$ Immobilized Bed for the Photodegradation of Bisphenol A under UV and Sunlight Irradiation. *React. Kinet. Mech. Catal.* **2015**, *114*, 753–766. [CrossRef]
28. Guo, W.; Guo, Y.; Dong, H.; Zhou, X. Tailoring the Electronic Structure of β-Ga$_2$O$_3$ by Non-Metal Doping from Hybrid Density Functional Theory Calculations. *Phys. Chem. Chem. Phys.* **2015**, *17*, 5817–5825. [CrossRef]
29. Raza, W.; Faisal, S.M.; Owais, M.; Bahnemann, D.; Muneer, M. Facile Fabrication of Highly Efficient Modified ZnO Photocatalyst with Enhanced Photocatalytic, Antibacterial and Anticancer Activity. *RSC Adv.* **2016**, *6*, 78335–78350. [CrossRef]
30. Patil, A.B.; Patil, K.R.; Pardeshi, S.K. Ecofriendly Synthesis and Solar Photocatalytic Activity of S-Doped ZnO. *J. Hazard. Mater.* **2010**, *183*, 315–323. [CrossRef]
31. Yuan, Z.; Jia, J.; Zhang, L. Influence of Co-Doping of Zn(II)+Fe(III) on the Photocatalytic Activity of TiO$_2$ for Phenol Degradation. *Mater. Chem. Phys.* **2002**, *73*, 323–326. [CrossRef]
32. Saad, A.M.; Abukhadra, M.R.; Abdel-Kader Ahmed, S.; Elzanaty, A.M.; Mady, A.H.; Betiha, M.A.; Shim, J.-J.; Rabie, A.M. Photocatalytic Degradation of Malachite Green Dye Using Chitosan Supported ZnO and Ce–ZnO Nano-Flowers under Visible Light. *J. Environ. Manag.* **2020**, *258*, 110043. [CrossRef]
33. Ran, L.; Zhao, D.; Gao, X.; Yin, L. Highly Crystalline Ti-Doped SnO$_2$ Hollow Structured Photocatalyst with Enhanced Photocatalytic Activity for Degradation of Organic Dyes. *CrystEngComm* **2015**, *17*, 4225–4237. [CrossRef]
34. Gnanamozhi, P.; Rajivgandhi, G.; Alharbi, N.S.; Kadaikunnan, S.; Khaled, J.M.; Almanaa, T.N.; Pandiyan, V.; Li, W.-J. Enhanced Antibacterial and Photocatalytic Degradation of Reactive Red 120 Using Lead Substituted ZnO Nanoparticles Prepared by Ultrasonic-Assisted Co-Precipitation Method. *Ceram. Int.* **2020**, *46*, 19593–19599. [CrossRef]
35. Sharma, H.K.; Archana, R.; Sankar ganesh, R.; Singh, B.P.; Ponnusamy, S.; Hayakawa, Y.; Muthamizhchelvan, C.; Raji, P.; Kim, D.Y.; Sharma, S.K. Substitution of Al^{3+} to Zn^{2+} Sites of ZnO Enhanced the Photocatalytic Degradation of Methylene Blue under Irradiation of Visible Light. *Solid State Sci.* **2019**, *94*, 45–53. [CrossRef]
36. Wang, P.; Zhou, T.; Wang, R.; Lim, T.-T. Carbon-Sensitized and Nitrogen-Doped TiO$_2$ for Photocatalytic Degradation of Sulfanilamide under Visible-Light Irradiation. *Water Res.* **2011**, *45*, 5015–5026. [CrossRef]
37. Yadav, S.; Jaiswar, G. Review on Undoped/Doped TiO$_2$ Nanomaterial; Synthesis and Photocatalytic and Antimicrobial Activity. *J. Chin. Chem. Soc.* **2017**, *64*, 103–116. [CrossRef]
38. Xu, J.; Chen, Z.; Zapien, J.A.; Lee, C.-S.; Zhang, W. Surface Engineering of ZnO Nanostructures for Semiconductor-Sensitized Solar Cells. *Adv. Mater.* **2014**, *26*, 5337–5367. [CrossRef]
39. Yaghoubi, H.; Li, Z.; Chen, Y.; Ngo, H.T.; Bhethanabotla, V.R.; Joseph, B.; Ma, S.; Schlaf, R.; Takshi, A. Toward a Visible Light-Driven Photocatalyst: The Effect of Midgap-States-Induced Energy Gap of Undoped TiO$_2$ Nanoparticles. *ACS Catal.* **2015**, *5*, 327–335. [CrossRef]
40. Özgür, Ü.; Alivov, Y.I.; Liu, C.; Teke, A.; Reshchikov, M.A.; Doğan, S.; Avrutin, V.; Cho, S.-J.; Morkoç, H. A Comprehensive Review of ZnO Materials and Devices. *J. Appl. Phys.* **2005**, *98*, 041301. [CrossRef]
41. Klingshirn, C. ZnO: Material, Physics and Applications. *ChemPhysChem* **2007**, *8*, 782–803. [CrossRef]
42. Klingshirn, C.; Hauschild, R.; Priller, H.; Decker, M.; Zeller, J.; Kalt, H. ZnO Rediscovered—Once Again!? *Superlattices Microstruct.* **2005**, *38*, 209–222. [CrossRef]
43. Thangavel, R.; Singh Moirangthem, R.; Lee, W.-S.; Chang, Y.-C.; Wei, P.-K.; Kumar, J. Cesium Doped and Undoped ZnO Nanocrystalline Thin Films: A Comparative Study of Structural and Micro-Raman Investigation of Optical Phonons. *J. Raman Spectrosc.* **2010**, *41*, 1594–1600. [CrossRef]
44. Ellmer, K.; Klein, A. ZnO and Its Applications. In *Transparent Conductive Zinc Oxide*; Springer: Berlin/Heidelberg, Germany, 2008; pp. 1–33. [CrossRef]
45. Wang, Z.L.; Song, J. Piezoelectric Nanogenerators Based on Zinc Oxide Nanowire Arrays. *Science* **2006**, *312*, 242–246. [CrossRef]
46. Gao, P.X.; Song, J.; Liu, J.; Wang, Z.L. Nanowire Piezoelectric Nanogenerators on Plastic Substrates as Flexible Power Sources for Nanodevices. *Adv. Mater.* **2007**, *19*, 67–72. [CrossRef]
47. Wang, Z.L. Nanopiezotronics. *Adv. Mater.* **2007**, *19*, 889–892. [CrossRef]
48. Song, J.; Wang, X.; Liu, J.; Liu, H.; Li, Y.; Wang, Z.L. Piezoelectric Potential Output from ZnO Nanowire Functionalized with P-Type Oligomer. *Nano Lett.* **2008**, *8*, 203–207. [CrossRef]
49. Lu, M.-P.; Song, J.; Lu, M.-Y.; Chen, M.-T.; Gao, Y.; Chen, L.-J.; Wang, Z.L. Piezoelectric Nanogenerator Using P-Type ZnO Nanowire Arrays. *Nano Lett.* **2009**, *9*, 1223–1227. [CrossRef] [PubMed]

50. Sharma, D.K.; Shukla, S.; Sharma, K.K.; Kumar, V. A Review on ZnO: Fundamental Properties and Applications. *Mater. Today Proc.* **2022**, *49*, 3028–3035. [CrossRef]
51. Ayanda, O.S.; Aremu, O.H.; Akintayo, C.O.; Sodeinde, K.O.; Igboama, W.N.; Oseghe, E.O.; Nelana, S.M. Sonocatalytic Degradation of Amoxicillin from Aquaculture Effluent by Zinc Oxide Nanoparticles. *Environ. Nanotechnol. Monit. Manag.* **2021**, *16*, 100513. [CrossRef]
52. Azeez, F.; Al-Hetlani, E.; Arafa, M.; Abdelmonem, Y.; Nazeer, A.A.; Amin, M.O.; Madkour, M. The Effect of Surface Charge on Photocatalytic Degradation of Methylene Blue Dye Using Chargeable Titania Nanoparticles. *Sci. Rep.* **2018**, *8*, 7104. [CrossRef]
53. Gautam, S.; Agrawal, H.; Thakur, M.; Akbari, A.; Sharda, H.; Kaur, R.; Amini, M. Metal Oxides and Metal Organic Frameworks for the Photocatalytic Degradation: A Review. *J. Environ. Chem. Eng.* **2020**, *8*, 103726. [CrossRef]
54. Yi, Z.; Wang, J.; Jiang, T.; Tang, Q.; Cheng, Y. Photocatalytic Degradation of Sulfamethazine in Aqueous Solution Using ZnO with Different Morphologies. *R. Soc. Open Sci.* **2018**, *5*, 171457. [CrossRef]
55. Mirzaei, A.; Chen, Z.; Haghighat, F.; Yerushalmi, L. Removal of Pharmaceuticals and Endocrine Disrupting Compounds from Water by Zinc Oxide-Based Photocatalytic Degradation: A Review. *Sustain. Cities Soc.* **2016**, *27*, 407–418. [CrossRef]
56. Qamar, M.; Muneer, M. A Comparative Photocatalytic Activity of Titanium Dioxide and Zinc Oxide by Investigating the Degradation of Vanillin. *Desalination* **2009**, *249*, 535–540. [CrossRef]
57. Gupta, D.; Chauhan, R.; Kumar, N.; Singh, V.; Srivastava, V.C.; Mohanty, P.; Mandal, T.K. Enhancing Photocatalytic Degradation of Quinoline by $ZnO:TiO_2$ Mixed Oxide: Optimization of Operating Parameters and Mechanistic Study. *J. Environ. Manag.* **2020**, *258*, 110032. [CrossRef]
58. He, S.; Hou, P.; Petropoulos, E.; Feng, Y.; Yu, Y.; Xue, L.; Yang, L. High Efficient Visible-Light Photocatalytic Performance of Cu/ZnO/RGO Nanocomposite for Decomposing of Aqueous Ammonia and Treatment of Domestic Wastewater. *Front. Chem.* **2018**, *6*, 219. [CrossRef]
59. Sansenya, T.; Masri, N.; Chankhanittha, T.; Senasu, T.; Piriyanon, J.; Mukdasai, S.; Nanan, S. Hydrothermal Synthesis of ZnO Photocatalyst for Detoxification of Anionic Azo Dyes and Antibiotic. *J. Phys. Chem. Solids* **2022**, *160*, 110353. [CrossRef]
60. Qu, Y.; Xu, X.; Huang, R.; Qi, W.; Su, R.; He, Z. Enhanced Photocatalytic Degradation of Antibiotics in Water over Functionalized N,S-Doped Carbon Quantum Dots Embedded ZnO Nanoflowers under Sunlight Irradiation. *Chem. Eng. J.* **2020**, *382*, 123016. [CrossRef]
61. Mohamed, R.M.; Ismail, A.A.; Alhaddad, M. A Novel Design of Porous Cr_2O_3@ZnO Nanocomposites as Highly Efficient Photocatalyst toward Degradation of Antibiotics: A Case Study of Ciprofloxacin. *Sep. Purif. Technol.* **2021**, *266*, 118588. [CrossRef]
62. Dworschak, D.; Brunnhofer, C.; Valtiner, M. Photocorrosion of ZnO Single Crystals during Electrochemical Water Splitting. *ACS Appl. Mater. Interfaces* **2020**, *12*, 51530–51536. [CrossRef]
63. Fang, Y.; Li, Z.; Xu, S.; Han, D.; Lu, D. Optical Properties and Photocatalytic Activities of Spherical ZnO and Flower-like ZnO Structures Synthesized by Facile Hydrothermal Method. *J. Alloy. Compd.* **2013**, *575*, 359–363. [CrossRef]
64. Pal, M.; Bera, S.; Sarkar, S.; Jana, S. Influence of Al Doping on Microstructural, Optical and Photocatalytic Properties of Sol–Gel Based Nanostructured Zinc Oxide Films on Glass. *RSC Adv.* **2014**, *4*, 11552–11563. [CrossRef]
65. Liang, S.; Xiao, K.; Mo, Y.; Huang, X. A Novel ZnO Nanoparticle Blended Polyvinylidene Fluoride Membrane for Anti-Irreversible Fouling. *J. Memb. Sci.* **2012**, *394–395*, 184–192. [CrossRef]
66. Meng, Y.; Lin, Y.; Yang, J. Synthesis of Rod-Cluster ZnO Nanostructures and Their Application to Dye-Sensitized Solar Cells. *Appl. Surf. Sci.* **2013**, *268*, 561–565. [CrossRef]
67. Jimenez-Cadena, G.; Comini, E.; Ferroni, M.; Vomiero, A.; Sberveglieri, G. Synthesis of Different ZnO Nanostructures by Modified PVD Process and Potential Use for Dye-Sensitized Solar Cells. *Mater. Chem. Phys.* **2010**, *124*, 694–698. [CrossRef]
68. Luo, J.; Ma, S.Y.; Sun, A.M.; Cheng, L.; Yang, G.J.; Wang, T.; Li, W.Q.; Li, X.B.; Mao, Y.Z.; GZ, D.J. Ethanol Sensing Enhancement by Optimizing ZnO Nanostructure: From 1D Nanorods to 3D Nanoflower. *Mater. Lett.* **2014**, *137*, 17–20. [CrossRef]
69. Srujana, S.; Bhagat, D. Chemical-Based Synthesis of ZnO Nanoparticles and Their Applications in Agriculture. *Nanotechnol. Environ. Eng.* **2022**, *7*, 269–275. [CrossRef]
70. Banerjee, P.; Chakrabarti, S.; Maitra, S.; Dutta, B.K. Zinc Oxide Nano-Particles—Sonochemical Synthesis, Characterization and Application for Photo-Remediation of Heavy Metal. *Ultrason. Sonochem.* **2012**, *19*, 85–93. [CrossRef]
71. Vafaee, M.; Ghamsari, M.S. Preparation and Characterization of ZnO Nanoparticles by a Novel Sol–Gel Route. *Mater. Lett.* **2007**, *61*, 3265–3268. [CrossRef]
72. Ciciliati, M.A.; Silva, M.F.; Fernandes, D.M.; de Melo, M.A.C.; Hechenleitner, A.A.W.; Pineda, E.A.G. Fe-Doped ZnO Nanoparticles: Synthesis by a Modified Sol–Gel Method and Characterization. *Mater. Lett.* **2015**, *159*, 84–86. [CrossRef]
73. El-Eskandarany, M.S. Introduction. In *Mechanical Alloying*; Elsevier: Amsterdam, The Netherlands, 2015; pp. 1–12. [CrossRef]
74. Özgür, Ü.; Avrutin, V.; Morkoç, H. Zinc Oxide Materials and Devices Grown by MBE. In *Molecular Beam Epitaxy*; Elsevier: Amsterdam, The Netherlands, 2013; pp. 369–416. [CrossRef]
75. Shen, J.-H.; Chiang, T.-H.; Tsai, C.-K.; Jiang, Z.-W.; Horng, J.-J. Mechanistic Insights into Hydroxyl Radical Formation of Cu-Doped $ZnO/g-C_3N_4$ Composite Photocatalysis for Enhanced Degradation of Ciprofloxacin under Visible Light: Efficiency, Kinetics, Products Identification and Toxicity Evaluation. *J. Environ. Chem. Eng.* **2022**, *10*, 107352. [CrossRef]
76. Bechambi, O.; Sayadi, S.; Najjar, W. Photocatalytic Degradation of Bisphenol A in the Presence of C-Doped ZnO: Effect of Operational Parameters and Photodegradation Mechanism. *J. Ind. Eng. Chem.* **2015**, *32*, 201–210. [CrossRef]

77. Li, Y.; Zhao, X.; Fan, W. Structural, Electronic, and Optical Properties of Ag-Doped ZnO Nanowires: First Principles Study. *J. Phys. Chem. C* **2011**, *115*, 3552–3557. [CrossRef]
78. Ahmad, M.; Ahmed, E.; Ahmed, W.; Elhissi, A.; Hong, Z.L.; Khalid, N.R. Enhancing Visible Light Responsive Photocatalytic Activity by Decorating Mn-Doped ZnO Nanoparticles on Graphene. *Ceram. Int.* **2014**, *40*, 10085–10097. [CrossRef]
79. Kaur, J.; Singhal, S. Facile Synthesis of ZnO and Transition Metal Doped ZnO Nanoparticles for the Photocatalytic Degradation of Methyl Orange. *Ceram. Int.* **2014**, *40*, 7417–7424. [CrossRef]
80. Polat, İ.; Yılmaz, E.; Altın, İ.; Bacaksız, E.; Sökmen, M. The Influence of Cu-Doping on Structural, Optical and Photocatalytic Properties of ZnO Nanorods. *Mater. Chem. Phys.* **2014**, *148*, 528–532. [CrossRef]
81. Fu, M.; Li, Y.; Wu, S.; Lu, P.; Liu, J.; Dong, F. Sol–Gel Preparation and Enhanced Photocatalytic Performance of Cu-Doped ZnO Nanoparticles. *Appl. Surf. Sci.* **2011**, *258*, 1587–1591. [CrossRef]
82. Mohan, R.; Krishnamoorthy, K.; Kim, S.-J. Enhanced Photocatalytic Activity of Cu-Doped ZnO Nanorods. *Solid State Commun.* **2012**, *152*, 375–380. [CrossRef]
83. Le, T.H.; Bui, A.T.; Le, T.K. The Effect of Fe Doping on the Suppression of Photocatalytic Activity of ZnONanopowder for the Application in Sunscreens. *Powder Technol.* **2014**, *268*, 173–176. [CrossRef]
84. Mardani, H.R.; Forouzani, M.; Ziari, M.; Biparva, P. Visible Light Photo-Degradation of Methylene Blue over Fe or Cu Promoted ZnO Nanoparticles. *Spectrochim. Acta A Mol. Biomol. Spectrosc.* **2015**, *141*, 27–33. [CrossRef]
85. Yu, Z.; Yin, L.-C.; Xie, Y.; Liu, G.; Ma, X.; Cheng, H.-M. Crystallinity-Dependent Substitutional Nitrogen Doping in ZnO and Its Improved Visible Light Photocatalytic Activity. *J. Colloid Interface Sci.* **2013**, *400*, 18–23. [CrossRef]
86. Alam, M.S.; Manzoor, U.; Mujahid, M.; Bhatti, A.S. Highly Responsive UV Light Sensors Using Mg-Doped ZnO Nanoparticles. *J. Sens.* **2016**, *2016*, 8296936. [CrossRef]
87. Mridha, S.; Basak, D. Aluminium Doped ZnO Films: Electrical, Optical and Photoresponse Studies. *J. Phys. D Appl. Phys.* **2007**, *40*, 6902–6907. [CrossRef]
88. Dai, W.; Pan, X.; Chen, C.; Chen, S.; Chen, W.; Zhang, H.; Ye, Z. Enhanced UV Detection Performance Using a Cu-Doped ZnO Nanorod Array Film. *RSC Adv.* **2014**, *4*, 31969. [CrossRef]
89. Al-Buriahi, M.S.; Hessien, M.; Alresheedi, F.; Al-Baradi, A.M.; Alrowaili, Z.A.; Kebaili, I.; Olarinoye, I.O. ZnO– Bi_2O_3 Nanopowders: Fabrication, Structural, Optical, and Radiation Shielding Properties. *Ceram. Int.* **2022**, *48*, 3464–3472. [CrossRef]
90. Wang, Y.; Hao, X.; Wang, Z.; Dong, M.; Cui, L. Facile Fabrication of Mn^{2+}-Doped ZnO Photocatalysts by Electrospinning. *R. Soc. Open Sci.* **2020**, *7*, 191050. [CrossRef]
91. Singh, J.; Soni, R.K. Controlled Synthesis of CuO Decorated Defect Enriched ZnO Nanoflakes for Improved Sunlight-Induced Photocatalytic Degradation of Organic Pollutants. *Appl. Surf. Sci.* **2020**, *521*, 146420. [CrossRef]
92. Wang, J.; Wang, G.; Jiang, J.; Wan, Z.; Su, Y.; Tang, H. Insight into Charge Carrier Separation and Solar-Light Utilization: RGO Decorated 3D ZnO Hollow Microspheres for Enhanced Photocatalytic Hydrogen Evolution. *J. Colloid Interface Sci.* **2020**, *564*, 322–332. [CrossRef] [PubMed]
93. Van Tuan, P.; Phuong, T.T.; Tan, V.T.; Nguyen, S.X.; Khiem, T.N. In-Situ Hydrothermal Fabrication and Photocatalytic Behavior of ZnO/Reduced Graphene Oxide Nanocomposites with Varying Graphene Oxide Concentrations. *Mater. Sci. Semicond. Process* **2020**, *115*, 105114. [CrossRef]
94. Eskandari Azar, B.; Ramazani, A.; TaghaviFardood, S.; Morsali, A. Green Synthesis and Characterization of $ZnAl_2O_4$@ZnO Nanocomposite and Its Environmental Applications in Rapid Dye Degradation. *Optik* **2020**, *208*, 164129. [CrossRef]
95. Długosz, O.; Szostak, K.; Krupiński, M.; Banach, M. Synthesis of Fe_3O_4/ZnO Nanoparticles and Their Application for the Photodegradation of Anionic and Cationic Dyes. *Int. J. Environ. Sci. Technol.* **2021**, *18*, 561–574. [CrossRef]
96. Zhang, Y.; Liu, B.; Chen, N.; Du, Y.; Ding, T.; Li, Y.; Chang, W. Synthesis of SnO_2/ZnO Flowerlike Composites Photocatalyst for Enhanced Photocatalytic Degradation of Malachite Green. *Opt. Mater.* **2022**, *133*, 112978. [CrossRef]
97. Trakulmututa, J.; Chuaicham, C.; Shenoy, S.; Srikhaow, A.; Sasaki, K.; Smith, S.M. Effect of Transformation Temperature toward Optical Properties of Derived CuO/ZnO Composite from Cu–Zn Hydroxide Nitrate for Photocatalytic Ciprofloxacin Degradation. *Opt. Mater.* **2022**, *133*, 112941. [CrossRef]
98. Hemnil, P.; Prapawasit, Y.; Karthikeyan, V.; Wongwuttanasatian, T.; Seithtanabutara, V. Novel Cubic Heterojunction Fe_2O_3/ZnO Composite for the Photocatalyst Application. *Mater. Today Proc.* **2022**, *in press*. [CrossRef]
99. Krishnan, U.; Kaur, M.; Kaur, G.; Singh, K.; Dogra, A.R.; Kumar, M.; Kumar, A. MoS_2/ZnO Nanocomposites for Efficient Photocatalytic Degradation of Industrial Pollutants. *Mater. Res. Bull.* **2019**, *111*, 212–221. [CrossRef]
100. Zhang, J.; Kuang, M.; Cao, Y.; Ji, Z. Environment-Friendly Ternary ZnO/$ZnFe_2O_4$/TiO_2 Composite Photocatalyst with Synergistic Enhanced Photocatalytic Activity under Visible-Light Irradiation. *Solid State Sci.* **2022**, *129*, 106913. [CrossRef]
101. Fatimah, I.; Fadillah, G.; Sahroni, I.; Kamari, A.; Sagadevan, S.; Doong, R.-A. Nanoflower-like Composites of ZnO/SiO_2 Synthesized Using Bamboo Leaves Ash as Reusable Photocatalyst. *Arab. J. Chem.* **2021**, *14*, 102973. [CrossRef]
102. Sabri, M.; Habibi-Yangjeh, A.; Ghosh, S. Novel ZnO/$CuBi_2O_4$ Heterostructures for Persulfate-Assisted Photocatalytic Degradation of Dye Contaminants under Visible Light. *J. Photochem. Photobiol. A Chem.* **2020**, *391*, 112397. [CrossRef]
103. Abdurahman, M.H.; Abdullah, A.Z.; Shoparwe, N.F. A Comprehensive Review on Sonocatalytic, Photocatalytic, and Sonophotocatalytic Processes for the Degradation of Antibiotics in Water: Synergistic Mechanism and Degradation Pathway. *Chem. Eng. J.* **2021**, *413*, 127412. [CrossRef]

104. Wei, Z.; Liu, J.; Shangguan, W. A Review on Photocatalysis in Antibiotic Wastewater: Pollutant Degradation and Hydrogen Production. *Chin. J. Catal.* **2020**, *41*, 1440–1450. [CrossRef]
105. Rizzi, V.; Lacalamita, D.; Gubitosa, J.; Fini, P.; Petrella, A.; Romita, R.; Agostiano, A.; Gabaldón, J.A.; ForteaGorbe, M.I.; Gómez-Morte, T.; et al. Removal of Tetracycline from Polluted Water by Chitosan-Olive Pomace Adsorbing Films. *Sci. Total Environ.* **2019**, *693*, 133620. [CrossRef]
106. Zhang, Z.; Ding, C.; Li, Y.; Ke, H.; Cheng, G. Efficient Removal of Tetracycline Hydrochloride from Aqueous Solution by Mesoporous Cage MOF-818. *SN Appl. Sci.* **2020**, *2*, 669. [CrossRef]
107. Zhao, Y.; Gu, X.; Li, S.; Han, R.; Wang, G. Insights into Tetracycline Adsorption onto Kaolinite and Montmorillonite: Experiments and Modeling. *Environ. Sci. Pollut. Res.* **2015**, *22*, 17031–17040. [CrossRef]
108. Zhang, M.; Xu, L.; Qi, C.; Zhang, M. Highly Effective Removal of Tetracycline from Water by Hierarchical Porous Carbon: Batch and Column Adsorption. *Ind. Eng. Chem. Res.* **2019**, *58*, 20036–20046. [CrossRef]
109. World Health Organization. *Ten Threats to Global Health in 2019*; World Health Organization: Geneva, Switzerland, 2019.
110. World Health Organization. *WHO Report on Surveillance of Antibiotic Consumption: 2016–2018 Early Implementation*; World Health Organization: Geneva, Switzerland, 2018.
111. Rodi, P.; Obermeyer, W.; Pablos-Mendez, A.; Gori, A.; Raviglione, M.C. Political Rationale, Aims, and Outcomes of Health-Related High-Level Meetings and Special Sessions at the UN General Assembly: A Policy Research Observational Study. *PLoS Med.* **2022**, *19*, e1003873. [CrossRef] [PubMed]
112. Lofrano, G.; Libralato, G.; Brown, J. (Eds.) *Nanotechnologies for Environmental Remediation*; Springer International Publishing: Cham, Switzerland, 2017. [CrossRef]
113. Rizzo, L.; Agovino, T.; Nahim-Granados, S.; Castro-Alférez, M.; Fernández-Ibáñez, P.; Polo-López, M.I. Tertiary Treatment of Urban Wastewater by Solar and UV-C Driven Advanced Oxidation with Peracetic Acid: Effect on Contaminants of Emerging Concern and Antibiotic Resistance. *Water Res.* **2019**, *149*, 272–281. [CrossRef] [PubMed]
114. Zammit, I.; Vaiano, V.; Ribeiro, A.; Silva, A.; Manaia, C.; Rizzo, L. Immobilised Cerium-Doped Zinc Oxide as a Photocatalyst for the Degradation of Antibiotics and the Inactivation of Antibiotic-Resistant Bacteria. *Catalysts* **2019**, *9*, 222. [CrossRef]
115. O'Neill, J. *Antimicrobial Resistance: Tackling a Crisis for the Health and Wealth of Nations*; Review on Antimicrobial Resistance: London, UK, 2014.
116. Brisson-Noël, A.; Trieu-Cuot, P.; Courvalin, P. Mechanism of Action of Spiramycin and Other Macrolides. *J. Antimicrob. Chemother.* **1988**, *22* (Suppl. B), 13–23. [CrossRef] [PubMed]
117. Bellino, A.; Lofrano, G.; Carotenuto, M.; Libralato, G.; Baldantoni, D. Antibiotic Effects on Seed Germination and Root Development of Tomato (*Solanum lycopersicum* L.). *Ecotoxicol. Environ. Saf.* **2018**, *148*, 135–141. [CrossRef]
118. Fan, X.; Gao, J.; Li, W.; Huang, J.; Yu, G. Determination of 27 Pharmaceuticals and Personal Care Products (PPCPs) in Water: The Benefit of Isotope Dilution. *Front. Environ. Sci. Eng.* **2020**, *14*, 8. [CrossRef]
119. Dehghan, A.; Zarei, A.; Jaafari, J.; Shams, M.; Mousavi Khaneghah, A. Tetracycline Removal from Aqueous Solutions Using Zeolitic Imidazolate Frameworks with Different Morphologies: A Mathematical Modeling. *Chemosphere* **2019**, *217*, 250–260. [CrossRef]
120. Ibrahim, F.A.; Al-Ghobashy, M.A.; Abd El-Rahman, M.K.; Abo-Elmagd, I.F. Optimization and in Line Potentiometric Monitoring of Enhanced Photocatalytic Degradation Kinetics of Gemifloxacin Using TiO_2 Nanoparticles/H_2O_2. *Environ. Sci. Pollut. Res.* **2017**, *24*, 23880–23892. [CrossRef]
121. Lindberg, R.; Jarnheimer, P.-Å.; Olsen, B.; Johansson, M.; Tysklind, M. Determination of Antibiotic Substances in Hospital Sewage Water Using Solid Phase Extraction and Liquid Chromatography/Mass Spectrometry and Group Analogue Internal Standards. *Chemosphere* **2004**, *57*, 1479–1488. [CrossRef]
122. Shokoohi, R.; Leili, M.; Dargahi, A.; Vaziri, Y.; Khamutian, R. Common Antibiotics in Wastewater of Sina and Besat Hospitals, Hamadan, Iran. *Arch. Hyg. Sci.* **2017**, *6*, 152–159. [CrossRef]
123. Hu, Y.; Song, C.; Liao, J.; Huang, Z.; Li, G. Water Stable Metal-Organic Framework Packed Microcolumn for Online Sorptive Extraction and Direct Analysis of Naproxen and Its Metabolite from Urine Sample. *J. Chromatogr. A* **2013**, *1294*, 17–24. [CrossRef]
124. Vitiello, G.; Iervolino, G.; Imparato, C.; Rea, I.; Borbone, F.; de Stefano, L.; Aronne, A.; Vaiano, V. F-Doped ZnO Nano- and Meso-Crystals with Enhanced Photocatalytic Activity in Diclofenac Degradation. *Sci. Total Environ.* **2021**, *762*, 143066. [CrossRef]
125. Bendtsen, L.; Jensen, R.; Olesen, J. A Non-Selective (Amitriptyline), but Not a Selective (Citalopram), Serotonin Reuptake Inhibitor Is Effective in the Prophylactic Treatment of Chronic Tension-Type Headache. *J. Neurol. Neurosurg. Psychiatry* **1996**, *61*, 285–290. [CrossRef]
126. Li, H.; Sumarah, M.W.; Topp, E. Persistence of the Tricyclic Antidepressant Drugs Amitriptyline and Nortriptyline in Agriculture Soils. *Environ. Toxicol. Chem.* **2013**, *32*, 509–516. [CrossRef]
127. Togola, A.; Budzinski, H. Multi-Residue Analysis of Pharmaceutical Compounds in Aqueous Samples. *J. Chromatogr. A* **2008**, *1177*, 150–158. [CrossRef]
128. Kasprzyk-Hordern, B.; Dinsdale, R.M.; Guwy, A.J. The Occurrence of Pharmaceuticals, Personal Care Products, Endocrine Disruptors and Illicit Drugs in Surface Water in South Wales, UK. *Water Res.* **2008**, *42*, 3498–3518. [CrossRef]
129. Sabourin, L.; Duenk, P.; Bonte-Gelok, S.; Payne, M.; Lapen, D.R.; Topp, E. Uptake of Pharmaceuticals, Hormones and Parabens into Vegetables Grown in Soil Fertilized with Municipal Biosolids. *Sci. Total Environ.* **2012**, *431*, 233–236. [CrossRef]

130. Snyder, S.A. Endocrine Disruptors as Water Contaminants: Toxicological Implications for Humans and Wildlife. *Southwest Hydrol.* **2003**, *2*, 14–15.
131. Vymazal, J.; Březinová, T.; Koželuh, M. Occurrence and Removal of Estrogens, Progesterone and Testosterone in Three Constructed Wetlands Treating Municipal Sewage in the Czech Republic. *Sci. Total Environ.* **2015**, *536*, 625–631. [CrossRef]
132. Han, J.; Qiu, W.; Cao, Z.; Hu, J.; Gao, W. Adsorption of Ethinylestradiol (EE2) on Polyamide 612: Molecular Modeling and Effects of Water Chemistry. *Water Res.* **2013**, *47*, 2273–2284. [CrossRef]
133. Limpiyakorn, T.; Homklin, S.; Ong, S.K. Fate of Estrogens and Estrogenic Potentials in Sewerage Systems. *Crit. Rev. Environ. Sci. Technol.* **2011**, *41*, 1231–1270. [CrossRef]
134. Nghiem, L.D.; Schäfer, A.I. Adsorption and Transport of Trace Contaminant Estrone in NF/RO Membranes. *Environ. Eng. Sci.* **2002**, *19*, 441–451. [CrossRef]
135. Liu, Y.; Gao, W. Photodegradation of Endocrine Disrupting Chemicals by ZnO Nanorod Arrays. *Mol. Cryst. Liq. Cryst.* **2014**, *603*, 194–201. [CrossRef]
136. Su, Y.; Zhao, X.; Bi, Y.; Li, C.; Feng, Y.; Han, X. High-Concentration Organic Dye Removal Using $Fe_2O_3 \cdot 3.9MoO_3$ Nanowires as Fenton-like Catalysts. *Environ. Sci. Nano* **2018**, *5*, 2069–2076. [CrossRef]
137. Jiang, W.-L.; Xia, X.; Han, J.-L.; Ding, Y.-C.; Haider, M.R.; Wang, A.-J. Graphene Modified Electro-Fenton Catalytic Membrane for in Situ Degradation of Antibiotic Florfenicol. *Environ. Sci. Technol.* **2018**, *52*, 9972–9982. [CrossRef]
138. Chen, T.; Zhu, Z.; Zhang, H.; Shen, X.; Qiu, Y.; Yin, D. Enhanced Removal of Veterinary Antibiotic Florfenicol by a Cu-Based Fenton-like Catalyst with Wide PH Adaptability and High Efficiency. *ACS Omega* **2019**, *4*, 1982–1994. [CrossRef]
139. Oberoi, A.S.; Jia, Y.; Zhang, H.; Khanal, S.K.; Lu, H. Insights into the Fate and Removal of Antibiotics in Engineered Biological Treatment Systems: A Critical Review. *Environ. Sci. Technol.* **2019**, *53*, 7234–7264. [CrossRef]
140. Shehu Imam, S.; Adnan, R.; MohdKaus, N.H. Photocatalytic Degradation of Ciprofloxacin in Aqueous Media: A Short Review. *Toxicol. Environ. Chem.* **2018**, *100*, 518–539. [CrossRef]
141. Cuerda-Correa, E.M.; Alexandre-Franco, M.F.; Fernández-González, C. Advanced Oxidation Processes for the Removal of Antibiotics from Water. An Overview. *Water* **2019**, *12*, 102. [CrossRef]
142. Boxi, S.S.; Paria, S. Visible Light Induced Enhanced Photocatalytic Degradation of Organic Pollutants in Aqueous Media Using Ag Doped Hollow TiO_2 Nanospheres. *RSC Adv.* **2015**, *5*, 37657–37668. [CrossRef]
143. Kushwaha, H.S.; Halder, A.; Jain, D.; Vaish, R. Visible Light-Induced Photocatalytic and Antibacterial Activity of Li-Doped $Bi_{0.5}Na_{0.45}K_{0.5}TiO_3$–$BaTiO_3$ Ferroelectric Ceramics. *J. Electron. Mater.* **2015**, *44*, 4334–4342. [CrossRef]
144. Djurišić, A.B.; Leung, Y.H.; Ching Ng, A.M. Strategies for Improving the Efficiency of Semiconductor Metal Oxide Photocatalysis. *Mater. Horiz.* **2014**, *1*, 400. [CrossRef]
145. Sacco, O.; Vaiano, V.; Han, C.; Sannino, D.; Dionysiou, D.D. Photocatalytic Removal of Atrazine Using N-Doped TiO_2 Supported on Phosphors. *Appl. Catal. B* **2015**, *164*, 462–474. [CrossRef]
146. Zhao, G.; Ding, J.; Zhou, F.; Chen, X.; Wei, L.; Gao, Q.; Wang, K.; Zhao, Q. Construction of a Visible-Light-Driven Magnetic Dual Z-Scheme $BiVO_4$/$g-C_3N_4$/$NiFe_2O_4$ Photocatalyst for Effective Removal of Ofloxacin: Mechanisms and Degradation Pathway. *Chem. Eng. J.* **2021**, *405*, 126704. [CrossRef]
147. Bai, X.; Chen, W.; Wang, B.; Sun, T.; Wu, B.; Wang, Y. Photocatalytic Degradation of Some Typical Antibiotics: Recent Advances and Future Outlooks. *Int. J. Mol. Sci.* **2022**, *23*, 8130. [CrossRef]
148. Yashni, G.; Al-Gheethi, A.; Mohamed, R.; Hossain, M.S.; Kamil, A.F.; AbiramaShanmugan, V. Photocatalysis of Xenobiotic Organic Compounds in Greywater Using Zinc Oxide Nanoparticles: A Critical Review. *Water Environ. J.* **2021**, *35*, 190–217. [CrossRef]
149. Chankhanittha, T.; Yenjai, C.; Nanan, S. Utilization of Formononetin and Pinocembrin from Stem Extract of Dalbergia Parviflora as Capping Agents for Preparation of ZnO Photocatalysts for Degradation of RR141 Azo Dye and Ofloxacin Antibiotic. *Catal. Today* **2022**, *384–386*, 279–293. [CrossRef]
150. Ravbar, M.; Kunčič, A.; Matoh, L.; SmoleMožina, S.; Šala, M.; Šuligoj, A. Controlled Growth of ZnO Nanoparticles Using Ethanolic Root Extract of Japanese Knotweed: Photocatalytic and Antimicrobial Properties. *RSC Adv.* **2022**, *12*, 31235–31245. [CrossRef]
151. Vignati, D.; Lofrano, G.; Libralato, G.; Guida, M.; Siciliano, A.; Carraturo, F.; Carotenuto, M. Photocatalytic ZnO-Assisted Degradation of Spiramycin in Urban Wastewater: Degradation Kinetics and Toxicity. *Water* **2021**, *13*, 1051. [CrossRef]
152. Bhuyan, B.; Paul, B.; Purkayastha, D.D.; Dhar, S.S.; Behera, S. Facile Synthesis and Characterization of Zinc Oxide Nanoparticles and Studies of Their Catalytic Activity towards Ultrasound-Assisted Degradation of Metronidazole. *Mater. Lett.* **2016**, *168*, 158–162. [CrossRef]
153. El-Kemary, M.; El-Shamy, H.; El-Mehasseb, I. Photocatalytic Degradation of Ciprofloxacin Drug in Water Using ZnO Nanoparticles. *J. Lumin.* **2010**, *130*, 2327–2331. [CrossRef]
154. Zhang, Z.; Zada, A.; Cui, N.; Liu, N.; Liu, M.; Yang, Y.; Jiang, D.; Jiang, J.; Liu, S. Synthesis of Ag Loaded ZnO/BiOCl with High Photocatalytic Performance for the Removal of Antibiotic Pollutants. *Crystals* **2021**, *11*, 981. [CrossRef]
155. Majumder, S.; Chatterjee, S.; Basnet, P.; Mukherjee, J. ZnO Based Nanomaterials for Photocatalytic Degradation of Aqueous Pharmaceutical Waste Solutions—A Contemporary Review. *Environ. Nanotechnol. Monit. Manag.* **2020**, *14*, 100386. [CrossRef]
156. Semeraro, P.; Bettini, S.; Sawalha, S.; Pal, S.; Licciulli, A.; Marzo, F.; Lovergine, N.; Valli, L.; Giancane, G. Photocatalytic Degradation of Tetracycline by ZnO/γ-Fe_2O_3 Paramagnetic Nanocomposite Material. *Nanomaterials* **2020**, *10*, 1458. [CrossRef]

157. Jalili-Jahani, N.; Rabbani, F.; Fatehi, A.; MusaviHaghighi, T. Rapid One-Pot Synthesis of Ag-Decorated ZnO Nanoflowers for Photocatalytic Degradation of Tetracycline and Product Analysis by LC/APCI-MS and Direct Probe ESI-MS. *Adv. Powder Technol.* **2021**, *32*, 3075–3089. [CrossRef]
158. Hashem, E.M.; Hamza, M.A.; El-Shazly, A.N.; Abd El-Rahman, S.A.; El-Tanany, E.M.; Mohamed, R.T.; Allam, N.K. Novel Z-Scheme/Type-II CdS@ZnO/g-C3N4 Ternary Nanocomposites for the Durable Photodegradation of Organics: Kinetic and Mechanistic Insights. *Chemosphere* **2021**, *277*, 128730. [CrossRef]
159. Chankhanittha, T.; Komchoo, N.; Senasu, T.; Piriyanon, J.; Youngme, S.; Hemavibool, K.; Nanan, S. Silver Decorated ZnO Photocatalyst for Effective Removal of Reactive Red Azo Dye and Ofloxacin Antibiotic under Solar Light Irradiation. *Colloids Surf. A Physicochem. Eng. Asp.* **2021**, *626*, 127034. [CrossRef]
160. Mccally, M.; Solomon, G.M.; Schettler, T. Environment and Health: 6. Endocrine Disruption and Potential Human Health Implications. *CMAJ* **2000**, *163*, 1471–1476.
161. Ji, B.; Zhang, J.; Zhang, C.; Li, N.; Zhao, T.; Chen, F.; Hu, L.; Zhang, S.; Wang, Z. Vertically Aligned ZnO@ZnS Nanorod Chip with Improved Photocatalytic Activity for Antibiotics Degradation. *ACS Appl. Nano Mater.* **2018**, *1*, 793–799. [CrossRef]
162. Jingyu, H.; Ran, Y.; Zhaohui, L.; Yuanqiang, S.; Lingbo, Q.; NtiKani, A. In-Situ Growth of ZnO Globular on g-C3N4 to Fabrication Binary Heterojunctions and Their Photocatalytic Degradation Activity on Tetracyclines. *Solid State Sci.* **2019**, *92*, 60–67. [CrossRef]
163. Ramasamy, B.; Jeyadharmarajan, J.; Chinnaiyan, P. Novel Organic Assisted Ag-ZnO Photocatalyst for Atenolol and Acetaminophen Photocatalytic Degradation under Visible Radiation: Performance and Reaction Mechanism. *Environ. Sci. Pollut. Res.* **2021**, *28*, 39637–39647. [CrossRef]
164. Piña-Pérez, Y.; Aguilar-Martínez, O.; Acevedo-Peña, P.; Santolalla-Vargas, C.E.; Oros-Ruíz, S.; Galindo-Hernández, F.; Gómez, R.; Tzompantzi, F. Novel ZnS-ZnO Composite Synthesized by the Solvothermal Method through the Partial Sulfidation of ZnO for H2 Production without Sacrificial Agent. *Appl. Catal. B Environ.* **2018**, *230*, 125–134. [CrossRef]
165. Costa, L.N.; Nobre, F.X.; Lobo, A.O.; de Matos, J.M.E. Photodegradation of Ciprofloxacin Using Z-Scheme TiO2/SnO2 Nanostructures as Photocatalyst. *Environ. Nanotechnol. Monit. Manag.* **2021**, *16*, 100466. [CrossRef]
166. Kumar, S.; Kaushik, R.D.; Purohit, L.P. RGO supported ZnO/SnO2 Z-scheme heterojunctions with enriched ROS production towards enhanced photocatalytic mineralization of phenolic compounds and antibiotics at low temperature. *J. Colloid Interface Sci.* **2023**, *632*, 196–215. [CrossRef]
167. Yu, Y.; Yao, B.; Cao, B.; Ma, W. Morphology-controlled Fabrication of SnO2/ZnO Nanocomposites with Enhanced Photocatalytic Performance. *Photochem. Photobiol.* **2019**, *95*, 1131–1141. [CrossRef]
168. Ciğeroğlu, Z.; Şahin, S.; Kazan, E.S. One-Pot Green Preparation of Deep Eutectic Solvent-Assisted ZnO/GO Nanocomposite for Cefixime Trihydrate Photocatalytic Degradation under UV-A Irradiation. *Biomass Convers. Biorefin.* **2022**, *12* (Suppl. 1), 73–86. [CrossRef]
169. Sayadi, M.H.; Sobhani, S.; Shekari, H. Photocatalytic Degradation of Azithromycin Using GO@Fe3O4/ZnO/SnO2 Nanocomposites. *J. Clean. Prod.* **2019**, *232*, 127–136. [CrossRef]
170. Mahmud, S. One-Dimensional Growth of Zinc Oxide Nanostructures from Large Micro-Particles in a Highly Rapid Synthesis. *J. Alloy. Compd.* **2011**, *509*, 4035–4040. [CrossRef]
171. Tran, M.L.; Nguyen, C.H.; Fu, C.-C.; Juang, R.-S. Hybridizing Ag-Doped ZnO Nanoparticles with Graphite as Potential Photocatalysts for Enhanced Removal of Metronidazole Antibiotic from Water. *J. Environ. Manag.* **2019**, *252*, 109611. [CrossRef]
172. Sabouni, R.; Gomaa, H. Photocatalytic Degradation of Pharmaceutical Micro-Pollutants Using ZnO. *Environ. Sci. Pollut. Res.* **2019**, *26*, 5372–5380. [CrossRef]
173. Yasir, M.; Šopík, T.; Ali, H.; Kimmer, D.; Sedlařík, V. Green synthesis of titanium and zinc oxide nanoparticles for simultaneous photocatalytic removal of estrogens in wastewater. In Proceedings of the NANOCON 2021—International Conference on Nanomaterials, Brno, Czech Republic, 20–22 October 2021; pp. 189–196. [CrossRef]
174. Finčur, N.; ŠojićMerkulov, D.; Putnik, P.; Despotović, V.; Banić, N.; Lazarević, M.; Četojević-Simin, D.; Agbaba, J.; Abramović, B. Environmental Photocatalytic Degradation of Antidepressants with Solar Radiation: Kinetics, Mineralization, and Toxicity. *Nanomaterials* **2021**, *11*, 632. [CrossRef] [PubMed]
175. Teixeira, S.; Gurke, R.; Eckert, H.; Kühn, K.; Fauler, J.; Cuniberti, G. Photocatalytic Degradation of Pharmaceuticals Present in Conventional Treated Wastewater by Nanoparticle Suspensions. *J. Environ. Chem. Eng.* **2016**, *4*, 287–292. [CrossRef]
176. Bohdziewicz, J.; Kudlek, E.; Dudziak, M. Influence of the Catalyst Type (TiO2 and ZnO) on the Photocatalytic Oxidation of Pharmaceuticals in the Aquatic Environment. *Desalination Water Treat.* **2016**, *57*, 1552–1563. [CrossRef]
177. Divband, B.; Khatamian, M.; Eslamian, G.R.K.; Darbandi, M. Synthesis of Ag/ZnO Nanostructures by Different Methods and Investigation of Their Photocatalytic Efficiency for 4-Nitrophenol Degradation. *Appl. Surf. Sci.* **2013**, *284*, 80–86. [CrossRef]

178. Cheng, J.; Shen, Y.; Chen, K.; Wang, X.; Guo, Y.; Zhou, X.; Bai, R. Flower-like Bi2WO6/ZnO Composite with Excellent Photocatalytic Capability under Visible Light Irradiation. *Chin. J. Catal.* **2018**, *39*, 810–820. [CrossRef]
179. Mills, A.; Wiser, R.; Millstein, D.; Carvallo, J.P.; Gorman, W.; Seel, J.; Jeong, S. The Impact of Wind, Solar, and Other Factors on the Decline in Wholesale Power Prices in the United States. *Appl. Energy* **2021**, *283*, 116266. [CrossRef]

Disclaimer/Publisher's Note: The statements, opinions and data contained in all publications are solely those of the individual author(s) and contributor(s) and not of MDPI and/or the editor(s). MDPI and/or the editor(s) disclaim responsibility for any injury to people or property resulting from any ideas, methods, instructions or products referred to in the content.

Review

Mitochondrial Dysfunction Induced by Zinc Oxide Nanoparticles

Leslie Patrón-Romero [1,2], Priscy Alfredo Luque-Morales [2,*], Verónica Loera-Castañeda [3], Ismael Lares-Asseff [3], María Ángeles Leal-Ávila [4], Jorge Arturo Alvelais-Palacios [5], Ismael Plasencia-López [6,7] and Horacio Almanza-Reyes [1,7,*]

[1] Faculty of Medicine and Psychology, Autonomous University of Baja California, Tijuana 22390, Baja California, Mexico
[2] Faculty of Engineering, Architecture and Design, Autonomous University of Baja California, Ensenada 22860, Baja California, Mexico
[3] Instituto Politécnico Nacional, CIIDIR-Unidad Durango, Durango 34220, Durango, Mexico
[4] University Center for Health Education, Autonomous University of Baja California, Tijuana 22010, Baja California, Mexico
[5] School of Health Sciences, Valle de Las Palmas, Autonomous University of Baja California, Tijuana 22260, Baja California, Mexico
[6] Faculty of Accounting and Administration, Autonomous University of Baja California, Tijuana 22390, Baja California, Mexico
[7] Bioeconomy Cluster of Baja California, A.C., Tijuana 22040, Baja California, Mexico
* Correspondence: pluque@uabc.edu.mx (P.A.L.-M.); almanzareyes@uabc.edu.mx (H.A.-R.)

Abstract: The constant evolution and applications of metallic nanoparticles (NPs) make living organisms more susceptible to being exposed to them. Among the most used are zinc oxide nanoparticles (ZnO-NPs). Therefore, understanding the molecular effects of ZnO-NPs in biological systems is extremely important. This review compiles the main mechanisms that induce cell toxicity by exposure to ZnO-NPs and reported in vitro research models, with special attention to mitochondrial damage. Scientific evidence indicates that in vitro ZnO-NPs have a cytotoxic effect that depends on the size, shape and method of synthesis of ZnO-NPs, as well as the function of the cells to which they are exposed. ZnO-NPs come into contact with the extracellular region, leading to an increase in intracellular [Zn^{2+}] levels. The mechanism by which intracellular ZnO-NPs come into contact with organelles such as mitochondria is still unclear. The mitochondrion is a unique organelle considered the "power station" in the cells, participates in numerous cellular processes, such as cell survival/death, multiple biochemical and metabolic processes, and holds genetic material. ZnO-NPs increase intracellular levels of reactive oxygen species (ROS) and, in particular, superoxide levels; they also decrease mitochondrial membrane potential (MMP), which affects membrane permeability and leads to cell death. ZnO-NPs also induced cell death through caspases, which involve the intrinsic apoptotic pathway. The expression of pro-apoptotic genes after exposure to ZnO-NPs can be affected by multiple factors, including the size and morphology of the NPs, the type of cell exposed (healthy or tumor), stage of development (embryonic or differentiated), energy demand, exposure time and, no less relevant, the dose. To prevent the release of pro-apoptotic proteins, the damaged mitochondrion is eliminated by mitophagy. To replace those mitochondria that underwent mitophagy, the processes of mitochondrial biogenesis ensure the maintenance of adequate levels of ATP and cellular homeostasis.

Keywords: mitochondria; apoptosis; zinc oxide; nanoparticles

1. Introduction

The almost universal applications of metallic nanoparticles (NP) in daily life increase the probability of being in constant contact with them; therefore, investigating the beneficial and harmful effects is of particular relevance. Due to their multiple areas of application, among the most relevant NPs are zinc oxide NPs (ZnO-NPs). Its usefulness in the biomedical filed is undoubtedly among the most relevant and promising advances. Within the nanotechnology industry, it is among the most synthesized NPs, ranking below silver and gold NPs [1,2]. Even though ZnO-NPs offer essential benefits in many areas, such as the pharmacological, medical, biochemical, and microbiological, some studies show that ZnO-NPs can cause harmful effects compared to other metallic NPs such as Fe_3O_4, Al_2O_3, and TiO_2 [3–6]. Unlike their therapeutic applications in the oncology field, toxicological studies have shown that ZnO-NPs can represent a severe danger to specific tissues and organs, such as the lungs, skin, or muscles, to name a few [6–8]. Therefore, investigating the potential health risk due to increased exposure has recently generated concern. ZnO-NPs can enter indirectly through contact or consumption of previously exposed water, plants, and animals, or they can enter directly through inhalation, dermal contact, ingestion, or absorption.

ZnO-NP morphology depends on the synthesis process. They may be nanospheres, nanoplates, nanowires, nanotubes, nanorings, nanocages, nanoflowers, nanoflakes, or hexagonal. ZnO-NPs penetrate the cell membrane by multiple mechanisms, the most studied are ion channels (located both in the cell membrane and in the mitochondrial membrane), specific receptors for metal ions, endocytosis and by direct uptake of ZnO-NPs. Size and shape play a crucial role. The smaller the ZnO-NPs, the more easily they enter the cells and the greater the cytotoxic effect (<50 nm), due to the higher surface area/volume ratio. Morphology plays another relevant role—for example, spherical and hexagonal shapes penetrate easier than tubular shapes. In terms of surface charge, positively charged ZnO showed high cellular uptake compared to ZnO with negative charge. Additionally, the toxicity of nanoparticles, releasing toxic ions, has been considered. Since zinc oxide is amphoteric in nature, it reacts with both acids and alkalis, giving Zn^{2+} ions.

Once ZnO-NPs are internalized in the cell, they are distributed in all the organelles, particularly in the mitochondria [9,10]. The mitochondria is a unique organelle considered the "energy supply center" of the cell, present in all eukaryotic organisms and all mammalian cells. Mitochondria are responsible for numerous cellular processes, such as β-oxidation of fatty acids, amino acid metabolism, pyridine synthesis, phospholipid modifications, generation of reactive oxygen species (ROS), oxidative stress homeostasis, cell death/survival, and senescence. It is also well known that, in multicellular organisms, the number of mitochondria is variable. The number of mitochondria will depend on energy demands, function, and stage of development [11].

At the structural level, mitochondria are composed of an outer membrane and an inner membrane; the latter contains proteins involved in the electron transfer chain (ETC). The integrity of the ETC is crucial for ATP generation; as a result of its activation, reactive oxygen species (ROS) are generated. ROS are responsible for causing cellular stress, mutations and induce apoptosis [4]. The mitochondria carries a unique genetic material, which is characterized by having a circular double-strand (without ends) of nucleic acids called mitochondrial DNA (mtDNA), containing information to encode 37 genes—of which 13 genes encode for messenger RNA (mRNA), from which 13 mitochondrial proteins are subsequently obtained. Two genes encode ribosomal RNA (rRNA), and 22 genes synthesize transference RNA (tRNA). This smaller but no less critical genome is replicated, transcribed, and translated within the mitochondrial matrix, independently of cell cycle phases. Like nuclear genetic material, mtDNA is susceptible to genotoxicity produced by chemical (ZnO-NPs) and physical agents [12]. This review compiles the most relevant mechanisms leading to mitochondrial dysfunction caused by ZnO-NP exposure.

This review compiles the multiple mitochondrial pathways affected by toxicity after exposure to ZnO-NPs, among which loss of mitochondrial membrane potential, depletion of ATP synthesis due to electron transport chain abnormality, mtDNA synthesis, mitochondrial biogenesis, cell survival and apoptosis are mentioned in detail. The cytotoxicity induced by ZnO-NPs depends on numerous factors, such as ZnO-NPs synthesis methodology, nanoparticle size, morphology, concentration, exposure time and, not least, the type of cell exposed. Factors associated with cellular response include the stage of differentiation, e.g., a differentiated cell versus a cell of embryonic origin and whether the cell is healthy or tumorigenic. In the case of differentiated human cells, liver, neuronal, cardiac, epidermal, fibroblast, blood and other cells were included; and for tumor models: hepatocyte cancer cell, breast cancer, multiple myeloma cell, cervical cancer, etc. The maximum inhibitory concentration (IC50) and biological methods to obtain it were investigated. Finally, a concise discussion was made on the most relevant factors involved in the mitochondrial and cellular response after exposure to ZnO-NPs found in the most updated literature.

ZnO-NPs penetrate the cell membrane via four mechanisms: the first mechanism involves receptors associated with effector proteins that activate multiple signaling pathways. In the second mechanism, ion channels located in the mitochondrial membrane and cell membrane facilitate entry into the cytoplasm. The third mechanism is endocytosis—in this process, a fragment of the lipid cell membrane covers ZnO-NPs, allowing the formation of vacuoles, which later merge with the membranes of other organelles such as the nucleus and the mitochondria. The fourth mechanism is direct absorption, in which ZnO-NPs (depending on their size and shape) penetrate the cellular lipid bilayer. The entry of ZnO-NPs is also facilitated by two specialized zinc transporters—ZnT1 and ZnT2. Zinc dissociation also occurs, and zinc can penetrate as zinc ions.

2. Impairment of Mitochondrial Biogenesis Induced by ZnO-NPs

Mitochondrial biogenesis is essential to maintain the proper mitochondrial population and cellular functioning. In humans, it is estimated that >50% of the cytoplasm volume is occupied by the mitochondrial mass, >30% in cardiomyocytes and >25% in fibroblasts [13]. Under physiological conditions, the induction of mitochondrial biogenesis is associated with the activation of transcription factors that act on nuclear DNA (nDNA) and mtDNA to control local (within mitochondria) or cytoplasmic protein synthesis [14]. For example, exercise increases mitochondrial protein synthesis, resulting in mitochondrial biogenesis in skeletal muscles [15]. The half-life of mitochondria depends on multiple factors, such as ATP demand and cellular function. Severely damaged mitochondria will be eliminated by mitophagy to prevent the release of pro-apoptotic proteins. The fine coordination between these two opposing processes will determine mitochondrial homeostasis [16,17]. Some of the factors involved in mitochondrial biogenesis are mitochondrial fission protein 1 (FIS1) and mitochondrial fission factor (MFF), which promote mitochondrial fission by recruiting GTPase dynamin-related protein 1 (DRP1) [18,19].

Other factors involved in regulating mitochondrial biogenesis include mitochondrial transcription factor A (mtTFA), which drives mtDNA transcription and replication. The expression of mtTFA is regulated by the peroxisome proliferator-activated receptor gamma-coactivator 1α (PGC-1α) and PGC-1β, which are the main regulatory proteins of mitochondrial biogenesis [20]. More recently, the involvement of NF-E2-p45-related factor 2 (NRF2), encoded by the NFE2L2 gene, has gained attention. Free NRF2 translocates to the nucleus, where it targets the promoters of genes that induce mitochondrial biogenesis [21]. Li et al., [22] in 2013, reported the adverse effects caused in mature human cardiomyocytes derived from human stem cells after exposure to ZnO-NPs. They proposed that ZnO-NPs could alter mitochondrial biogenesis, as shown by decreased mitochondrial density, altered mtDNA copy number, and inhibition of the PGC-1α pathway that eventually lead to mitochondrial depletion. The mtDNA replication and transcription system is activated to counteract oxidative damage inflicted on the mitochondrial respiratory chain under mild oxidative stress. When cellular stress increases, mitochondrial biogenesis intervenes to maintain cellular homeostasis.

Yousef et al., [23] reported similar results in 2019, where they performed a quantitative analysis of mtFTA and PGC-1α expression in rat-derived liver and kidney cells. The results showed significant suppression of hepatic mtTFA expression by approximately 62% relative to the control value and, similarly, suppression in PGC-1α expression by approximately 51% relative to the control group, following ZnO-NP exposure. In addition, increased expression of the tumor suppressor gene p53, which is involved in the induction of apoptosis, was documented. On the other hand, mitochondria are highly dynamic organelles, constantly undergoing fission and fusion processes in response to changing energy demands and cellular stress environments. Mitochondrial fusion is mediated by proteins such as mitofusin 1 (MFN1) and mitofusin 2 (MFN2), located on the outer mitochondrial membrane [24]. Phosphatidylserine decarboxylase (PSD1), located in the inner mitochondrial membrane, is involved in the biosynthesis of phosphatidylcholine, which in turn regulates mitochondrial fusion. Another protein involved is optic atrophy 1 (OPA1), which participates in the inner membrane fusion and apoptosis [25,26].

In 2018, Babele et al., [27] reported that exposure to ZnO-NPs conditioned the release of mitochondrial PSD1 into the cytosol. This indicated loss of integrity in the inner mitochondrial membrane, decreased levels in biogenesis, and defects in lipid metabolism. A year later, to expand on the toxic effects of ZnO-NPs, Babele et al. combined a proteome and metabolome analysis. He identified 46 metabolites (including seven unknown). Among those affecting the oxidative pathway were Super Oxide Dismutase 1 (SOD1), Super Oxide Dismutase 2 (SOD2), and yeast AP-1 (YAP1). The metabolites involved in lipid biosynthetic pathways were transcription factor 2 requiring INOsitol (INO2), transcription factor 4 requiring INOsitol (INO4), CHOline requiring CHO1, cardiolipin synthase (CRD1), and PSD1. The aforementioned metabolites are involved in crucial mitochondrial processes, i.e., glycolysis, the tricarboxylic acid cycle (TCA), pentose phosphate pathway, and central carbon metabolism (CCM) in the Saccharomyces cerevisiae model [28]. Up to 10% of the proteins encoded by the human genome interact with zinc ions. Approximately 3000 proteins interact with zinc ions, acting mainly as a cofactor [29]. Therefore, determining how ZnO-NP are internalized, transported, and distributed within the cell and mitochondria is of particular interest. Zinc transporter 2 (ZnT2) was the first zinc transporter found. In 2020, Chevallet et al., [30] reported a 30- to 40-fold dose-dependent overexpression of ZnT2 in human hepatocytes (HepG2) with a dose-dependent effect. This proposed active sequestration of ZnO-NPs into mitochondria by ZnT2. Abnormalities in mitochondrial morphology were also reported, suggesting disruptions in mitochondrial fission and fusion dynamics after exposure to ZnO-NPs. Skin cells are among the most exposed tissues due to the use of sunscreens and beauty products. In 2013, Yu et al. showed in normal mouse-derived skin the alteration of mitochondria; ZnO-NP exposure negatively affects mitochondrial network and biogenesis after 48 h of exposure [31]. In 2020, Khan et al. proposed that dose-dependent abnormalities in mitochondrial morphology and a significant increase in apoptotic behavior in human skin carcinoma A431 cells compared to normal renal epithelial NRK-52E cells may be due to a higher expression of anionic membrane phospholipids in cancer cells [32]. Table 1, presents numerous cytotoxicity studies in various biological systems, both in embryonic cell lines and in differentiated human and non-human cells, healthy and tumor cells, exposed to ZnO-NPs. Morphological characteristics are described, as well as dose, time and molecular analysis, to facilitate comparative analysis.

Figure 1, shows a compilation of the main molecular pathways involved by exposure to ZnO-NPs, including their entry into the membrane, effect on the cell nucleus, mitochondria and, finally, activation of apoptosis.

Table 1. Cell models exposed to dose-time-dependent ZnO-NPs and the mitochondrial pathway analyzed.

Cell Type	Size and Shape of ZnO-NP	Synthesis and Characterization of ZnO-NPs	Time of Exposition (h)	Doses mg/L	IC50	Biological Characterization	Mitochondrial Pathway	Reference
Human embryonic kidney cells (HEK293T) Chicken embryo (cranial neural crest cells HH10)	<50 nm Spherical shape	Synthesis method ND/OC, TEM, SEM-EDS, DLS	12 h	12.5, 25, 50, 100, 200 µg/mL	50 µg/mL ND *	Cell culture, CCK-8, CNCCs, ARS, IHC, RNA-seq, q-PCR, DAPI, FM/IFA	Ion release triggered ROS production, which further induced cell toxicity, inflammation and apoptosis, which are mediated by NF-κB signaling cascades and mitochondria dysfunction. Increase in the expression of Nrf2, HO-1, NQO1, Cat, GLXR and NOS. Increase in Pax7 and Casp-3 expression.	[8]
Human cardiomyocytes (hiPSC-CMs)	40 nm–60 nm Spheroid and rod shaped	Synthesis method ND/OC, TEM, DLS	0, 2, 6 h	0–200 µg/mL	62.5 µg/mL	Cell culture, LDH, HCA, CCK-8, MMP, ROS, q-PCR, WB, MEA	ROS generation and induced mitochondrial dysfunction. Impair mitochondrial biogenesis and inhibit the PGC-1α pathway. Cardiac electrophysiological alterations.	[22]
Human epidermoid carcinoma cell A431 Normal kidney epithelial NRK-52E cells	<40 nm Hexagonal structure	Chemical synthesis by reduction Zinc acetate dyhidrate XRD, TEM, SEM, UV-Vis	24 h 24 h	0–25 µg/mL	24 µg/mL 15 µg/mL NS*	Cell culture, MTT, FM, ROS, DAPI, Casp-3 and cell morphology	The anti-proliferative activity, morphological changes, ROS generation, nuclear apoptosis and caspase-3 in a dose-dependent manner.	[32]
Human aortic endothelial cells (HAECs)	70 nm Rod shaped	Synthesis method ND/OC, TEM, XRD	12 or 24 h	8–50 µg/mL	50 µg/mL	Cell culture, MTT, LDH, ROS, FM, MMP, LA, CACA, Cas-3, Cas-9, Cyt-c, ICC, IFA	Apoptosis was confirmed using reactive oxygen species (ROS). Decrease in MMP. Increased release of Cyt-c, Caspase-3, caspase-9, BAX, BCL-2 and FAS receptor expression. The antioxidant LA was able to protect HAECs from apoptosis induced by ZnO-NPs.	[33]
Human tenon fibroblast (HTF)	56 nm ND	Synthesis method ND/OC, TEM	24 h 48 h 72 h	0–16.0 µg/mL	1.51 µg/mL 1.03 µg/mL 0.57 µg/mL	Cell culture, CCK-8, RT-CES, ROS, FM, MMP, qPCR, FC, Apaf-1, Cas-3, Cas-9, FSP-1	Inhibit the viability of HTFs and decrease MMP. Elevated ROS, caspase-3, caspase-9, and apoptotic Apaf-1 expression. Decrease the levels of FSP-1, collagen III, and E-cadherin expression, leading to HTF apoptosis.	[34]
Human erythrocytes	47.8–52.5 nm Rod shaped	Chemical synthesis by Pechini method XRD, TEM, FTIR	1 h	50–500 µg/mL	200 µg/mL	Cell culture, H, SOD, CAT, LPO, GST, GSH, ROS, OH, O2, CA, IFA/FM	Concentration-dependent hemolytic activity to human erythrocytes. ROS generation. Depletion of glutathione and GST levels. Increased SOD, CAT and lipid peroxidation in dose-dependent manner.	[35]

Table 1. Cont.

Cell Type	Size and Shape of ZnO-NP	Synthesis and Characterization of ZnO-NPs	Time of Exposition (h)	Doses mg/L	IC50	Biological Characterization	Mitochondrial Pathway	Reference
Human monocytes U-937 Human promyelocytes HL-60 Human B lymphocytes COLO-720L Human T lymphocytes HUT-78	15 nm ND	Synthesis method ND/OC. TEM, DLS	24 h	1.6–25 µg/mL	1.5625 µg/mL 3.125 µg/mL 6.25 µg/mL 12.5 µg/mL 25 µg/mL	Cell culture, MTT, LDH, MDA	ZnO-NPs caused lipid peroxidation of all cells and correlated with apoptosis. The level of cholesterol in membranes strongly modifies the effect exerted by ZnO-NPs.	[36]
Human monocytes U-937 Human promyelocytes HL-60 Human B lymphocytes COLO-720L Human T lymphocytes HUT-78	100–130 nm ND	Synthesis method ND/OC. TEM	24 h	0–25 µg/mL	1.56 µg/mL 3.12 µg/mL 6.25 µg/mL 12.5 µg/mL 25 µg/mL	Cell culture, MTT, LDH, NO, MDA, Casp-9, IL-6, TNF-α	Activation of the mitochondrial apoptosis pathway. TNF-α concentration increased for both cell lines. IL-6 concentration rose on average 5-fold in HL-60 cells in all experimental variants, the opposite trend was observed for COLO-720L. ZnO-NPs are cytotoxic to immune system and cause peroxidation od membrane lipids.	[37]
Human colon carcinoma (HCT116) Human myelogenous leukemic (K562)	30–48.5 nm Hexagonal structure	Green Synthesis method (Spondias pinnata) UV-Vis, FTIR, XRD, FESEM, HRTEM, EDX	24 and 48 h	0.25–200 µg/mL	82 and 60µg/mL 55 and 35µg/mL	Cell culture, MTT, IFA/FM, VWH, MMP, FC, CA, DNAf, RT-qPCR, WB, H	Upregulation of pro-apoptotic (PUMA, Bax, Cyt-c, cas-9, cas-3, and PARP) and downregulation of anti-apoptotic (Bcl-2 and survivin) genes. Elevation in the expression of Cyt-c, cas-9, and cas-3 genes involved in mitochondrial (intrinsic) apoptotic pathway.	[38]
Human keratinocytes (HaCaT) Human gingival fibroblast (HGF-1) Human gingival carcinomas (Ca9-22) Human gingival carcinomas (OECM-1)	100 nm ND	Synthesis method ND/OC. SEM	24 h	0–100 µg/mL	100 µg/mL 100 µg/mL 17.4 ± 0.6 µg/mL 51.0 ± 0.6 µg/mL	Cell culture, MTT, ROS, MMP, WB, ZXA, FC	Mitochondrial oxidative damage and p70S6K signaling pathway inhibition, caspase-3, caspase-8, caspase-9 and PARP activation. Induce sub-G1 arrest of the cell cycle followed by apoptosis in human GSCC. ROS is essential for the anti-cancer activity. Did not affect the expression of pro-survival Bcl-2 members (Bcl-2, Bcl-xl, and Mcl-1) as well as pro-apoptosis Bcl-2 members (Bax, Bad, and Bid).	[39]
Human oral caner (CAL 27)	50 nm Hexagonal prism	Synthesis method ND/OC. TEM, XRD and DLS	24 h	0–100 µg/mL	25 µg/mL	Cell culture, CCK-8, LC3, P62, GAPDH, PINK1, Parkin, IHC, ROS, MDC, MMP, WB, JC-1, ROS	Increased ROS levels, decreased MMP and mitochondrial dysfunction time-dependent manner. Increased levels of LC3-II, PINK1, mito-Parkin and decreased P62 and Parkin.	[40]

Table 1. *Cont.*

Cell Type	Size and Shape of ZnO-NP	Synthesis and Characterization of ZnO-NPs	Time of Exposition (h)	Doses mg/L	IC50	Biological Characterization	Mitochondrial Pathway	Reference
Human hepatocytes (HepG2)	237 nm 79 nm Rod shaped and spherical	Synthesis method ND/OC. TEM, ICP-AES, EDX and DLS	24 h	0–300 µg/mL	>150 µg/mL	Cell culture, qRT-PCR, AVO, MMP, TBEA, SOD, CAT	Increased mitochondrial zinc transporter 1 and 2 expression levels. Response and the upregulation of MET, ZnT1, ZnT2, no change in CAT, SOD1 and SOD2 expression, nor in the enzymatic activities of catalase and SOD, implying a minimal activation of oxidative stress. Abnormal mitochondria morphologies and autophagy vesicles in response to ZnO-NPs.	[30]
Human hepatocytes (HL-7702) Human colorectal (Caco-2)	<100 nm ND	Synthesis method ND/OC. ICP-MS, TEM, SEM	3 h	0.665 µg/mL	NS* NS*	Cell culture, MTT, LDH, MMP, ROS, qRT-PCR	Increased ROS levels due to NP exposure, overexpression of HMOX1 in response to the increased oxidative stress. The cytotoxicity induced by high ROS levels, oxidative stress and depolarization of mitochondrial membrane.	[41]
Huh7 hepatocytes cancer cell line	18 nm Polycrystalline nature	Green synthesis (*Luffa acutangula*) UV-Vis, HRTEM, SAED	8 h	0–60 µg/mL	40 µg/mL	Cell culture, MTT, ROS, MMP, DNAf, FM	Huh 7 liver cancer cells undergo apoptosis as a result of generation of ROS molecules. Stimulation of apoptotic signaling pathway in a dose-dependent manner. Condensed chromatin and fragmented DNA.	[42]
Human hepatocytes (HepG2) Human breast cancer (MCF-7)	~13 ± 2 nm ND	Chemical synthesis by reduction Zinc acetate dihydrate FESEM, TEM, XRD, FTIR, UV-Vis	24 and 48 h	0–100 µg/mL	25 and 10 µg/mL 25 µg/mL	FC, MTT, FACS, qRT-PCR, cell morphology, Casp-3, p53, Bax, Bcl-2	Significant upregulation of mRNA expression levels of Bax, p53, and caspase-3 and the downregulation of the anti-apoptotic gene Bcl-2.	[43]
Human breast cancer (MCF-7)	41 nm flakes Hexagonal	Chemical synthesis Triethanolamine XRD, FESEM, EDS	24 h	1–5 µg/mL	3 µg/mL	Cell culture, MTT, DPPH, AO/EB, PI	Dose-dependent loss of cell viability. ZnO-NP exposure increases necrosis and apoptosis of MCF cells.	[44]
Human breast cancer (MCF-7)	43 nm Spherical	Synthesis using sol-gel method. XRD, HR-TEM, EDS, TEM, XRD	24 h	0–100 µg/mL	44 µg/mL	Cell culture, MTT, LDH, ROS, GSH, TSH, MMP, IFA/FM, q-PCR	Upregulation of apoptotic genes (p53, Bax/Bcl2 ratio, caspase-3 and caspase-9). Loss of MMP and apoptosis in MCF-7 cells through the mitochondrial pathway.	[45]

Table 1. Cont.

Cell Type	Size and Shape of ZnO-NP	Synthesis and Characterization of ZnO-NPs	Time of Exposition (h)	Doses mg/L	IC50	Biological Characterization	Mitochondrial Pathway	Reference
Human breast cancer (MCF-7) Human Non-tumorigenic human mammary epithelial cell line (MCF-10A)	63.7 ±6.5 nm Spherical	Synthesis method ND/OC. DLS, AFM	24 and 48 h	0.5–20 µg/mL	20 and 10 µg/mL >90% IC50 with max doses	MTT, DAPI, qRT-PCR, IFA/FM and cell morphology	The increased expression of Bax/Bcl-2 ratio confirmed the induction of apoptosis in MCF-7 cells. ZnO-NPs induced casp3- and casp-8 upregulation.	[46]
Human multiple myeloma cell (RPMI8226) Peripheral blood mononuclear cell	30 nm	Synthesis method ND/OC. TEM	24 and 48 h	0–60 µg/mL	34 and 29 µg/mL 120 and 79 µg/mL	ROS, FC, q-PCR, ATP assay, cell morphology	Increase expression and proteins levels of caspase-3, caspase-9, Apaf-1, and Cyt-c, decrease ATP levels in time- and dose-dependent manner, ROS generation.	[47]
Human cervical cancer cell (SiHa)	20–50 nm	Green synthesis (*Gracilaria edulis*) UV-Vis, FTIR, XRD, FESEM, XPS, SEM, EDS, EDX and HRTEM	24 h	0–200 µg/mL	35 µg/mL	MTT, FC, ROS, MMP AO/EB, IFA/FM, CA, JC-1	Apoptotic and necrotic effect. ROS elevation, DNA damage, activation of mitochondrial intrinsic pathway.	[48]
Human cervical cancer cell line (HeLa)	10–70 nm	Green synthesis (*Aspergillus terreus*) UV-Vis, XRD, EDX, FTIR and TEM.	24 h	5–80 µg/mL	20 µg/mL	MTT, WB, SOD, CAT, ROS, MMP, p53, BAX, casp-3, casp-9, Cyt-c. Cell morphology.	Reduced SOD, CAT and GPO, increased ROS, diminished MMP. Upregulation of p53, casp-9, BAX, RAD51 and downregulation of BCL-2.	[49]
Human cervical cancer (HeLa) Human colon cancer (HT-29)	30 nm	Green synthesis (*Bergenia ciliate*) UV-Vis, FTIR, DLS, EDX, XDR and SEM	24 h	25–200 µg/mL	101.7 µg/mL 124.3 µg/mL	MTT, DPPH and ABTS assay	Increased ROS and cytotoxicity against cancer cells.	[50]
Human ovarian cancer (SKOV3)	20 nm	Synthesis method ND/OC. UV-Vis, TEM, XRD, FTIR and AFM	24 h	0–30 µg/mL	20–30 µg/mL	Cell morphology, MTT, ICC, JC-1 assay, TUNEL, WB, ROS levels and IFA/FM	Genotoxicity, double DNA strand breaks and apoptosis. Loss of MMP, ROS increased levels. Upregulation of p53, casp-9, BAX, RAD51 and downregulation of BCL-2.	[51]
AGS gastric cancer cells	100 nm	Green synthesis (*Morus nigra*) UV-Vis, TEM, SEM, FT-IR, XRD and XDR	24 h	0–25 µg/mL	10 µg/mL	MTT, MMP, AO/EB, ROS, FC, CAT, TBARS, GSH SOD, RT-PCR	MMP decreased. Apoptosis induced by ROS generation and gene expressions of apoptosis markers effects, increased lipid peroxidation, cell cycle arrest.	[52]
Human osteosarcoma cell line (MG-63)	10–12 nm	Green synthesis (*Radix Rehmanniae*) UV-Vis, TEM, XRD and FTIR	24 h	5–80 µg/mL	30 µg/mL	MTT, MMP, WB, ROS, BAX, casp-3 and casp-9 levels. Morphology assay AO/EB	Increased levels of BAX, casp-3, casp-9. MMP decreased and ROS generation.	[53]

Table 1. Cont.

Cell Type	Size and Shape of ZnO-NP	Synthesis and Characterization of ZnO-NPs	Time of Exposition (h)	Doses mg/L	IC50	Biological Characterization	Mitochondrial Pathway	Reference
Human lymphocytes Mouse fibroblast cell (C2C12) Mouse myoblast cell (L929)	16–24 nm	Green synthesis (*Beta vulgaris*) HRTEM, FTIR, XRD	24 and 48 h	1×10^{-6} 100 µg/mL	37 and 30 µg/mL 32 and 23 µg/mL 28 and 20 µg/mL	MTT, TBEA, cell kinetic and morphology assay	Abnormal cell morphology, adherence and viability manner. Mitochondrial function. Increased doses-dependent mortality rate.	[54]
Mouse macrophage line (J774A1)	140 ± 16 nm	Synthesis method ND/OC. TEM, SEM DLS	24 h	200 µg/mL	8 µg/mL	IFA/FM. MMP, GAPDH assay, mass spectrometry, HPLC, 2D and SDS electrophoresis.	Mitochondrial proteomics abnormalities, MMP proteins loss, nuclear DNA damage, genotoxic effect. GAPDH inhibition, increased levels od methylglyoxal-associated DNA damage.	[55]
Chicken embryo HH9 Neural stem cells Neuroblastoma SH-SY5Y cell	50 m	Synthesis method ND/OC. TEM, SEM, EDS and DLS	24 h 12 h 12 h	12.5–50 µg/mL 0–50 µg/mL 0–50 µg/mL	ND * 25 µg/mL 25 µg/mL	Cell/embryo IFA, FC, WB, RNA-seq, q-PCR, IFA/FM	Endoplasmic reticulum stress, increased Ca2+ levels, Casp-3, increased levels, abnormal mitochondrial morphology. ZnO-NPs induced failure of neural tube closure.	[56]
Neuroblastoma SH-SY5Y cell	24–30 nm	Green synthesis (*Clausena lansium*) UV-Vis, XRD, FT-IR and TEM	24 h	10–20 µg/mL	15 µg/mL	Cell viability, MTT, ROS levels, WB, CA (DNA damage)	Increased ROS levels, increased levels of BAX, caspase-3, and BCL-2 proteins. Autophagy (beclin-1, LC3-I, LCEII and ATG4B) increased levels. DNA loss and damage. NAC prevents ROS in SH-SY5Y.	[57]
Mouse Leydig cells Mouse Sertoli cells	70 nm	Synthesis method ND/OC. TEM, AFM, XRD, FTIR, UV-Vis, DLS	24 h 24 h	0, 5, 10, 15, and 20 µg/mL	15 µg/mL 15 µg/mL	MTT, LDH, ROS levels, TUNEL and JC-1 assay, WB, sperm morphology	Increased ROS levels, MMP abnormalities, increased apoptotic proteins, loss of MMP, nuclear DNA damage and breakage.	[58]
Mouse ovarian germ cells (CHO-K1)	80 nm	Co-precipitation method SEM, TEM	24 h (1 day) 168 h (7day)	10, 20 and 30 µg/mL	30 µg/mL 10 µg/mL	IFA/FM; FC. Quantification of ROS. qRT-PCR of Prdm1, Dppa3, Ifitm3, Ddx4 and Dazl	ROS level, cell membrane integrity. Significant increase in expression of premeiotic germ cells markers but a decrease in meiotic and post-meiotic markers.	[59]
Murine microglial BV-2 cells	20 nm	Synthesised by using the wet chemical method. TEM, SEM, XRD, XPS	6 h 24 h	0–80 µg/mL	20 µg/mL 10 µg/mL	MTT, TBEA, FC, MMP, ROS levels, PMP levels, cell cycle analysis, WB	Absence of caspase-3 cleavage and PARP fragmentation. Mitochondrial dysfunction and lysosomal alteration increased PMP. Accumulation of ROS. Non-apoptotic hallmarks, necrotic markers in BV-2 cells.	[60]

Table 1. Cont.

Cell Type	Size and Shape of ZnO-NP	Synthesis and Characterization of ZnO-NPs	Time of Exposition (h)	Doses mg/L	IC50	Biological Characterization	Mitochondrial Pathway	Reference
S. Cerevisiae (yeast BY4743)	20–80 nm	Synthesis method ND/OC. SEM, HRFESEM, UV-Vis and DLS analysis.	3 h	5–20 µg/mL	10 µg/mL	Cell death assay, WB, ROS levels, chitin and lipid droplets measurement. Vacuolar organelle morphology. RT-PCR	Mitochondrial morphology and lipid homeostasis abnormalities, ROS elevation. Perturbations in peripheral endoplasmic reticulum. ZnO-NPs do not affect histone epigenetic marks. Increased levels of Hog1. Inhibition of cell growth due alterations in CWI and HOG signaling pathway.	[27]
S. Cerevisiae (yeast BY4741)	20–80 nm	Synthesis method ND/OC. SEM, HRFESEM, UV-Vis	3 h	10 µg/mL	10 µg/mL	Cell culture, WB, HNMR, RT-PCR. RNA and metabolites identification/quantification	Mitochondrial expression of PSD1, SOD1, SOD2, KGD1. TCA cycle genes (ACO1 and KDG1) Oxidative pathway abnormalities (SOD1, SOD2 and YAP1) Lipid biosynthesis pathway abnormalities (INO2, INO4, CHO1, PSD1 and CRD1)	[28]

ND/OC: not determined/obtained commercially; NS *: not significant at the maximum doses of ZnO-NP exposure; TEM: transmission electron microscopy; XRD: x-ray diffraction; MTT: [3-(4,5-dimethylthiazoyl-2-yl)-2,5 e diphenylte-trazolium bromide]; LDH: lactate dehydrogenase; LA: antioxidant alpha-lipoic acid; ROS: reactive oxygen species; FM: fluorescence microscopy; MMP: mitochondrial membrane potential; CACAK: Caspase Apoptosis Colorimetric Assay Kit; Cyt-c: measurements of Cyt-c release; ICC: immunocytochemistry; IFA: immunofluorescence; SEM-EDS: scanning electron microscope-energy dispersive spectroscopy; DLS: Dynamic Light Scattering; CNCCs: primary culture of cranial neural crest cells; ARS: Alizarin red staining; IHC: immunohistochemistry; RNA-seq: RNA sequencing transcript profiling; q-PCR: quantitative polymerase chain reaction; DHE: dihydroethidium staining; HCA: High-Content Analysis; CCK-8: cell viability was analyzed by using the Cell Counting Kit-8; WB: Western Blot; MEA: Microelectrode Assay; RT-CES: real-time cell electronic sensing system; ELISA: enzyme-linked immunosorbent assay; FC: flow cytometry; FSP-1: fibroblast-specific protein-1; Apaf-1: protease-activating factor-1; FTIR: Fourier transform infrared; SOD: superoxide dismutase; CAT: catalase; TBARS: thiobarbituric acid reactive substances; LPO: lipid peroxidation; GST: glutathione-S-transferase; GSH: glutathione estimation; OH: hydroxyl radicals; O2: superoxide radical; UV-Vis: UV-visible spectroscopy; DAPI: 4′,6′-diamidino-2 phenylindole; Cas-3: caspase-3; Cas-9: caspase-9; H: hemolysis; MDA: determination of lipid peroxidation; NO: nitric oxide production; IL-6: interleukin 6; TNF-α: tumor necrosis factor α; HRTEM: high-resolution transmission electron microscopy; FESEM: field-emission scanning electron microscope; VWH: Vitro Wound Healing; CA: Comet Assay; DNAf: DNA Fragmentation; RT-qPCR: Reverse Transcription-Quantitative Polymerase Chain Reaction; FACScan: fluorescence associated cell sorter scan; ZXA: Zebrafish Xenograft Assay; GAPDH: glyceraldehyde 3-phosphate dehydrogenase; MDC: Autophagic Vacuole Indicator Monodansylcadaverine; ICP-AES: inductively coupled plasma atomic emission spectroscopy; AVO: acidic vesicular organelles; ICP-MS: Inductively Coupled Plasma Mass Spectroscopy; SAED: Selected Area Electron Diffraction; AO/EB: acridine orange/ethidium bromide; PI: propidium iodide; HNMR: hydrogen-1 nuclear magnetic resonance spectroscopy; HR-FESEM: high-resolution field-emission scanning electron microscope; CWI: cell wall integrity; TCA: tricarboxylic acid cycle; XPS: X-ray photoelectron spectroscopy; TBEA: trypan blue exclusion assay; PMP: plasma membrane permeability; HPLC: high-performance liquid chromatography; DPPH: 2,2-diphenyl-1-picrylhydrazyl; ABTS: 2,2′-Azino-Bis-3-Ethylbenzothiazoline-6-Sulfonic Acid.

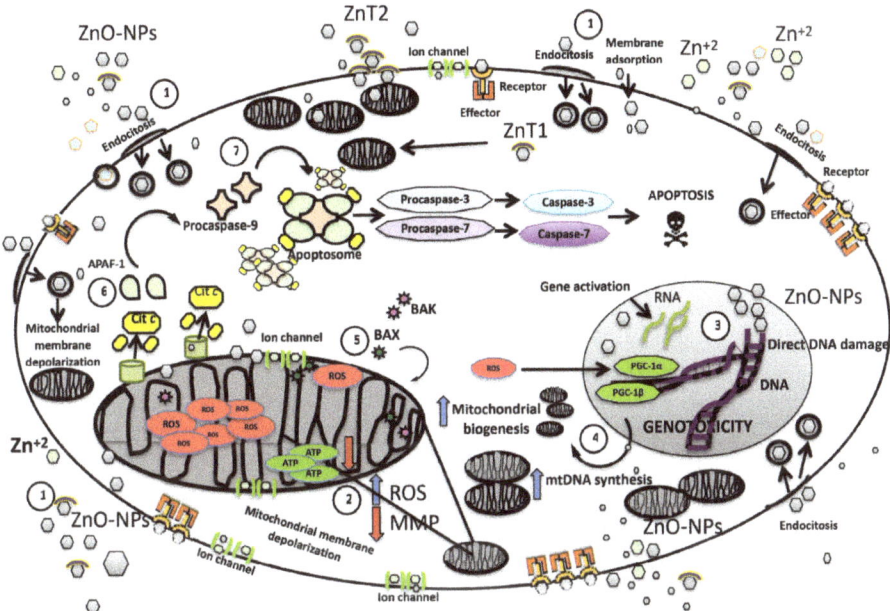

Figure 1. An integrated overview of the molecular pathways activated by ZnO-NP exposure inside the cell, while also summarizing the initial mitochondrial effects such as increased mitochondrial biogenesis, the potential depolarization of the mitochondrial membrane, increased ROS production, and the cumulative effects of ZnO-NPs leading to activation of the intrinsic (mitochondrial) pathway of programmed cell death. 1. ZnO-NPs penetrate the cell membrane via four mechanisms: the first mechanism occurs when ZnO-NPs interact with receptors embedded in the cell membrane, which, in association with the effector, function as second messengers, activating other signaling pathways. The second mechanism occurs when ion channels facilitate entry into the cytoplasm. The third mechanism is endocytosis—in this process, a fragment of the lipid cell membrane covers ZnO-NPs, allowing the formation of vacuoles, which later merge with the membranes of other organelles such as the nucleus and the mitochondria. The fourth mechanism is direct absorption, in which ZnO-NPs (depending on their size and shape) penetrate the cellular lipid bilayer. The entry of ZnO-NPs is also facilitated by two specialized zinc transporters, called ZnT1 and ZnT2. Zinc dissociation also occurs, and zinc can penetrate as zinc ions. The effects of ZnO-NPs depend on the dose, the exposure time, the size of the nanoparticles—as mentioned above, the smaller the size, the greater the cytotoxic effect—and also the shape (spherical, hexagonal, tubular, etc.) [9,10,36] 2. ZnO-NPs, such as Zn ions, readily enter all cellular compartments, including mitochondria, mainly through mitochondrial ion channels, conditioning the voltage change and depolarization of the inner mitochondrial membrane. Within the inner membrane, there is also the protein complex that constitutes the electron transport chain responsible for the synthesis of ATP. ZnO-NPs not only induce a decrease in MMP due to ion channel damage but also cause a decrease in ATP production and an increase in ROS production [13]. 3. The increase in ROS levels in the cell nucleus causes direct damage to the genetic material (double-strand breaks and the oxidation of nitrogenous bases), induces a greater expression of the PGC-1α and PGC-1β factors in the nucleus, and promotes mitochondrial biogenesis, to compensate for the increased energy demands [15,16,21] 4. mtDNA replication and transcription pathways are simultaneously activated. mtDNA number is significantly increased in cells exposed to ZnO-NP to counteract mitochondrial oxidative damage under mild oxidative stress [31]. 5. As cellular stress increases, the pathways to compensate for the damage trigger the release of pro-apoptotic proteins, such as BCL-associated protein X-2 (BAX) and/or BCL antagonist/killer-2 (BAK). Activated BAX or BAK

forms pores in the mitochondrial membrane, leading to the release of apoptogenic molecules [34,35,47] 6. leads to the release of pro-apoptotic proteins, including cytochrome-c and Apaf-1. 7. Finally, pro-caspase-9 is added, to form the apoptosome, which promotes the activation of pro-caspase-3 and pro-caspase-7 and conversion to caspase-3 and caspase-7, respectively; this protein complex will initiate cell dismantling and lead to apoptosis. ROS: reactive oxygen species; MMP: mitochondrial membrane potential; ZnO-NPs: zinc oxide nanoparticles; Cit-c: cytochrome-c; APAF-1: apoptosis protease-activating factor-1; ZnT1, ZnT2: Zinc transporter 1/2; mtDNA: mitochondrial DNA; BAX: BCL-2-associated X protein; BAK: BCL-2 antagonist/killer; PGC-1α: peroxisome proliferator-activated receptor gamma-coactivator 1α and 1β [61–63].

3. Apoptosis Induction ZnO-NPs

To maintain normal tissue function, damaged or aged cells are constantly eliminated by apoptosis. There are two routes to initiate apoptosis: the extrinsic pathway, conditioned by the binding of extracellular factors to the receptors present in the plasma membrane, and the intrinsic pathway, mediated by the action of mitochondrial factors [64]. The intrinsic apoptosis pathway is activated by physical (radiation), chemical (ZnO-NPs), or biological (viruses) stimuli. The BCL-2 family proteins orchestrate the regulation and execution of this pathway, which includes both pro-apoptosis and pro-survival proteins [65]. To initiate apoptosis, cellular stress or damage signals turn on the cascade of pro-apoptotic proteins, such as BCL-2-associated protein X (BAX) and/or BCL-2 antagonist/killer (BAK). Once activated, BAX or BAK induces mitochondrial pore formation. Provoking the release of apoptogenic molecules, which includes the second mitochondria-derived caspase activator (SMAC), serine proteases called caspases, and cytochrome-c derived from the mitochondrial intermembrane space. The mitochondrial release of cytochrome-c into the cytoplasm leads to the formation of apoptosomes. These events promote the activation of the initiator caspase (caspase-9) and executioner caspases (caspases-3, 6, and 7) for the orderly dismantling of the cell [61,62,66]. Considering the complexity of cancer treatment and that ZnO-NPs activate the intrinsic apoptosis pathway and caspases, the study of the cytotoxic response to ZnO-NP exposure in tumor cells is promising. A study performed on gingival squamous cell carcinomas (GSCC) in two human cell lines (Ca9-22 and OECM-1). Showed that ZnO-NPs induced growth inhibition of GSCCs, and did not cause damage to normal human keratinocytes (HaCaT cells) or gingival fibroblasts (HGF-1 cells). ZnO-NPs caused apoptotic death of GSCC cells in a concentration-dependent manner, increased intracellular ROS and specifically superoxide levels. Importantly, antioxidant and caspase inhibitors prevent ZnO-NP-induced cell death, indicating that humans' superoxide-induced mitochondrial dysfunction is related to caspase-mediated apoptosis. On the other hand, treatment of ZnO-NPs in a dose-dependent condition triggered the expression of p53, BAX, and cytochrome-c in ovarian and cervical tumor cells. The expression of caspase-9, caspase-3, cleaved caspase-9 and cleaved caspase-3 in HeLa cells was increased in cells treated with ZnO-NPs, compared with control cells. Multiple experiments in carcinogenic human hepatocytes HepG2 and HL-7702 cells, human hepatocytes cancer Huh7 cells, chronic myelogenous leukemia cells, Human colorectal Caco-2, colon carcinoma HCT116 cells, human cervical cancer SiHa cells and HeLa cells, cell models such as human ovarian cancer SKOV3, and neuroblastoma SH-SY5Y cells, breast cancer MDAMB-23 and MCF-7 cells, human head, and neck squamous cell carcinoma FaDu cells, showed a dose-dependent effect after treatment with ZnO-NPs [38,53,67–69], resulting in stimulation of the intrinsic pathway of apoptosis with very similar patterns.

ZnO-NPs induced similar changes in transcript levels of diverse proteins, such as p53, BAX, BCL-2, CYT-c, CAS-9, CAS-8 and CAS-3, with significant differences versus control cell models. They also documented the induction of apoptosis through oxidative stress-mediated Ca^{2+} release, ROS generation, and loss of mitochondrial membrane potential. Tumor cells have an increased susceptibility to ZnO-NP exposure. ZnO-NPs also induce apoptosis in germ ovarian germ cells CHO-K1, male germ Leydig and Sertoli cells (mouse model), causing time-dose-dependent damage, genotoxicity caused by increased ROS, and loss of mitochondrial membrane potential was reported. In addition, injection of ZnO-NPs into male mice caused structural alterations in the seminiferous epithelium and sperm abnormalities [58,59]. In 2020, Sruthi et al. [60], proposed a non-apoptotic mode of cell death following exposure of ZnO-NPs in murine microglial BV-2 cells. By triggering an accumulation of ROS favoring altered and increased plasma membrane permeability and also loss of mitochondrial membrane potential, leading to the formation of "ghost cells" without apoptotic features, there was no caspase-3 or PARP fragmentation in BV-2 cells. However, an adaptive response involving increased mitochondrial biogenesis cannot be excluded.

4. Caspases Expression and ZnO-NPs

The caspases are at the core of executing apoptosis by orchestrating cellular destruction with a proteolytic mechanism. Caspases are a family of at least a dozen cysteine proteases; they are expressed as inactive proteases [70–72] Caspase activation occurs by cleavage of the pro-domain and inter-subunit linker, the latter of which is most critical. The mature enzyme is formed by a dimer containing a small and a large subunit that generates an active site on each side. Apoptotic caspases are functionally subdivided into initiator (caspases-8, 9, and 10) and effector (caspases-3, 6, and 7), Apaf-1 attaches to cytochrome-c and procaspase-9 to form a proteic complex called the apoptosome that activates the caspase-3, which finally initiates apoptosis. Because caspase-3 is involved in the final molecular steps, it is used as a molecular marker to confirm apoptosis. It is well documented that mRNA levels of caspase-3, caspase-9, and Apaf-1 increase dose-dependent mode after ZnO-NP exposure. Thus, ZnO-NPs induce caspase signaling-related protein expression. In a study by Cheng et al. in bone cancer cells, there is a substantial increase in the protein levels of BAX, caspase-3 and caspase-9 when treated with ZnO-NPs, compared with control cells. Control cells displayed a decreased rate of apoptosis.

Cells administered with 50 µg/mL of ZnO-NPs exhibited a higher rate of apoptosis than cells supplemented with 30 µg/mL of Zn-ONPs [53,73–75]. Tumor cells show greater susceptibility to apoptosis after exposure to ZnO-NPs. Considering that the molecular changes in cancer cells include resistance to apoptosis, which can lead to immortality and a high rate of cellular division, ZnO-NPs undoubtedly constitute a very promising line of therapeutic research. Pro-apoptotic and anti-apoptotic molecules, such as p53 and the B-cell lymphoma 2 (BCL-2) family, closely regulate apoptosis. ZnO-NPs treatment causes a significant increase in mRNA expression of cell cycle checkpoint proteins p53, BAX, and caspase-3 and downregulates anti-apoptotic BCL-2 proteins in epidermoidal carcinoma cells. A similar effect in the overexpression of p53 was reported in hepatocytes and kidney tissue after oral administration in rats with ZnO-NPs. ROS induces the translocation, phosphorylation, and cleavage of pro-apoptosis BCL-2 members, leading to the induction of apoptosis. Caspase-9 and caspase-3 activation can be completely abolished by the presence of antioxidants, which confirms that is a ROS-dependent pathway in ZnO-NP-induced apoptosis [33,76]. On the other hand, there are reported differences in apoptotic gene expression between fresh and aged ZnO-NPs. Wang et al., in 2020, reported at least a dozen apoptosis genes that were significantly overexpressed in fresh ZnO-NP-exposed cells, compared to only half of the genes overexpressed in human peripheral blood T lymphocyte cells after addition of aged ZnO-NPs; interestingly, the expression of caspase-3 was not significantly changed [77]. ZnO-NPs are used in numerous dermatological products, cosmetics and sunscreens. Application of ZnO to the skin increases the level of zinc ions in the epidermis, then in the systemic circulation, and finally in the urine; although the amount that penetrates the skin and enters the circulation is low, it can be detected in urine and blood. Therefore, in vivo studies on blood cells, including erythrocytes, lymphocytes, monocytes, eosinophils, and platelets, are essential. In a study using 100–130 nm ZnO-NPs, lymphocytes were found to be the most sensitive to the action of the NPs. Human B lymphocyte and human monocyte cell lines were the most susceptible to the activity of 100 nm diameter nanoparticles. In contrast, human T lymphocytes and human promyelocytes were slightly more sensitive to 130 nm ZnO-NPs. Human promyelocytes are the most resistant to ZnO nanoparticles, while human monocytes are the most sensitive. When exposed to ZnO-NPs with a diameter of 15–24 nm, human lymphocytes display a dose-dependent decrease in mitochondrial activity. Khan et al., [35] in 2015, found ROS elevation and hemolytic action conditioned by ZnO-NPs in human erythrocytes. In contrast, in human eosinophils, ZnO-NPs delays apoptosis through caspase suppression and de novo protein synthesis. This evidence demonstrates that although caspases are ubiquitous proteins in all multicellular systems, cell type, energetic demands and their functions play an important role and influence susceptibility to apoptosis [35–37,78].

5. Effect of ZnO-NPs in Mitochondrial Membrane Potential

Mitochondrial membrane potential (MMP) maintains efficient ATP synthesis and an appropriate proton gradient across its lipid bilayer; also its integrity plays a vital role in inducing apoptosis. The effects of ZnO-NPs on MMP are widely studied. Pro-apoptosis BCL-2 family proteins also induce the permeabilization of the outer mitochondrial membrane, as well as modulate mitochondrial homeostasis for contributing to the loss of MMP. There is a strong correlation between MMP and ROS production and it is widely accepted that mitochondria produce more ROS at a high membrane potential. Depolarization of MMP produced either by a closure of the mitochondrial permeability transition pore or inhibition of ATP synthase is associated with increased ROS production. Additionally, the mitochondrial permeability transition pore has been demonstrated to induce depolarization of MMP, release of apoptogenic factors and loss of oxidative phosphorylation. Wang et al., in 2018, reported that ZnO-NPs increased the intracellular ROS level and decreased MMP in CAL 27 oral cancer cell lines. Similar results were reported in AGS gastric cancer cells, with significantly reduced MMP levels; this diminution of MMP after inner mitochondrial membrane permeabilization stimulates the release of several apoptotic factors [40,52,78].

6. Conclusions

This systematic review aimed to include the most recent scientific evidence concerning the cytotoxic effect of ZnO-NPs in biological models and in particular on mitochondrial damage in multiple biological models. Since mitochondria are among the most complex and relevant organelles for cellular homeostasis, it is indispensable to define the most relevant mechanisms leading to cell dysfunction/death. Although the applications of ZnO-NPs have revolutionized modern life with unprecedented advances in the medical field, their harmful and cytotoxic effects cannot be neglected. The unique properties of ZnO-NPs are dependent on morphology, size, concentration, and exposition period.

The contribution and usefulness of ZnO-NPs in medical oncology is of great interest; in particular, their contribution in the therapeutic area is increasingly relevant due to the immense potential in the public health sector because cancer remains among the leading causes of death worldwide. Tumor cells show a different cytotoxic effect compared to healthy cells of the same lineage, and the response also varies depending on the exposure time, size and shape of ZnO-NPs. Due to the heterogeneity in research design, more in vitro investigations are required, to determine the exact mechanisms of cytotoxicity, and further clarify the anti-cancer effect of ZnO-NPs. Therefore, it is imperative to perform further in vitro trials with different tumor lines from the same tissue—for example, the human cervical cancer cell SiHa and the cervical cancer cell line HeLa—at different concentrations to properly examine the cytotoxic properties of ZnO-NPs. More emphasis should be given to the correlation between the size (5–100 nm) and structural shapes, including spherical, oval, elongated, irregular, and hexagonal, of ZnO-NPs and their mitochondrial toxicity. Because ZnO-NPs can act as anti-cancer agents against different tumor lines resistant to conventional chemotherapeutic treatments, they provide a talented substitute approach to chemotherapies.

On the other hand, other important characteristic of ZnO-NPs, with the same impact on public health, are the possible teratogenic effects in embryonic cell models, affecting migration, cell–cell interaction and cell differentiation as well as intervening in mitochondrial pathways, as already mentioned.

Since we are certainly in daily contact with ZnO-NPs, more attention needs to be addressed to them, since bioaccumulation of these elements can occur in plants, food and aquatic species, with a direct impact on human and environmental health. Table 2 aims to summarize the most relevant factors involved in the mitochondrial and cellular response after exposure to ZnO-NPs found in the most relevant literature. We believe that this systematic review provides information on correlation and impact on future research.

Table 2. Effects triggered after exposure to ZnO-NPs.

Mitochondrial Pathway	ZnO-NPs Effect
Mitochondrial biogenesis	↑ mDNA replication and transcription systems. ↑ PSD1 to mitochondrial maintenance. ↑ Mitochondrial biogenesis by PGC-1. ↑ Mitophagy.
Apoptosis induction	↑ p53, BAX, BAK, SMAC and cytochrome-c levels to induce apoptosis. ↑ DNA fragmentation and cytoplasmic reduction. ↑ Anti-BCL2.
Caspases induction	↑ Pro-caspases-9 and pro-caspase-3 to convert in activated caspase-9. ↑ Apaf-1. ↑ Apoptosome.
Mitochondrial membrane potential	↑ Mitochondrial membrane potential. ↑ ATP synthase. ↑ Reactive oxygen species. ↑ Mitochondrial pore abnormalities with delivery of pro-apoptotic proteins.

↑: Increase; mDNA: mitochondrial DNA; PSD1: phosphatidylserine decarboxylase 1; PGC-1: peroxisome proliferator-activated receptor γ coactivator 1α; p53: p53 protein; BAX; Bax protein; BAK: Bak protein; SMAC: second mitochondria-derived activator of caspase; BCL2: B-cell lymphoma-2 proteins; Apaf-1: Apoptotic Peptidase-Activating Factor 1; ATP synthase: adenosine triphosphate synthase.

Author Contributions: L.P.-R., V.L.-C., I.L.-A., M.Á.L.-Á., J.A.A.-P. and I.P.-L. gathered the research data. L.P.-R., P.A.L.-M. and H.A.-R. analyzed these data and wrote this review paper. All authors have read and agreed to the published version of the manuscript.

Funding: This research received no external funding.

Institutional Review Board Statement: Not applicable.

Informed Consent Statement: Not applicable.

Data Availability Statement: All data is in this review.

Conflicts of Interest: The authors declare no conflict of interest.

References

1. Mathew, E.N.; Hurst, M.N.; Wang, B.; Murthy, V.; Zhang, Y.; De Long, R.K. Interaction of Ras Binding Domain (RBD) by chemotherapeutic zinc oxide nanoparticles: Progress towards RAS pathway protein interference. *PLoS ONE* **2020**, *15*. [CrossRef] [PubMed]
2. Bharathi, D.; Bhuvaneshwari, V. Synthesis of zinc oxide nanoparticles (ZnO NPs) using pure bioflavonoid rutin and their biomedical applications: Antibacterial, antioxidant and cytotoxic activities. *Res. Chem. Intermed.* **2019**, *45*, 2065–2078. [CrossRef]
3. Danielsen, P.H.; Cao, Y.; Roursgaard, M.; Møller, P.; Loft, S. Endothelial cell activation, oxidative stress and inflammation induced by a panel of metal-based nanomaterials. *Nanotoxicology* **2015**, *9*, 813–824. [CrossRef] [PubMed]
4. Liu, N.; Tang, M. Toxic effects and involved molecular pathways of nanoparticles on cells and subcellular organelles. *J. Appl. Toxicol.* **2020**, *40*, 16–36. [CrossRef]

5. Hu, X.; Cook, S.; Wang, P.; Hwang, H.-M. In vitro evaluation of cytotoxicity of engineered metal oxide nanoparticles. *Sci. Total Environ.* **2009**, *407*, 3070–3072. [CrossRef] [PubMed]
6. Sharma, V.; Shukla, R.K.; Saxena, N.; Parmar, D.; Das, M.; Dhawan, A. DNA damaging potential of zinc oxide nanoparticles in human epidermal cells. *Toxicol. Lett.* **2009**, *185*, 211–218. [CrossRef] [PubMed]
7. Donaldson, K.; Stone, V.; Clouter, A.; Renwick, L.; Macnee, W. Ultrafine particles. *Occup. Environ. Med.* **2001**, *58*, 211–216. [CrossRef] [PubMed]
8. Yan, Y.; Wang, G.; Huang, J.; Zhang, Y.; Cheng, X.; Chuai, M.; Brand-Saberi, B.; Chen, G.; Jiang, X.; Yang, X. Zinc oxide nanoparticles exposure-induced oxidative stress restricts cranial neural crest development during chicken embryogenesis. *Ecotoxicol. Environ. Saf.* **2020**, *194*, 110415. [CrossRef] [PubMed]
9. Wang, M.; Wang, J.; Liu, Y.; Wang, J.; Nie, Y.; Si, B.; Liu, Y.; Wang, X.; Chen, S.; Hei, T.K.; et al. Subcellular Targets of Zinc Oxide Nanoparticles during the Aging Process: Role of Cross-Talk between Mitochondrial Dysfunction and Endoplasmic Reticulum Stress in the Genotoxic Response. *Toxicol. Sci.* **2019**, *171*, 159–171. [CrossRef]
10. Limo, M.J.; Sola-Rabada, A.; Boix, E.; Thota, V.; Westcott, Z.C.; Puddu, V.; Perry, C.C. Interactions between Metal Oxides and Biomolecules: From Fundamental Understanding to Applications. *Chem. Rev.* **2018**, *118*, 11118–11193. [CrossRef] [PubMed]
11. Popov, L.D. Mitochondrial biogenesis: An update. *J. Cell. Mol. Med.* **2020**, *24*, 4892–4899. [CrossRef] [PubMed]
12. Zhao, L.; Sumberaz, P. Mitochondrial DNA Damage: Prevalence, Biological Consequence, and Emerging Pathways. *Chem. Res. Toxicol.* **2020**, *33*, 2491–2502. [CrossRef] [PubMed]
13. Yu, S.B.; Pekkurnaz, G. Mechanisms Orchestrating Mitochondrial Dynamics for Energy Homeostasis. *J. Mol. Biol.* **2018**, *430*, 3922–3941. [CrossRef] [PubMed]
14. Wang, F.; Zhang, D.; Zhang, D.; Li, P.; Gao, Y. Mitochondrial Protein Translation: Emerging Roles and Clinical Significance in Disease. *Front. Cell Dev. Biol.* **2021**, *9*, 675465. [CrossRef] [PubMed]
15. Yokokawa, T.; Kido, K.; Suga, T.; Isaka, T.; Hayashi, T.; Fujita, S. Exercise-induced mitochondrial biogenesis coincides with the expression of mitochondrial translation factors in murine skeletal muscle. *Physiol. Rep.* **2018**, *6*, e13893. [CrossRef]
16. Holt, A.G.; Davies, A.M. The significance of mitochondrial DNA half-life to the lifespan of post-mitotic cells. *bioRxiv* **2020**. [CrossRef]
17. Byrne, J.J.; Soh, M.S.; Chandhok, G.; Vijayaraghavan, T.; Teoh, J.S.; Crawford, S.; Cobham, A.E.; Yapa, N.M.B.; Mirth, C.K.; Neumann, B. Disruption of mitochondrial dynamics affects behaviour and lifespan in Caenorhabditis elegans. *Cell. Mol. Life Sci.* **2019**, *76*, 1967–1985. [CrossRef]
18. Yu, R.; Lendahl, U.; Nistér, M.; Zhao, J. Regulation of Mammalian Mitochondrial Dynamics: Opportunities and Challenges. *Front. Endocrinol.* **2020**, *11*, 374. [CrossRef]
19. Zhang, C.S.; Lin, S.C. AMPK promotes autophagy by facilitating mitochondrial fission. *Cell Metab.* **2016**, *23*, 399–401. [CrossRef]
20. Wu, Z.; Puigserver, P.; Andersson, U.; Zhang, C.; Adelmant, G.; Mootha, V.; Troy, A.; Cinti, S.; Lowell, B.; Scarpulla, R.C.; et al. Mechanisms controlling mitochondrial biogenesis and respiration through the thermogenic coactivator PGC-1. *Cell* **1999**, *98*, 115–124. [CrossRef]
21. Gureev, A.P.; Shaforostova, E.A.; Popov, V.N. Regulation of mitochondrial biogenesis as a way for active longevity: Interaction between the Nrf2 and PGC-1α signaling pathways. *Front. Genet.* **2019**, *10*, 435. [CrossRef] [PubMed]
22. Li, Y.; Li, F.; Zhang, L.; Zhang, C.; Peng, H.; Lan, F.; Peng, S.; Liu, C.; Guo, J. Zinc oxide nanoparticles induce mitochondrial biogenesis impairment and cardiac dysfunction in human ipsc-derived cardiomyocytes. *Int. J. Nanomed.* **2020**, *15*, 2669–2683. [CrossRef] [PubMed]
23. Yousef, M.I.; Mutar, T.F.; Kamel, M.A.E.N. Hepato-renal toxicity of oral sub-chronic exposure to aluminum oxide and/or zinc oxide nanoparticles in rats. *Toxicol. Rep.* **2019**, *6*, 336–346. [CrossRef] [PubMed]
24. Liu, Y.J.; McIntyre, R.L.; Janssens, G.E.; Houtkooper, R.H. Mitochondrial fission and fusion: A dynamic role in aging and potential target for age-related disease. *Mech. Ageing Dev.* **2020**, *186*, 111212. [CrossRef]
25. Li, J.; Zhang, B.; Chang, X.; Gan, J.; Li, W.; Niu, S.; Kong, L.; Wu, T.; Zhang, T.; Tang, M.; et al. Silver nanoparticles modulate mitochondrial dynamics and biogenesis in HepG2 cells. *Environ. Pollut.* **2020**, *256*, 113430. [CrossRef] [PubMed]
26. Aung, L.H.H.; Jumbo, J.C.C.; Wang, Y.; Li, P. Therapeutic potential and recent advances on targeting mitochondrial dynamics in cardiac hypertrophy: A concise review. *Mol. Ther. Nucleic Acids* **2021**, *25*, 416–443. [CrossRef]
27. Babele, P.K.; Thakre, P.K.; Kumawat, R.; Tomar, R.S. Zinc oxide nanoparticles induce toxicity by affecting cell wall integrity pathway, mitochondrial function and lipid homeostasis in *Saccharomyces cerevisiae*. *Chemosphere* **2018**, *213*, 65–75. [CrossRef]
28. Kumar Babele, P. Zinc oxide nanoparticles impose metabolic toxicity by de-regulating proteome and metabolome in *Saccharomyces cerevisiae*. *Toxicol. Rep.* **2019**, *6*, 64–73. [CrossRef] [PubMed]
29. Bird, A.J.; Wilson, S. Zinc homeostasis in the secretory pathway in yeast. *Curr. Opin. Chem. Biol.* **2020**, *55*, 145–150. [CrossRef] [PubMed]
30. Chevallet, M.; Gallet, B.; Fuchs, A.; Jouneau, P.H.; Um, K.; Mintz, E.; Michaud-Soret, I. Metal homeostasis disruption and mitochondrial dysfunction in hepatocytes exposed to sub-toxic doses of zinc oxide nanoparticles. *Nanoscale* **2016**, *8*, 18495–18506. [CrossRef]
31. Yu, K.N.; Yoon, T.J.; Minai-Tehrani, A.; Kim, J.E.; Park, S.J.; Jeong, M.S.; Ha, S.W.; Lee, J.K.; Kim, J.S.; Cho, M.H. Zinc oxide nanoparticle induced autophagic cell death and mitochondrial damage via reactive oxygen species generation. *Toxicol. Vitr.* **2013**, *27*, 1187–1195. [CrossRef] [PubMed]

32. Khan, M.F.; Siddiqui, S.; Zia, Q.; Ahmad, E.; Jafri, A.; Arshad, M.; Jamal, A.; Alam, M.M.; Banawas, S.; Alshehri, B.A.; et al. Characterization and in vitro cytotoxic assessment of zinc oxide nano-particles in human epidermoid carcinoma cells. *J. Environ. Chem. Eng.* **2021**, *9*, 105636. [CrossRef]
33. Liang, S.; Sun, K.; Wang, Y.; Dong, S.; Wang, C.; Liu, L.X.; Wu, Y.H. Role of Cyt-C/caspases-9,3, Bax/Bcl-2 and the FAS death receptor pathway in apoptosis induced by zinc oxide nanoparticles in human aortic endothelial cells and the protective effect by alpha-lipoic acid. *Chem. Biol. Interact.* **2016**, *258*, 40–51. [CrossRef] [PubMed]
34. Wang, L.; Guo, D.; Wang, Z.; Yin, X.; Wei, H.; Hu, W.; Chen, R.; Chen, C. Zinc oxide nanoparticles induce human tenon fibroblast apoptosis through reactive oxygen species and caspase signaling pathway. *Arch. Biochem. Biophys.* **2020**, *683*, 108324. [CrossRef] [PubMed]
35. Khan, M.; Naqvi, A.H.; Ahmad, M. Comparative study of the cytotoxic and genotoxic potentials of zinc oxide and titanium dioxide nanoparticles. *Toxicol. Rep.* **2015**, *2*, 765–774. [CrossRef]
36. Czyżowska, A.; Dyba, B.; Rudolphi-Szydło, E.; Barbasz, A. Structural and biochemical modifications of model and native membranes of human immune cells in response to the action of zinc oxide nanoparticles. *J. Appl. Toxicol.* **2021**, *41*, 458–469. [CrossRef] [PubMed]
37. Czyżowska, A.; Barbasz, A. Cytotoxicity of zinc oxide nanoparticles to innate and adaptive human immune cells. *J. Appl. Toxicol.* **2021**, *41*, 1425–1437. [CrossRef] [PubMed]
38. Ahlam, A.A.; Shaniba, V.S.; Jayasree, P.R.; Manish Kumar, P.R. *Spondias pinnata* (L.f.) Kurz Leaf Extract Derived Zinc Oxide Nanoparticles Induce Dual Modes of Apoptotic-Necrotic Death in HCT 116 and K562 Cells. *Biol. Trace Elem. Res.* **2021**, *199*, 1778–1801. [CrossRef]
39. Wang, S.W.; Lee, C.H.; Lin, M.S.; Chi, C.W.; Chen, Y.J.; Wang, G.S.; Liao, K.W.; Chiu, L.P.; Wu, S.H.; Huang, D.M.; et al. ZnO nanoparticles induced caspase-dependent apoptosis in gingival squamous cell carcinoma through mitochondrial dysfunction and p70s6K signaling pathway. *Int. J. Mol. Sci.* **2020**, *21*, 1612. [CrossRef]
40. Wang, J.; Gao, S.; Wang, S.; Xu, Z.; Wei, L. Zinc oxide nanoparticles induce toxicity in CAL 27 oral cancer cell lines by activating PINK1/Parkin-mediated mitophagy. *Int. J. Nanomed.* **2018**, *13*, 3441–3450. [CrossRef]
41. Li, J.; Song, Y.; Vogt, R.D.; Liu, Y.; Luo, J.; Li, T. Bioavailability and cytotoxicity of Cerium- (IV), Copper- (II), and Zinc oxide nanoparticles to human intestinal and liver cells through food. *Sci. Total Environ.* **2020**, *702*, 134700. [CrossRef] [PubMed]
42. Ananthalakshmi, R.; Rathinam, S.R.X.R.; Sadiq, A.M. Apoptotic Signalling of Huh7 Cancer Cells by Biofabricated Zinc Oxide Nanoparticles. *J. Inorg. Organomet. Polym. Mater.* **2021**, *31*, 1764–1773. [CrossRef]
43. Wahab, R.; Siddiqui, M.A.; Saquib, Q.; Dwivedi, S.; Ahmad, J.; Musarrat, J.; Al-Khedhairy, A.A.; Shin, H.S. ZnO nanoparticles induced oxidative stress and apoptosis in HepG2 and MCF-7 cancer cells and their antibacterial activity. *Colloids Surfaces B Biointerfaces* **2014**, *117*, 267–276. [CrossRef] [PubMed]
44. Arasu, M.V.; Madankumar, A.; Theerthagiri, J.; Salla, S.; Prabu, S.; Kim, H.S.; Al-Dhabi, N.A.; Arokiyaraj, S.; Duraipandiyan, V. Synthesis and characterization of ZnO nanoflakes anchored carbon nanoplates for antioxidant and anticancer activity in MCF7 cell lines. *Mater. Sci. Eng. C* **2019**, *102*, 536–540. [CrossRef] [PubMed]
45. Akhtar, M.J.; Alhadlaq, H.A.; Alshamsan, A.; Majeed Khan, M.A.; Ahamed, M. Aluminum doping tunes band gap energy level as well as oxidative stress-mediated cytotoxicity of ZnO nanoparticles in MCF-7 cells. *Sci. Rep.* **2015**, *5*, 13876. [CrossRef] [PubMed]
46. Farasat, M.; Niazvand, F.; Khorsandi, L. Zinc oxide nanoparticles induce necroptosis and inhibit autophagy in MCF-7 human breast cancer cells. *Biologia* **2020**, *75*, 161–174. [CrossRef]
47. Li, Z.; Guo, D.; Yin, X.; Ding, S.; Shen, M.; Zhang, R.; Wang, Y.; Xu, R. Zinc oxide nanoparticles induce human multiple myeloma cell death via reactive oxygen species and Cyt-C/Apaf-1/Caspase-9/Caspase-3 signaling pathway in vitro. *Biomed. Pharmacother.* **2020**, *122*, 109712. [CrossRef] [PubMed]
48. Gowdhami, B.; Jaabir, M.; Archunan, G.; Suganthy, N. Anticancer potential of zinc oxide nanoparticles against cervical carcinoma cells synthesized via biogenic route using aqueous extract of *Gracilaria edulis*. *Mater. Sci. Eng. C* **2019**, *103*, 109840. [CrossRef]
49. Chen, H.; Luo, L.; Fan, S.; Xiong, Y.; Ling, Y.; Peng, S. Zinc oxide nanoparticles synthesized from *Aspergillus terreus* induces oxidative stress-mediated apoptosis through modulating apoptotic proteins in human cervical cancer HeLa cells. *J. Pharm. Pharmacol.* **2021**, *73*, 221–232. [CrossRef]
50. Dulta, K.; Koşarsoy Ağçeli, G.; Chauhan, P.; Jasrotia, R.; Chauhan, P.K. A Novel Approach of Synthesis Zinc Oxide Nanoparticles by Bergenia ciliata Rhizome Extract: Antibacterial and Anticancer Potential. *J. Inorg. Organomet. Polym. Mater.* **2021**, *31*, 180–190. [CrossRef]
51. Bai, D.P.; Zhang, X.F.; Zhang, G.L.; Huang, Y.F.; Gurunathan, S. Zinc oxide nanoparticles induce apoptosis and autophagy in human ovarian cancer cells. *Int. J. Nanomed.* **2017**, *12*, 6521–6535. [CrossRef]
52. Tang, Q.; Xia, H.; Liang, W.; Huo, X.; Wei, X. Synthesis and characterization of zinc oxide nanoparticles from Morus nigra and its anticancer activity of AGS gastric cancer cells. *J. Photochem. Photobiol. B Biol.* **2020**, *202*, 111698. [CrossRef] [PubMed]
53. Cheng, J.; Wang, X.; Qiu, L.; Li, Y.; Marraiki, N.; Elgorban, A.M.; Xue, L. Green synthesized zinc oxide nanoparticles regulates the apoptotic expression in bone cancer cells MG-63 cells. *J. Photochem. Photobiol. B Biol.* **2020**, *202*, 111644. [CrossRef] [PubMed]
54. Patrón-Romero, L.; Luque, P.A.; Soto-Robles, C.A.; Nava, O.; Vilchis-Nestor, A.R.; Barajas-Carrillo, V.W.; Martínez-Ramírez, C.E.; Chávez Méndez, J.R.; Alvelais Palacios, J.A.; Leal Ávila, M.; et al. Synthesis, characterization and cytotoxicity of zinc oxide nanoparticles by green synthesis method. *J. Drug Deliv. Sci. Technol.* **2020**, *60*, 101925. [CrossRef]

55. Aude-Garcia, C.; Dalzon, B.; Ravanat, J.L.; Collin-Faure, V.; Diemer, H.; Strub, J.M.; Cianferani, S.; Van Dorsselaer, A.; Carrière, M.; Rabilloud, T. A combined proteomic and targeted analysis unravels new toxic mechanisms for zinc oxide nanoparticles in macrophages. *J. Proteomics* **2016**, *134*, 174–185. [CrossRef]
56. Yan, Y.; Wang, G.; Luo, X.; Zhang, P.; Peng, S.; Cheng, X.; Wang, M.; Yang, X. Endoplasmic reticulum stress-related calcium imbalance plays an important role on Zinc oxide nanoparticles-induced failure of neural tube closure during embryogenesis. *Environ. Int.* **2021**, *152*, 106495. [CrossRef] [PubMed]
57. Li, F.; Song, L.; Yang, X.; Huang, Z.; Mou, X.; Syed, A.; Bahkali, A.H.; Zheng, L. Anticancer and genotoxicity effect of (*Clausena lansium* (Lour.) Skeels) Peel ZnONPs on neuroblastoma (SH-SY5Y) cells through the modulation of autophagy mechanism. *J. Photochem. Photobiol. B Biol.* **2020**, *203*, 111748. [CrossRef]
58. Han, Z.; Yan, Q.; Ge, W.; Liu, Z.G.; Gurunathan, S.; De Felici, M.; Shen, W.; Zhang, X.F. Cytotoxic effects of ZnO nanoparticles on mouse testicular cells. *Int. J. Nanomed.* **2016**, *11*, 5187–5203. [CrossRef]
59. Saber, M.; Hayaei-Tehrani, R.S.; Mokhtari, S.; Hoorzad, P.; Esfandiari, F. In vitro cytotoxicity of zinc oxide nanoparticles in mouse ovarian germ cells. *Toxicol. Vitr.* **2021**, *70*, 105032. [CrossRef]
60. Sruthi, S.; Nury, T.; Millot, N.; Lizard, G. Evidence of a non-apoptotic mode of cell death in microglial BV-2 cells exposed to different concentrations of zinc oxide nanoparticles. *Environ. Sci. Pollut. Res.* **2021**, *28*, 12500–12520. [CrossRef]
61. Singh, R.; Letai, A.; Sarosiek, K. Regulation of apoptosis in health and disease: The balancing act of BCL-2 family proteins. *Nat. Rev. Mol. Cell Biol.* **2019**, *20*, 175–193. [CrossRef] [PubMed]
62. Ladokhin, A.S. Regulation of Apoptosis by the Bcl-2 Family of Proteins: Field on a Brink. *Cells* **2020**, *9*, 2121. [CrossRef] [PubMed]
63. Morrish, E.; Brumatti, G.; Silke, J. Future Therapeutic Directions for Smac-Mimetics. *Cells* **2020**, *9*, 406. [CrossRef] [PubMed]
64. Carneiro, B.A.; El-Deiry, W.S. Targeting apoptosis in cancer therapy. *Nat. Rev. Clin. Oncol.* **2020**, *17*, 395–417. [CrossRef]
65. Wanner, E.; Thoppil, H.; Riabowol, K. Senescence and Apoptosis: Architects of Mammalian Development. *Front. Cell Dev. Biol.* **2021**, *8*, 1–16. [CrossRef] [PubMed]
66. Rajashekara, S.; Shrivastava, A.; Sumhitha, S.; Kumari, S. Biomedical Applications of Biogenic Zinc Oxide Nanoparticles Manufactured from Leaf Extracts of *Calotropis gigantea* (L.) Dryand. *Bionanoscience* **2020**, *10*, 654–671. [CrossRef]
67. Shobha, N.; Nanda, N.; Giresha, A.S.; Manjappa, P.; Sophiya, P.; Dharmappa, K.K.; Nagabhushana, B.M. Synthesis and characterization of Zinc oxide nanoparticles utilizing seed source of Ricinus communis and study of its antioxidant, antifungal and anticancer activity. *Mater. Sci. Eng. C* **2019**, *97*, 842–850. [CrossRef]
68. Moratin, H.; Scherzad, A.; Gehrke, T.; Ickrath, P.; Radeloff, K.; Kleinsasser, N.; Hackenberg, S. Toxicological characterization of ZnO nanoparticles in malignant and non-malignant cells. *Environ. Mol. Mutagen.* **2018**, *59*, 247–259. [CrossRef] [PubMed]
69. Julien, O.; Wells, J.A. Caspases and their substrates. *Cell Death Differ.* **2017**, *24*, 1380–1389. [CrossRef] [PubMed]
70. Boice, A.; Bouchier-Hayes, L. Targeting apoptotic caspases in cancer. *Biochim. Biophys. Acta Mol. Cell Res.* **2020**, *1867*, 118688. [CrossRef] [PubMed]
71. Nakajima, Y.I.; Kuranaga, E. Caspase-dependent non-apoptotic processes in development. *Cell Death Differ.* **2017**, *24*, 1422–1430. [CrossRef] [PubMed]
72. McComb, S.; Chan, P.K.; Guinot, A.; Hartmannsdottir, H.; Jenni, S.; Dobay, M.P.; Bourquin, J.P.; Bornhauser, B.C. Efficient apoptosis requires feedback amplification of upstream apoptotic signals by effector caspase-3 or -7. *Sci. Adv.* **2019**, *5*, 1–12. [CrossRef]
73. Kopeina, G.S.; Prokhorova, E.A.; Lavrik, I.N.; Zhivotovsky, B. Alterations in the nucleocytoplasmic transport in apoptosis: Caspases lead the way. *Cell Prolif.* **2018**, *51*, e12467. [CrossRef] [PubMed]
74. Shalini, S.; Dorstyn, L.; Dawar, S.; Kumar, S. Old, new and emerging functions of caspases. *Cell Death Differ.* **2015**, *22*, 526–539. [CrossRef] [PubMed]
75. Giordo, R.; Nasrallah, G.K.; Al-Jamal, O.; Paliogiannis, P.; Pintus, G. Resveratrol inhibits oxidative stress and prevents mitochondrial damage induced by zinc oxide nanoparticles in zebrafish (*Danio rerio*). *Int. J. Mol. Sci.* **2020**, *21*, 3838. [CrossRef] [PubMed]
76. Wang, J.; Wang, L. Aged Zinc Oxide Nanoparticles Did Not Induce Cytotoxicity Through Apoptosis Signaling Pathway as Fresh NPs. *Res. Sq.* **2020**. [CrossRef]
77. Van Opdenbosch, N.; Lamkanfi, M. Caspases in Cell Death, Inflammation, and Disease. *Immunity* **2019**, *50*, 1352–1364. [CrossRef]
78. Suski, J.; Lebiedzinska, M.; Bonora, M.; Pinton, P.; Duszynski, J.; Wieckowski, M.R. Relation between mitochondrial membrane potential and ROS formation. In *Methods in Molecular Biology*; Humana Press Inc.: Totowa, NJ, USA, 2018; Volume 1782, pp. 357–381.

Review

Review of the Nanostructuring and Doping Strategies for High-Performance ZnO Thermoelectric Materials

Suraya Sulaiman [1,2,*], Izman Sudin [1], Uday M. Basheer Al-Naib [1,3] and Muhammad Firdaus Omar [4]

[1] School of Mechanical Engineering, Faculty of Engineering, Universiti Teknologi Malaysia, Skudai 81310, Johor, Malaysia; izman@utm.my (I.S.); uday@utm.my (U.M.B.A.-N.)
[2] Faculty of Manufacturing and Mechatronic Engineering Technology, Universiti Malaysia Pahang, Pekan 26600, Pahang, Malaysia
[3] Centre for Advanced Composite Materials (CACM), Institute for Vehicle Systems and Engineering, Universiti Teknologi Malaysia, Skudai 81310, Johor, Malaysia
[4] Physics Department, Faculty of Science, Universiti Teknologi Malaysia, Skudai 81310, Johor, Malaysia; firdausomar@utm.my
* Correspondence: surayas@ump.edu.my

Abstract: Unique properties of thermoelectric materials enable the conversion of waste heat to electrical energies. Among the reported materials, Zinc oxide (ZnO) gained attention due to its superior thermoelectric performance. In this review, we attempt to oversee the approaches to improve the thermoelectric properties of ZnO, where nanostructuring and doping methods will be assessed. The outcomes of the reviewed studies are analysed and benchmarked to obtain a preliminary understanding of the parameters involved in improving the thermoelectric properties of ZnO.

Keywords: thermoelectric material; nanostructure approach; doping approach; nanoparticle; zinc oxides

1. Introduction

Pioneering studies on thermoelectric materials began in 1822. Thermoelectric was first proposed in the 1820s to convert waste energies from automobiles and manufacturing to useful electrical energies [1–4], which can potentially bring lucrative economic profits [1,5,6]. Thermoelectric materials can be applied in woodstove and diesel power plants for optimal power efficiency [5]. The ideal heat-to-electricity conversion rate in a typical thermoelectric device ranges from 15% to 20% [7]. These rates can significantly reduce the usage of non-renewable energy sources [8]. Despite the promising potential of thermoelectric conversion, the current energy conversion rate is only around 7% to 10% [1,9] due to the unsatisfactory performances of the existing thermoelectric materials, which do not meet the commercial requirement. Thus, studies have focused on synthesising high-quality and excellent-performance thermoelectric materials.

In the 1950s, researchers focused on conventional materials such as Bismuth telluride (Bi_2Te_3,) lead telluride (PbTe) and Silicon Germanium (Si-Ge) [10,11], which contain heavy elements and covalent bonding characteristics, aiming to reduce the thermal conductivity and increase the electron mobility, respectively [12,13]. The main figure of merit (ZT) of these materials ranged between one and two [13–15], which are sufficient for practical applications in thermoelectric fields [5]. However, the poor durability and chemical stability at high temperatures of these materials limit their feasibilities in thermoelectric applications. At the same time, these materials have high synthesis complexity and are toxic, further limiting their feasibilities in practical thermoelectric applications [5,12–14,16]. Thus, in the mid-1990s, the investigation paradigm switched from optimising conventional thermoelectric materials to exploratory ventures of novel thermoelectric materials. Of the investigated materials, metal oxides offered excellent stability and oxidation resistance and are primarily safe [12–14,17]. These properties complement the limitations of the conventional thermoelectric materials (Bi_2Te_3, PbTe, and Si-Ge), which were widely studied in the 1950s. Unfortunately, the major drawback of metal oxides is the lower ZT values. The low carrier mobility and high thermal conductivity in metal oxides [12,14] result in smaller ZT values than that of the conventional non-oxides materials. However, one characteristic of the metal oxides is opposite to that of conventional thermoelectric materials, whereby the ZT value of metal oxide can be significantly increased at higher temperatures. Therefore, suggesting various possibilities of elevated temperature operations in metal oxides [12]. Having said that, many studies have been reported on the excellent thermoelectric properties of metal oxides. For instance, zinc oxide (ZnO) exhibited a high ZT value of 0.44 at 1000 K [16] and a ZT value of 0.52 at 1100 K [18]. In order to achieve this ZT value, a nanostructured approach and chemical doping strategy were employed.

The ZT value of thermoelectric materials is determined using an equation $ZT = (S^2 \sigma T)\kappa^{-1}$, where S is the Seebeck coefficient, σ is the electrical conductivity, κ is thermal conductivity, and T is temperature [15,19]. According to the literature, the ZT value should be at least 1 to 1.25 for thermoelectric usage [15]. From the equation, the term $S^2\sigma$ is also known as the power factor (PF) [2,20–28]. Larger the ZT value leads to a higher heat-to-energy conversion rate, where the materials exhibit better thermoelectric performances. To achieve that, the PF should be maximised, and the thermal conductivity value should be kept to the lowest possible [29]. However, various parameters are involved in optimising the PF [30]. The parameters include the Seebeck coefficient and thermal conductivity, which are theoretically interdependent [13,24,29,30]. The alteration of these parameters involves complex, fundamental physics, including modification of carrier concentration, band structures, vacancies, defects, etc. [29]. These complex insights significantly disrupted the development of ZnO thermoelectric materials [5,31,32].

2. Zinc Oxide (ZnO)

ZnO is a promising n-type thermoelectric material [2,33–37] and an iconic representation of group II-IV compound semiconductor material [6,15,23,38,39]. The n-type semiconductor behaviour originated from the ionisation of excess zinc atoms in interstitial positions and the presence of oxygen vacancies [36]. ZnO is naturally found in the mineral called zincite [35], appears as a white powder and is soluble in water and alcohol. Due to its chemical structure, it has excellent physical and chemical properties, high electron mobility and can be easily synthesised [20,33,34,39–41]. Moreover, due to its abundance, ZnO can be obtained at a low cost [2,35,37]. It is also safer and of low toxicity [1,2,34,37,42], compared to other telluride-based nanomaterials such as PbTe and tin telluride (SnTe) [3,25,33,35,43,44].

ZnO crystallises in hexagonal wurtzite (Wz) structure with two lattice parameters, a and c (a = 3.249 and c = 5.206 Å) and space group P63mc (186) [6,15,20,23,38,45,46]. Each anion is surrounded by four cations at the corners of a tetrahedron and vice versa [45,47,48] (Figure 1a). Wz structure contributes to the high stability of ZnO [39], whereby it can also act as a benchmarking index in determining the doping effects in ZnO [20]. As indicated in Figure 1b, the X-Ray diffraction of ZnO exhibited unique and iconic peaks for ZnO crystals [47,49]. Apart from that, ZnO has a wide and direct bandgap semiconductor with 3.3 eV at room temperature [38,46] and ~3.3–3.37 eV at 300 K [23,44,50,51] with a large exciton binding energy of 60 meV at room temperature [15,22,35,42,44]. The bandgap reduces electronic noise and demonstrates superior electrical durability [40]. Moreover, the chemical and mechanical stability make ZnO an ideal candidate for high-temperature, high voltage thermoelectric applications [22,38,43,44,52].

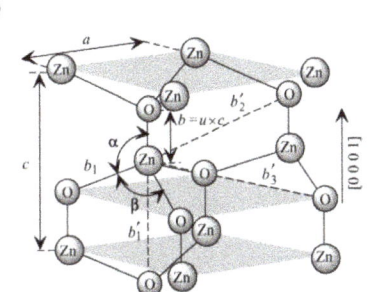 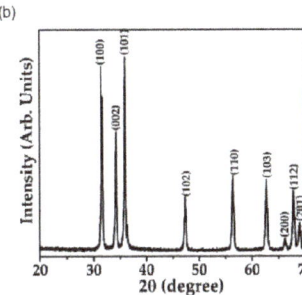

Figure 1. (a) A schematic of the hexagonal Wz ZnO crystal structure with lattice parameters a and c, bond length u, the nearest neighbour distance b, and three types of second-nearest-neighbour distances b′$_1$, b′$_2$, and b′$_3$. [48], (b) A powder diffraction pattern for ZnO [49].

Despite the promising potential of ZnO thermoelectric materials, the drawbacks in realising ZnO-based thermoelectric materials include poor electrical conductivity and high thermal conductivity. The thermal conductivity of ZnO is about 49 W/mK at 300 K and 10 W/mK at 1000 K, attributed to the stable covalent bond [16]. This disadvantage limits the ZT value of ZnO, which affects its feasibility in thermoelectric applications. Therefore, to enhance the thermoelectric properties of ZnO, various approaches, including foreign-particle doping and nanostructuring, are performed.

3. Enhancing the Thermoelectric Performance of ZnO

ZnO has a high Seebeck coefficient [34] between −350 and −430 µV/K [40,53]. These drawbacks lead to a poor ZT value of less than 0.01, which is undesirable for thermoelectric applications [14–16,54]. Many strategies have been attempted to obtain higher ZT values and lower thermal conductivity in ZnO ceramics. The strategies used include nanostructuring approach [2,50] focusing on the reduction of thermal conductivity [2,50,51,55], doping [2,21,50,51,55–57] and nanostructuring [58–61]. The following sections discuss the common approaches to enhancing the PF and lowering the thermal conductivity of ZnO, which include nanostructuring and doping approaches. This review seeks to provide a comprehensive compilation of the available strategies for enhancing the thermoelectric properties of ZnO, which can serve as a useful preliminary guide for the further development of ZnO in thermoelectric applications.

3.1. Nanostructure Approach

To date, substantial research activities have been conducted on nanostructuring thermoelectric materials. The nanostructuring approach has been reported to effectively reduce the thermal conductivity of many thermoelectric material systems to improve their thermoelectric properties. Fundamentally, the nanostructuring approach is the growth of superfine nanostructured materials [36,62]. The size of nanostructures/nanoparticles ranges between 1 and 100 nm, while non-nanostructured materials fall between 100 and 1000 nm, commonly referred to as submicron [63,64]. The altering of the synthesis method produces ZnO materials with improved thermoelectric properties.

Figure 2 indicates the effects of the nanostructuring and non-nanostructuring approaches on ZT values of ZnO ceramics over temperature. The dashed lines in Figure 2 [1,2,16,18,50,65] represent the nanostructuring approach and the solid lines represent the non-nanostructuring approach [21–23,34,51,53,55,66,67]. In the nanostructuring approach, strong phonon scattering is applied at grain boundaries, and the grain sizes are refined to lower the thermal conductivity. The correlation between ZnO grain size and its corresponding thermal conductivity can be explained using the Callaway model [15,68,69], where the thermal conductivity of nanostructured ZnO can be expressed as:

$$\kappa = CT^3 \int_0^{\theta_D/T} \frac{x^4 e^x (e^x - 1)^{-2}}{\alpha T^4 x^4 + (\beta_1 + \beta_1) T^5 x^2 + v/L} dx + \kappa_2 \quad (1)$$

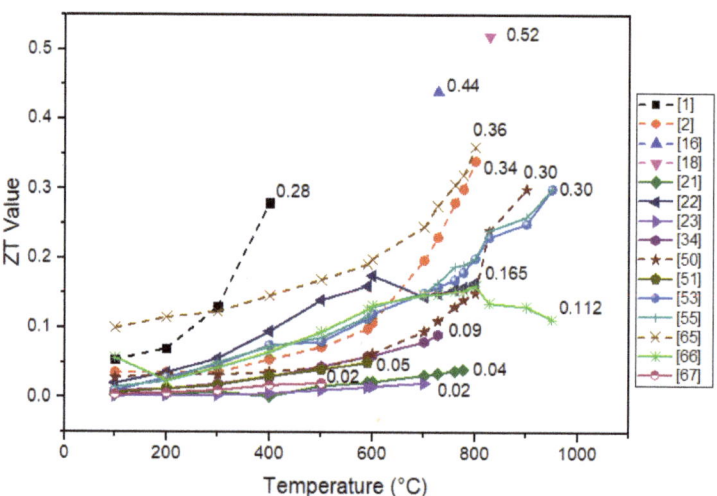

Figure 2. Comparison of nanostructuring and non-nanostructuring approach on the ZT value of ZnO ceramics over temperature.

C is an arbitrary constant in the Callaway equation which can be expressed as:

$$C = \frac{k_\beta}{2\pi^2 v} \left(\frac{k_\beta}{h} \right) \quad (2)$$

and x can be expressed as:

$$x = \frac{h\omega}{k_\beta T} \quad (3)$$

The key parameters can be found in Refs. [15,68,69]. For ZnO nanostructured thermoelectric materials, L can be deduced as the grain size of the nanostructures; thus, the smaller the grain size (L values), the lower the thermal conductivity (κ) of the nanostructure.

In the case of small, refined ZnO nanograins, a simultaneous increase in both phonon scattering and electron scattering can reduce the value of PF [2,65,70]. Nevertheless, the nanostructuring approach can successfully achieve higher ZT values compared to the non-nanostructuring approach, as illustrated in Figure 2. The highest reported ZT value from the nanostructuring approach was 0.52 at 827 °C [18], which is 42% higher than the best ZT value reported by its non-nanostructuring counterparts, 0.30 at 1000 °C [53,55]. At a lower temperature of 727 °C, the nanostructuring approach demonstrated a 62.5% higher ZT value compared to nanoparticles obtained from non-nanostructuring approaches [55]. Furthermore, the nanostructuring approach is capable of reaching a higher ZT value at a lower temperature than the non-nanostructuring approach. For instance, nanostructuring approach yielded a ZT value of 0.44 at 727 °C [16] and 0.3 at 777 °C [2], 950 °C [50] and 760 °C [65] compared to non-nanostructured approach which required 1000 °C [53,55]. This observation is in agreement with the Callaway model, where the larger grain size obtained from the non-nanostructuring approach resulted in higher thermal conductivity, whereby the corresponding ZT value was reduced, affecting its overall thermoelectric properties [71]. It can also be observed that ZnO nanoparticles demonstrated increasing ZT values at higher temperatures attributed to the temperature-dependency of ZnO grain size. Higher temperatures yielded smaller ZnO grains, favouring the ZT value and thermoelectric properties [72].

At present, the high thermal conductivity of ZnO remains a major challenge in restricting its usage in thermoelectric applications. As mentioned earlier, higher thermal conductivity in ZnO ceramics is caused by higher phonon frequency, attributing to the ionic bonding and low atomic mass of ZnO lattice [70]. Moreover, the parameters affecting ZnO's thermal conductivity values are complex (e.g. carrier concentration, electrical conductivity, grain size, etc.). Such complexity remains an obstacle in realising ZnO-based thermoelectric materials. Fundamentally, the thermal conductivity (κ) of ZnO nanoparticles can be expressed as the sum of two portions, phonons and electric charge carriers, as referred to in equation $\kappa = \kappa_{lat} + \kappa_{ele}$ [24], where κ_{lat} is the lattice thermal conductivity that arises from heat transport phonon conducted through the crystal lattice and κ_{ele} is the electron thermal conductivity that arises from a heat-carrying charge carrier moving through the crystal structure [24]. The lattice thermal conductivity is influenced by the mass of the unit cell and the grain size [13]. Based on the equation, reducing the lattice thermal conductivity will enhance the ZT value, but reducing the electron thermal conductivity will exhibit little or no improvement in ZT values [50,73]. Thus, lowering the lattice thermal conductivity of ZnO can achieve outstanding thermoelectric properties.

On the other hand, the lower thermal conductivity of ZnO is due to the phonon scattering at grain boundaries [74]. Thus, refining grain sizes and the control over nano-microstructures of ZnO can reduce its thermal conductivity and align well with the Callaway model [8,15,21,27,75]. However, refining grain size also increases the electron scattering and reduces the value of the PF [70,76]. Since smaller grain size yields a higher abundance of grain boundaries, lattice misalignment at grain boundaries will induce phonon scattering. Consequently, lower thermal conductivity is achieved with more electron scattering; hence, PF is reduced [77,78]. Nevertheless, several studies suggested that the reduction of grain size down to <100 nm reduces the thermal conductivity and improves the ZT values [2,15,16]. In addition, other strategies, such as the development of novel materials with unique lattice vibrational modes, have also been reported [27,79].

Despite reports on novel material development [79], the research community has emphasised refining grain size to lower the thermal conductivity of ZnO. Figure 3 illustrates the relationship between grain size and thermal conductivity on ZT value using non-nanostructuring and nanostructuring approaches. The solid lines in the graph represent the non-nanostructuring approach [21,34,51,67,80], while the dashed lines [1,2,16,18,50] the nanostructuring approach. Based on the graph, ZnO nanoparticles demonstrated lower thermal conductivity and smaller grain size when synthesised using the nanostructuring approach compared to their non-nanostructuring counterparts. According to Kinemuchi et al. [70], it can be computed using the Callaway formula that the thermal conductivity of ZnO nanocomposite was >5 W/mK at high temperature when the grain size was reduced to <100 nm [74]. This observation is in line with Figure 3, where the grain size of >100 nm resulted in lower thermal conductivity and higher ZT values.

Biswas et al. synthesised 1 at% Al with 1.5 wt % RGO into ZnO using the solvothermal method [18]. It has s been found to show significant improvement in ZT value of 0.52 at 1100 K, which is an order of magnitude larger compared to that of bare undoped ZnO. The minimum crystallite size obtained is ~20 nm, and the thermal conductivity was 2.96 W/mK at 1100 K. Jood et al. [16] reported a high ZT value of 0.44 at 727 °C for $Zn_{0.975}Al_{0.025}O$ via rapid and scalable microwave-activated aminolytic decomposition of zinc and aluminium salts. This method produces a ZnO grain size of 15 nm and a particle size of $\leq$ 25 nm. The thermal conductivity measured at 300 K was 1.5 W/mK and was 2 W/mK at 1000 K. The obtained thermal conductivity was ~96% lower than the 49 W/mK in the bulk ZnO sample. This thermal conductivity can be attributed to the increase in phonon scattering at ZnO nanograin promoted by Al-induced grain refinement and Zinc Aluminate ($ZnAl_2O_4$) nano precipitates. Additionally, Zhang et al. [65] reported that micro/nanostructured

$Zn_{0.98}Al_{0.02}O$ could be successfully fabricated via hydrothermal synthesis and spark plasma sintering. The effective scattering on the boundaries and interfaces reduced the thermal conductivity value to κ = 2.1 W/mK at 1073 K, leading to a high ZT value, 0.36 at 1073 K.

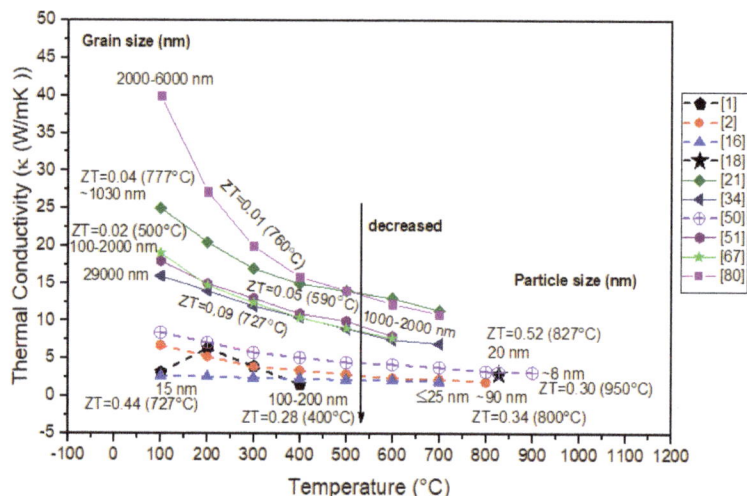

Figure 3. Effects of grain size on the thermal conductivity of ZnO ceramics at different temperatures.

Meanwhile, Nam et al. [2] reported a ZT value of 0.34 at 1073 K from the synthesised $Zn_{0.98}Al_{0.02}O$. The synthesis method used in the study was the nanostructuring approach involving a hybrid solution method and spark plasma sintering. The average ZnO grain size of ~90 nm was obtained. In addition to the ZT value of 0.34, the thermal conductivity was successfully reduced to 7.6 W/mK at room temperature, as it decreased with increasing temperature (κ_{700K} was ~3.2 W/mK and κ_{1000K} was < 2.0 W/mK). The thermal conductivity is reduced due to increasing phonon scattering at nanograin boundaries and nano precipitates. It also substantiated the observations in Figures 2 and 3, where the nanostructuring approach yielded ZnO nanocomposites with higher ZT values compared to the non-nanostructuring approach.

On the other hand, Han et al. [50] reported on the Al-doped ZnO nanoparticles using a forced-hydrolysis method. According to the study, ZnO ceramics were successfully produced with ZT values of 0.3 at 1223 K. The high ZT value was attributed to the low thermal conductivity of 3.2 W/mK at 1223 K, resulting from the nanostructuring of ZnO ceramics. The ZnO grain size that was obtained was ~8 nm, which was significantly smaller than the other Al-doped ZnO thermoelectrics reported by Jood et al. [16] and Nam et al. [2]. Meanwhile, Jantrasee et al. [1] reported a ZT value of 0.28 at 673 K of $Zn_{0.97}Al_{0.03}O$ synthesised using the hydrothermal method of zinc and aluminium salts. The nanostructuring approach yielded a ZT value of 1.59 W/mK with 100–200 nm grain size.

Based on the discussion above, it can be deduced that smaller grain size yields lower thermal conductivity in ZnO, leading to higher ZT values and favourable thermoelectric properties. Moreover, the nanostructuring approach yielded a smaller ZnO grain size than the non-nanostructuring approach, which resulted in lower thermal conductivity and higher ZT values. Additionally, doping of foreign particles (e.g. Al particles) can also effectively refine ZnO grain growth, resulting in smaller grain size. The next section discusses the effects of particle doping on ZnO's thermal conductivity and thermoelectric properties.

3.2. Doping Approach

Doping is another approach used to increase the ZT value of ZnO by increasing the carrier density [14,53,81] and refining its grain size. It is a process of adding impurities in minimal quantities to intrinsic semiconductors to alter their behaviour or physical properties. Fundamentally, the Seebeck coefficient can be expressed by Joker relation as:

$$S = ATm^* \left(\frac{\pi}{3n}\right)^{\frac{2}{3}} \qquad (4)$$

Meanwhile, the electrical conductivity of the thermoelectric materials can be expressed as:

$$\sigma = ne\mu \qquad (5)$$

The definitions of the key parameters can be found in Ref. [82]. As the electrical conductivity (σ) and the Seebeck coefficient (S) are both a function of the carrier concentration (n), it can be deduced that higher carrier concentration leads to higher electrical conductivity but lower Seaback coefficient. Thus, an optimal doping condition is required to obtain optical thermoelectric properties from ZnO nanoparticles [83].

In achieving optimal doping conditions, selecting the appropriate doping elements to improve the carrier concentration by altering their electronic band structure [59] and electron transport properties [20] are important. At the same time, its Seebeck coefficient for optimal thermoelectric performances should also be retained. Meanwhile, undoped ZnO usually contains various intrinsic defects [84]. These intrinsic defects can significantly affect the electrical properties and thermoelectric performances of ZnO [84]. In this section, the dopings of ZnO and their resulting thermoelectric properties are briefly discussed. Based on the review, it can be observed that Al-doped ZnO and Gallium-doped ZnO demonstrated the best thermoelectric performances due to the high stability and distinguished thermopower [1]. The aluminium (Al) and gallium (Ga) dopants in ZnO increased the electrical conductivity of ZnO, thereby enhancing its thermoelectric performances [40].

3.2.1. Al-doped ZnO

Al is a cheap, durable, abundant, non-toxic and a common n-type donor dopant for ZnO [40]. The incorporation of Al into ZnO can reduce grain size and thermal conductivity without deteriorating its electrical conductivity and Seebeck coefficient. When Al is doped into ZnO, the additional enhancements in electrical properties on the existing high electron transport properties of ZnO result in outstanding thermoelectric performance [84]. Since the radius of Al atoms is smaller than Zn atoms, Zn atoms in ZnO lattice are easily displaced by Al dopant. As the name implied, Al atoms are n-type donor dopants. Thus, an additional free electron is obtained per lattice, which enhances the overall electrical conduction of Al-doped ZnO. Apart from the dopant effects, the covalent bond of the metal-to-oxygen bond in the oxide also contributes to the outstanding thermoelectric performance of Al/ZnO by improving carrier density and carrier mobility [40,53,55,66,85].

Table 1 summarises the thermoelectric properties of Al-doped ZnO. It can be deduced that Al-doped ZnO demonstrated improved thermoelectric performance [1,21,53,66,67,80]. The addition of small atomic percentages (1 to 3 at.%) of Al significantly improved the ZT value of ZnO compared with the undoped ZnO. However, the ZT value was still low and unsatisfactory [67]. The highest ZT value obtained was 0.44, which is insufficient for thermoelectric applications [16]. Furthermore, due to the presence of n-type dopant, the Seebeck coefficient exhibited a negative value within the investigated temperature range [16,80]. Apart from the improvements in the Seebeck coefficient, some results exhibited lower Seebeck coefficients upon Al doping [53,66,67]. This observation aligns with the Joker relation, where additional charge carriers in Al-doped ZnO reduce the Seebeck coefficient. Nevertheless, Al-doped ZnO improved electrical conductivity by introducing additional charge carriers. It is reflected from the metallic behaviour [21,53,66] of Al-doped Zn, which improves its electrical conductivity [16,53,66,67]. As ZnO is doped with Al at >3 at.% (near solubility limit), its resistivity is further reduced [22]. Despite the reduction in the Seebeck coefficient, the increase in electrical conductivity indicated readily allowable electric charge movement favouring the thermoelectric performance of ZnO [86].

Table 1. Thermoelectric properties of Al-doped ZnO.

Composition/Synthesis Method	Thermoelectric Properties	Ref
$(Zn_{0.99}Al_{0.01})O$/Solution combustion	ZT = 0.05, S = ~−148 µV/K, σ = ~21,000 S/m, PF = ~48,000 W/mK^2, κ = 8 W/mK at 590 °C	[51]
ZnO/RF plasma powder	ZT = ~0.007, S = ~−386 µV/K, σ = ~369 S/m, PF = ~0.55 × 10^{-4} W/mK^2, κ = 8 W/mK at 777 °C	[21]
$(Zn_{0.99}Al_{0.01})O$/RF plasma powder	ZT = 0.04, S = −68 µV/K, σ = ~86,505 S/m, PF = ~4 × 10^{-4} W/mK^2, κ = 10.6 W/mK at 777 °C	
$(Zn_{0.98}Al_{0.02})O$/Solid-state reaction	ZT = 0.3, S = −180 µV/K, σ = 40,000 S/m, κ = 5.4 W/mK at 1000 °C	[55]
ZnO/Nylon-lined ball mill	ZT = 0.028, S = −318 µV/K, σ = ~2037 S/m, PF = ~2.06 × 10^{-4} W/mK2, κ = 8 W/mK at 800 °C	[53,66]
$(Zn_{0.98}Al_{0.02})O$/Nylon-lined ball mill	ZT = 0.3, S = −182 µV/K, σ = ~39,971 S/m, PF = ~13.24 × 10^{-4} W/mK^2, κ = 5 W/mK at 1000 °C	
$(Zn_{0.98}Al_{0.02})O$/Solution method	ZT = 0.34, S = ~−148 µV/K, σ = ~29,000 S/m, κ = ~2 W/mK at 800 °C	[2]
$(Zn_{0.98}Al_{0.02})O$/Forced hydrolysis method	ZT = ~0.3, S = ~−205 µV/K, σ = ~18,798 S/m, κ = ~3.2 W/mK at 950 °C	[50]
ZnO/Solid-state reaction	ZT = ~0.009, S = ~−107 µV/K, σ = ~2500 S/m, κ = ~3.1 W/mK at 700 °C	[80]
$(Zn_{0.98}Al_{0.02})O$/Solid-state reaction	ZT = ~0.031, S = ~−133 µV/K, σ = 5400 S/m, κ = ~3 W/mK at 700 °C	
$(Zn_{0.975}Al_{0.025})O$/microwave synthesis	ZT = 0.44, S = ~−300 µV/K, σ = ~10,000 S/m κ = ~2 W/mK at 727 °C	[16]
ZnO/Sol-gel method	ZT = ~0.007, S = −240 µV/K, σ = ~1563 S/m, κ = ~9.8 W/mK at 500 °C	[67]
$(Zn_{0.97}Al_{0.03})O$/Sol-gel method	ZT = ~0.02, S = ~−81 µV/K, σ = ~37,500 S/m, κ = ~9.5 W/mK at 500 °C	
ZnO/Chemical co-deposition method	ZT = ~0.009, S = ~−250 µV/K, σ = ~1333 S/m, κ = ~8.3 W/mK at 600 °C	[22]
$(Zn_{0.97}Al_{0.03})O$/Chemical co-deposition method	ZT = 0.15, S = ~−240 µV/K, σ = ~15,000 S/m, κ = ~5 W/mK at 600 °C	

The thermal conductivity of ZnO is higher at room temperature (40 W/mK) and decreases with the increase of temperature, 5 W/mK at 1000 °C [53]. The high thermal conductivity of ZnO can be attributed to its lattice structure [53]. Upon Al doping, the thermal conductivity decreases with the increasing amount of Al added due to the enhancement of the phonon scattering at nanograin boundaries and nano precipitates [2,67]. Moreover, the increase in the Al contents may cause random disorder phonon scattering induced by Al deficient sites [67]. Hence, thermal conductivity is lowered, resulting in enhanced thermoelectric performances [51].

3.2.2. Ga-doped ZnO

Ga is in group III of the periodic table, akin to aluminium and indium. Through Ga doping, the carrier concentration and electron and point defect concentration are increased, subsequently increasing electrical conductivity and decreasing thermal conductivity [78]. Table 2 summarises the thermoelectric properties of ZnO doped with Ga at different temperatures. Based on Table 2, it can be deduced that the addition of Ga into ZnO improved the electrical conductivity. Simultaneously, the Seebeck coefficient of Ga-doped ZnO also increased due to the presence of the voids, despite the additional charge carriers, according to Joker relation. Apart from that, as suggested by Jood et al. [16], the voids reduce the thermal conductivity of ZnO by increasing phonon scattering centres, enhancing the thermoelectric performances of ZnO.

Table 2. Thermoelectric properties of Ga-doped ZnO.

Composition/Synthesis Method	Thermoelectric Properties	Ref
ZnO/Nylon-lined ball mill	ZT = ~0.076, S = ~−430 µV/K, σ = ~1622 S/m, PF = ~3 × 10^{-4} W/mK^2, κ = 5 W/mK at 1000 °C	[66]
($Zn_{0.98}Ga_{0.02}$)O/Nylon-lined ball mill	ZT = ~0.13, S = ~−180 µV/K, σ = ~16,600 S/m, PF = ~5.38 × 10^{-4} W/mK^2, κ = 5 W/mK at 1000 °C	
($Zn_{0.995}Ga_{0.005}$)O/high energy wet milling	ZT = ~0.0022, S = ~−133 µV/K, σ = ~10,989 S/m, κ = 26.9 W/mK at 27 °C	[87]
ZnO/Wet chemistry gel combustion method	ZT = ~0.0022, S = ~−312.5 µV/K, σ = ~197 S/m, κ = ~9.4 W/mK at 800 °C	[78]
($Zn_{0.98}Ga_{0.02}$)O/Wet chemistry gel combustion method	ZT = ~0.026, S = ~−562.5 µV/K, σ = ~268 S/m, κ = ~3.5 W/mK at 800 °C	
($Zn_{0.99}Ga_{0.01}$)O/Atomic layer deposition	S = 60 µV/K, σ = ~180,832 S/m, PF = ~6.6 × 10^{-4} W/mK^2	[88]

3.2.3. Ni-doped ZnO

Nickel (Ni) belongs to group 10 (transition metals) in the periodic table. It is very hard, ductile and has high thermal conductivity (90.9 W/mK). Ni dopant acts as an excellent electron donor, as their presence decreases the charge carrier concentration in ZnO [23]. The primary material used in Ni-doping is Nickel oxide (NiO), which has a cubic structure [34,46]. Table 3 indicates the thermoelectric properties of Ni-doped ZnO. The addition of Ni to ZnO with the composition $Zn_{1-x}Ni_xO$ (x = 0.03) indicates the optimal value of ZT at the higher temperature. ZT value reduces as the composition increases above x = 0.03. The average size of samples is smaller (17.1 to 5.1 µm) with higher Ni content [46]. Nevertheless, this method produced an average ZnO grain size of 28 nm for intrinsic ZnO, where the grain size was reduced to 26 nm for Ni-doped ($Zn_{0.97}Ni_{0.03}$)O powder [34]. Meanwhile, Ni-doped ZnO with the composition of x < 0.03 demonstrated a dense microstructure of a single-phase wurtzite ZnO structure [23,34] and a secondary phase as cubic when the composition x > 0.0325. Despite the reduction in thermal conductivity associated with grain size reduction, the charge mobility in the grain boundaries decreases due to the pinning effects. This observation exhibits a trade-off in Ni-doped ZnO thermoelectric materials [46].

Table 3. Thermoelectric properties of Ni-doped ZnO.

Composition/Synthesis Method	Thermoelectric Properties	Ref
ZnO/Precipitation	ZT = 0.002, S = ~−517 µV/K, σ = ~50 S/m, κ = 8 W/mK at 700 °C	[23]
($Zn_{0.97}Ni_{0.03}$)O/Precipitation	ZT = 0.02, S = ~−306 µV/K, σ = ~1790 S/m, κ = 8 W/mK at 700 °C	
ZnO/Tape casting method	PF = 0.62 × 10^{-4} W/mK^2, S = ~−203 µV/K, σ = ~1500 S/m at 800 °C	[46]
($Zn_{0.97}Ni_{0.03}$)O/Tape casting method	PF = 17.6 × 10^{-4} W/mK^2, S = −503 µV/K, σ = 6970 S/m at 800 °C	
ZnO/Liquid route synthesis	ZT = 0.0034, S = ~−310 µV/K, σ = ~250 S/m, PF = ~0.24 × 10^{-4} W/mK^2, κ = 7 W/mK at 727 °C	[34]
($Zn_{0.97}Ni_{0.03}$)O/Liquid route synthesis	ZT = 0.09, S = ~−420 µV/K, σ = ~3401 S/m, PF = 6 × 10^{-4} W/mK^2, κ = 6.5 W/mK at 727 °C	

Among the reported ZT values of Ni-doped ZnO thermoelectric materials, Colder et al. [34] reported the highest value of 0.09, which was 75% higher compared to the value reported by Koresh and Amouyal [23]. Similar to Al-doped ZnO, the Seebeck coefficients reported here were also negative values, indicating n-type conduction [23,34,46]. According to Koresh and Amouyal, higher Ni composition resulted in higher Seebeck coefficient values at x < 0.03 [34,46], whereby increased Ni composition will result in reduced Seebeck coefficient, x > 0.03 [23]. Furthermore, the electrical conductivity of ZnO was also increased upon Ni doping, primarily due to the increasing electron carrier concentration [34]. Similar to the Seebeck coefficient, when the Ni composition reaches $x > 0.03$, the electrical conductivity begins to decrease [46]. This phenomenon could be due to the occurrences of the secondary phase in the vicinity of the grain boundary, which in turn, decreases the grain size. Smaller grain sizes yield a higher abundance of high resistivity grain boundary, resulting in low electrical conductivity [34,46]. Apart from the electrical conductivity, the Ni-doped element is capable of reducing the thermal conductivity at room temperature [25]. However, despite the changes in electrical conductivity, the overall value of thermal conductivity does not differ much with the addition of Ni in ZnO samples at higher temperatures [23,34,46].

3.2.4. Bi-doped ZnO

Bismuth (Bi) belongs to group 15 in the periodic table. It has relatively lower toxicity among heavy metals and low thermal conductivity (7.97 W/mK). Guan et al. [80] investigated the effects of bismuth (III) oxide (Bi_2O_3) doped in ZnO. The samples were prepared using a solid-state reaction method that yielded grain size within the range of 2 to 6 μm. The Bi_2O_3 particles in ZnO were segregated at the grain boundaries level resulting in their low solubility. It was also reported that the low melting point of Bi_2O_3 in ZnO at grain boundaries affects the grain growth in Bi-doped ZnO samples by transport accelerations.

The addition of Bi in ZnO demonstrated a dramatic increase in the Seebeck coefficient values (-533 μV/K). However, the electrical conductivity decreased beyond this value owing to higher electrical resistivity ($>10^8$ Ω). Moreover, it is worth noting that the carrier concentration is also lower than that of Al-doped ZnO. Apart from the electrical conductivity, the addition of Bi in ZnO significantly improved its lattice thermal conductivity intrinsic ZnO. Based on the significant increase in grain size due to the segregation of Bi_2O_3, it can be suggested that Bi is not an ideal dopant to improve the ZT value of ZnO for thermoelectric applications. Table 4 illustrates the thermoelectric properties of Bi-doped in ZnO.

Table 4. Thermoelectric properties of Bi-doped ZnO.

Composition/Synthesis Method	Thermoelectric Properties	Ref
ZnO/Solid-state reaction	ZT = ~0.009, S = ~−107 μV/K, σ = ~2500 S/m, κ = ~3.1 W/mK at 700 °C	[80]
$(Zn_{0.98}Bi_{0.02})O$/Solid-state reaction	ZT = 0.006, S = −533 μV/K, σ = ~230 S/m, κ = 10.5 W/mK at 700 °C	

3.2.5. Sn-doped ZnO

Tin (Sn) belongs to group 14 of the periodic table, where it has low toxicity and is soft and ductile with a thermal conductivity value of 66.8 W/mK. Table 5 indicates the effects of tin (IV) oxide (SnO_2) doped in ZnO ($Zn_{1-x}Sn_xO$) on its thermoelectric properties. The average grain size obtained ranged from 13.9 to 3.5 μm. However, with higher Sn doping concentrations, the grain size is reduced [38]. This reduction can be attributed to the pinning effect caused by Zn_2SnO_4 particles present at the grain boundaries of ZnO, along with the dragging effects between the added SnO_2 and grain boundaries. These effects result in lower electron mobility by reducing grain size [38]. Meanwhile, Guan et al. [80] also reported lower mobility in Sn-doped ZnO samples caused by the increased ionised impurities scattering within the samples.

Table 5. Thermoelectric properties of Sn-doped ZnO.

Composition/Synthesis Method	Thermoelectric Properties	Ref
$(Zn_{0.99}Sn_{0.01})O$/Solid-state reaction	PF = 1.25 × 10^{-3} W/mK^2, S = ~−160 μV/K, σ = ~15 S/m at 800 °C	[38]
ZnO/Solid-state reaction	ZT = ~0.009, S = ~−107 μV/K, σ = ~2500 S/m, κ = ~3.1 W/mK at 700 °C	[80]
$(Zn_{0.98}Sn_{0.02})O$/Solid-state reaction	ZT = 0.012, S = −93 μV/K, σ = 900 S/m, κ = 7 W/mK at 700 °C	

The electrical conductivity for undoped ZnO increases slightly with the increment of temperature, indicating its inherent semiconductor behaviour [38]. However, the addition of Sn in ZnO modified the semiconductor conduction behaviour of intrinsic ZnO, where the electric conductivity of Sn-doped ZnO decreases with increasing temperature [38]. The electrical conductivity behaviour of Sn-doped ZnO observed by Guan et al. [80] exhibited the opposite trend from other dopants (Ni, Al, etc.). As the electrical conductivity increases, the Seebeck coefficient value reduces, leading to a lower ZT value attributed to the lattice structure of Sn dopants. Since Sn has twice the ionised donor impurity that provides carriers (electrons) [80], the addition of Sn content increases the carrier concentration compared to Al, Bi, Ni and undoped ZnO [80]. This observation aligned with the Joker relation, where a higher number of charge carriers result in higher electrical conductivity and a lower Seebeck coefficient.

The addition of small amounts of SnO_2 leads to a strong donor effect [38,80]. Zn atoms (Zn^{2+}) in the ZnO lattice can be easily substituted by Sn atoms (Sn^{4+}), whereby the lattice distortion is not affected. This differs from the Al dopant discussed in the previous section. Since Zn and Sn have similar radii, Sn dopant can substitute vacancies in ZnO lattice without creating significant voids or defects [38,80]. In terms of the alteration of thermoelectric properties, the addition of Sn increases the thermal conductivity due to the larger grain size leading to decreased phonon scattering.

4. Concluding Remarks

ZnO possesses promising potential in thermoelectric application due to its high physicochemical stability, tunable properties and high abundance, which subsequently yields low synthesis cost and complexity. The overall effect is a ZnO thermoelectric device can operate at higher temperatures, has higher conversion efficiency, has higher reliability and costs less to produce. With this research, millions of money in savings or in new opportunities to recover waste heat from high-temperature processes could be made available. In order to enhance its thermoelectric properties, several approaches can be employed: (i) lowering the thermal conductivity, (ii) increasing the Seebeck coefficient, and (iii) increasing the electrical conductivity. Based on the Hallaway model, smaller grain size ZnO is desired to achieve lower thermal conductivity. In order to improve the electrical conductivity of ZnO, a higher number of charge carriers is needed. However, based on the Joker relation on the Seebeck coefficient of ZnO, a high number of charge carriers result in lowered Seebeck coefficient in ZnO, negatively affecting the overall thermoelectric properties of ZnO. Therefore, optimisation of charge carriers in ZnO is needed to obtain the desired thermoelectric properties.

According to the extant literature, several strategies have been reported to enhance the thermoelectric properties of ZnO, including nanostructuring synthesis technique and doping. Between the nanostructuring and non-nanostructuring synthesis approaches, the nanostructuring approach yielded a smaller grain size corresponding to lower thermal conductivity and improved Seebeck coefficient corresponding to its thermoelectric properties. As for doping strategies, doping of foreign particles in ZnO improves the overall thermoelectric properties via several mechanisms. Firstly, the inclusion of foreign particles (such as Al) inhibits the grain growth, and the combination in the ZnO lattice results in low thermal conductivity. Meanwhile, the addition of charged particles increased the electrical conductivity of ZnO. These mechanisms effectively improved the thermoelectric properties of ZnO. However, above the optimal doping concentration, the excessively high number of charge carriers will reduce the Seebeck coefficient of ZnO, affecting its overall thermoelectric properties. Thus, optimal doping conditions are called for.

5. Way Forward

This section is not mandatory but may be added if there are patents resulting from the work reported in this manuscript. The review addressed the preliminary understanding of the effects of doping and nanostructuring on the thermoelectric properties of ZnO. The reported state-of-the-art strategy demonstrated the individual impact of doping concentration, doping species and nanostructuring strategies on the Seebeck coefficient, electrical conductivity and thermal conductivity of ZnO thermoelectric materials. Therefore, several technical gaps should be focused on to fully utilise ZnO materials for commercial thermoelectric devices. First are the effects of hybrid doping species. Each doping species possess individualised advantages and drawbacks. For instance, Al dopant introduces increased carrier concentration without trading off the Seebeck coefficient. Meanwhile, Sn dopant resulted in a relatively higher enhancement in electrical conductivity, but the presence of excessive charge carriers lowered the Seebeck coefficient. A hybridised doping of Al and Sn species can yield promising outcomes in the thermoelectric properties of ZnO.

Apart from the mixed doping species, another approach which can be attempted is the mixture of nanostructuring and doping strategies. Nanostructuring refines the grain size of ZnO to achieve low thermal conductivity and to enhance its corresponding thermoelectric properties. Meanwhile, doping of ZnO suppresses the growth of ZnO grains, enhancing its electrical properties by reducing the Seebeck coefficient. It can be speculated that these two strategies can complement each other to potentially enhance the thermoelectric performance of ZnO to another level. In order to achieve the desired properties, this review provides a preliminary understanding of the nanostructuring and doping of ZnO thermoelectric materials, which also serves as a useful fundamental for further developments of ZnO-based thermoelectric technologies.

Author Contributions: Conceptualisation (idea and realization), S.S.; review of the state of the art, S.S.; writing—original draft preparation, S.S.; visualisation, S.S.; writing—review and editing, S.S., I.S., U.M.B.A.-N. and M.F.O.; formal analysis, S.S.; supervision, S.S.; funding acquisition, S.S. All authors have read and agreed to the published version of the manuscript.

Funding: This research was funded by UMP-IIUM-UiTM Sustainable Research Collaboration 2020 grant number RDU200744.

Acknowledgments: The authors would like to acknowledge generous support from the School of Mechanical Engineering, Faculty of Engineering, Universiti Teknologi Malaysia, Faculty of Science, Universiti Teknologi Malaysia, Faculty of Manufacturing and Mechatronic Engineering Technology, Universiti Malaysia Pahang and UMP-IIUM-UiTM Sustainable Research Collaboration 2020 (RDU200744).

Conflicts of Interest: The authors declare no conflict of interest.

References

1. Jantrasee, S.; Moontragoon, P.; Pinitsoontorn, S. Thermoelectric properties of Al-doped ZnO: Experiment and simulation. *J. Semicond.* **2016**, *37*, 092002-1–092002-8. [CrossRef]
2. Nam, W.H.; Lim, Y.S.; Choi, S.-M.; Seo, W.-S.; Lee, J.Y. High-temperature charge transport and thermoelectric properties of a degenerately Al-doped ZnO nanocomposite. *J. Mater. Chem.* **2012**, *22*, 14633–14638. [CrossRef]
3. Ihns, M. Structural Engineering of $Ca_3Co_4O_9$ Thermoelectric Thin Films. Master's Thesis, University of Twente, Enschede, The Netherlands, 2013.
4. Zaferani, S.H.; Jafarian, M.; Vashaee, D.; Ghomashchi, R. Thermal Management Systems and Waste Heat Recycling by Thermoelectric Generators—An Overview. *Energies* **2021**, *14*, 5646. [CrossRef]
5. Fergus, J.W. Oxide materials for high temperature thermoelectric energy conversion. *J. Eur. Ceram. Soc.* **2012**, *32*, 525–540. [CrossRef]
6. Zhu, B.B.; Li, D.; Zhang, T.S.; Luo, Y.B.; Donelson, R.; Zhang, T.; Zheng, Y.; Du, C.F.; Wei, L.; Hng, H.H. The improvement of thermoelectric property of bulk ZnO via ZnS addition: Influence of intrinsic defects. *Ceram. Int.* **2018**, *44*, 6461–6465. [CrossRef]
7. Funahashi, R.; Barbier, T.; Combe, E. Thermoelectric materials for middle and high temperature ranges. *J. Mater. Res.* **2015**, *30*, 2544–2557. [CrossRef]
8. Koumoto, K.; Wang, Y.; Zhang, R.; Kosuga, A.; Funahashi, R. Oxide Thermoelectric Materials: A Nanostructuring Approach. *Annu. Rev. Mater. Res.* **2010**, *40*, 363–394. [CrossRef]
9. Jouharaa, H.; Zabnienska-Góra, A.; Khordehgaha, N.; Doraghia, Q.; Ahmada, L.; Normana, L.; Axcella, B.; Wrobela, L.; Daid, S. Thermoelectric generator (TEG) technologies and applications. *Int. J. Thermofluids.* **2021**, *9*, 100063. [CrossRef]
10. Huang, S.; Ning, S.; Xiong, R. First-Principles Study of Silicon–Tin Alloys as a High-Temperature Thermoelectric Material. *Materials* **2022**, *15*, 4107. [CrossRef] [PubMed]
11. Wang, H.; Liang, X.; Wang, J.; Jiao, S.; Xue, D. Multifunctional inorganic nanomaterials for energy applications. *Nanoscale* **2020**, *12*, 14–42. [CrossRef]

12. Ohtaki, M. Recent aspects of oxide thermoelectric materials for power generation from mid-to-high temperature heat source. *J. Ceram. Soc. Jpn.* **2011**, *119*, 770–775. [CrossRef]
13. Lourdes, M.G. Preparation and Characterisation of Nanostructured Bulk Bi_2Te_3 Thermoelectric Materials Using Ultrasound Milling. Ph.D. Thesis, Cardiff University, Cardiff, UK, 2016.
14. Yin, Y.; Tudu, B.; Tiwari, A. Recent advances in oxide thermoelectric materials and modules. *Vacuum* **2017**, *146*, 356–374. [CrossRef]
15. Sulaiman, S.; Izman, S.; Uday, M.B.; Omar, M.F. Review on grain size effects on thermal conductivity in ZnO thermoelectric materials. *RSC Adv.* **2022**, *12*, 5428–5438. [CrossRef]
16. Jood, P.; Mehta, R.J.; Zhang, Y.; Peleckis, G.; Wang, X.; Siegel, R.W.; Borca-tasciuc, T.; Dou, S.X.; Ramanath, G. Al-Doped Zinc Oxide Nanocomposites with Enhanced Thermoelectric Properties. *Nano. Lett.* **2011**, *11*, 4337–4342. [CrossRef] [PubMed]
17. Mohammed, M.A.; Sudin, I.; Noor, A.M.; Rajoo, S.; Uday, M.B.; Obayes, N.H.; Omar, M.F. A review of thermoelectric ZnO nanostructured ceramics for energy recovery. *Int. J. Eng. Technol.* **2018**, *7*, 27–30. [CrossRef]
18. Biswas, S.; Singh, S.; Singh, S.; Chattopadhyay, S.; Silva, K.K.H.D.; Yoshimura, M.; Mitra, J.; Kamble, V.B. Selective Enhancement in Phonon Scattering Leads to a High Thermoelectric Figure-of-Merit in Graphene Oxide-Encapsulated ZnO Nanocomposites. *ACS Appl. Mater. Interfaces* **2021**, *13*, 23771–23786. [CrossRef] [PubMed]
19. Chen, Z.; Shi, X.; Zhao, L.; Zou, J. High-performance SnSe thermoelectric materials: Progress and future challenge. *Prog. Mater. Sci.* **2018**, *97*, 283–346. [CrossRef]
20. Wiff, J.P.; Kinemuchi, Y.; Kaga, H.; Ito, C.; Watari, K. Correlations between thermoelectric properties and effective mass caused by lattice distortion in Al-doped ZnO ceramics. *J. Eur. Ceram. Soc.* **2008**, *29*, 1413–1418. [CrossRef]
21. Cheng, H.; Xu, X.J.; Hng, H.H.; Ma, J. Characterization of Al-doped ZnO thermoelectric materials prepared by RF plasma powder processing and hot press sintering. *Ceram. Int.* **2009**, *35*, 3067–3072. [CrossRef]
22. Qu, X.; Wang, W.; Lv, S.; Jia, D. Thermoelectric properties and electronic structure of Al-doped ZnO. *Solid State Commun.* **2011**, *151*, 332–336. [CrossRef]
23. Koresh, I.; Amouyal, Y. Effects of microstructure evolution on transport properties of thermoelectric nickel-doped zinc oxide. *J. Eur. Ceram. Soc.* **2017**, *37*, 3541–3550. [CrossRef]
24. Liu, W.; Hu, J.; Zhang, S.; Deng, M.; Han, C.-G.; Liu, Y. New trends, strategies and opportunities in thermoelectric materials: A perspective. *Mater. Today Phys.* **2017**, *1*, 50–60. [CrossRef]
25. Park, Y.; Cho, K.; Kim, S. Thermoelectric characteristics of glass fibers coated with ZnO and Al-doped ZnO. *Mater. Res. Bull.* **2017**, *96*, 246–249. [CrossRef]
26. Kedia, S.K.; Singh, A.; Chaudhary, S. Design, development, and testing of a thermopower measurement system by studying the electron transport properties on indium and nitrogen co-doped sputtered ZnO films. *Measurement* **2018**, *117*, 49–56. [CrossRef]
27. Zhou, X.; Yan, Y.; Lu, X.; Zhu, H.; Han, X.; Chen, G.; Ren, Z. Routes for high-performance thermoelectric materials. *Mater. Today* **2018**, *21*, 974–988. [CrossRef]
28. Zhou, B.; Chen, L.; Li, C.; Qi, N.; Chen, Z.; Su, X.; Tang, X. Significant Enhancement in the Thermoelectric Performance of Aluminum-Doped ZnO Tuned by Pore Structure. *ACS Appl. Mater. Interfaces* **2020**, *12*, 51669–51678. [CrossRef]
29. Wu, N. Development and Processing of p-Type Oxide Thermoelectric Materials. Ph.D. Thesis, Technical University of Denmark, Copenhagen, Denmark, 2014.
30. Tritt, T.M.; Subramanian, M. Thermoelectric Materials, Phenomena, and Applications: A Bird's Eye View. *MRS Bull.* **2006**, *31*, 188–198. [CrossRef]
31. Markov, M.; Rezaei, S.E.; Sadeghi, S.N.; Esfarjani, K.; Zebarjadi, M. Thermoelectric properties of semimetals. *Phys. Rev. Mater.* **2019**, *3*, 095401. [CrossRef]
32. Kim, H.J.; Skuza, J.R.; Park, Y.; King, G.C.; Choi, S.H.; Nagavalli, A. System to Measure Thermal Conductivity and Seebeck Coefficient for Thermoelectrics. *NasaTM* **2012**, *217791*, LF99-15831.
33. Yamaguchi, H.; Chonan, Y.; Oda, M.; Komiyama, T.; Aoyama, T.; Sugiyama, S. Thermoelectric properties of ZnO ceramics Co-doped with Al and transition metals. *J. Electron. Mater.* **2011**, *40*, 723–727. [CrossRef]
34. Colder, H.; Guilmeau, E.; Harnois, C.; Marinel, S.; Retoux, R.; Savary, E. Preparation of Ni-doped ZnO ceramics for thermoelectric applications. *J. Eur. Ceram. Soc.* **2011**, *31*, 2957–2963. [CrossRef]
35. Brintha, S.R.; Ajitha, M. Synthesis and characterization of ZnO nanoparticles via aqueous solution, sol-gel and hydrothermal methods. *J. Appl. Chem.* **2015**, *8*, 66–72. [CrossRef]
36. Jurablu, S.; Farahmandjou, M.; Firoozabadi, T.P. Sol-Gel Synthesis of Zinc Oxide (ZnO) Nanoparticles: Study of Structural and Optical Properties. *J. Sci. Islamic Repub. Iran* **2015**, *26*, 281–285.
37. Wu, Z.-H.; Xie, H.-Q.; Zhai, Y.-B. Preparation and Thermoelectric Properties of Co-Doped ZnO Synthesized by Sol–Gel. *J. Nanosci. Nanotechnol.* **2015**, *15*, 3147–3150. [CrossRef] [PubMed]
38. Park, K.; Seong, J.K.; Kwon, Y.; Nahm, S.; Cho, W.S. Influence of SnO_2 addition on the thermoelectric properties of $Zn_{1-x}Sn_xO$ ($0.01 \leq x \leq 0.05$). *Mater. Res. Bull.* **2008**, *43*, 54–61. [CrossRef]
39. Kumar, M.; Sahu, S.S. Zinc Oxide Nanostructures Synthesized by Oxidization of Zinc. Master's Thesis, National Institute of Technology Rourkela, Rourkela, India, 2010.
40. Ilican, S.; Gorgun, K.; Aksoy, S.; Caglar, Y.; Caglar, M. Fabrication of p-Si/n-ZnO:Al heterojunction diode and determination of electrical parameters. *J. Mol. Struct.* **2018**, *1156*, 675–683. [CrossRef]

41. Kołodziejczak-Radzimska, A.; Jesionowski, T. Zinc Oxide—From Synthesis to Application: A Review. *Materials* **2014**, *7*, 2833–2881. [CrossRef] [PubMed]
42. Baghdadi, N.; Salah, N.; Alshahrie, A.; Koumoto, K. Microwave Irradiation to Produce High Performance Thermoelectric Material Based on Al Doped ZnO Nanostructures. *Crystals* **2020**, *10*, 610. [CrossRef]
43. Patel, N. Characterization of Electrical Performance of Aluminum-Doped Zinc Oxide Pellets. *DePaul Discov.* **2014**, *3*, 6.
44. Feng, Q.; Shi, X.; Xing, Y.; Li, T.; Li, F.; Pan, D.; Liang, H. Thermoelectric microgenerators using a single large-scale Sb doped ZnO microwires. *J. Alloy. Compd.* **2018**, *739*, 298–304. [CrossRef]
45. Morkoç, H.; Özgür, Ü. *Zinc Oxide: Fundamentals, Materials and Device Technology*; WILEY-VCH Verlag GmbH & Co. KGaA: Weinheim, Germany, 2009; ISBN 978-3-527-40813-9.
46. Park, K.; Seong, J.K.; Kim, G.H. NiO added Zn1-xNixO ($0 \leq x \leq 0.05$) for thermoelectric power generation. *J. Alloy. Compd.* **2009**, *473*, 423–427. [CrossRef]
47. Janotti, A.; Walle, C.G.V.D. Fundamentals of zinc oxide as a semiconductor. *Rep. Prog. Phys.* **2009**, *72*, 126501. [CrossRef]
48. Wojnarowicz, J.; Chudoba, T.; Lojkowski, W. A Review of Microwave Synthesis of Zinc OxideNanomaterials: Reactants, Process Parametersand Morphologies. *Nanomaterials* **2020**, *10*, 1086. [CrossRef] [PubMed]
49. Kumar, P.; Saini, M.; Singh, M.; Chhillar, N.; Dehiya, B.S.; Kishor, K.; Alharthi, F.A.; Al-Zaqri, N.; Alghamdi, A.A. Micro-Plasma Assisted Synthesis of ZnO Nanosheets for theEfficient Removal of Cr^{6+} from the Aqueous Solution. *Crystals* **2021**, *11*, 2. [CrossRef]
50. Han, L. High Temperature Thermoelectric Properties of ZnO Based Materials. Ph.D. Thesis, Technical University of Denmark, Copenhagen, Denmark, 2014.
51. Zhang, L.; Tosho, T.; Okinaka, N.; Akiyama, T. Thermoelectric Properties of Solution Combustion Synthesized Al-Doped ZnO. *Mater. Trans.* **2008**, *49*, 2868–2874. [CrossRef]
52. Sayari, A.; El Mir, L. Structural and optical characterization of Ni and Al co-doped ZnO nanopowders synthesized via the sol-gel process. *Kona Powder Part. J.* **2015**, *32*, 154–162. [CrossRef]
53. Tsubota, T.; Ohtaki, M.; Eguchi, K.; Arai, H. Thermoelectric properties of Al-doped ZnO as a promising oxide material for high-temperature thermoelectric conversion. *J. Mater. Chem.* **1997**, *7*, 85–90. [CrossRef]
54. Virtudazo, R.V.R.; Srinivasan, B.; Guo, Q.; Wu, R.; Takei, T.; Shimasaki, Y.; Wada, H.; Kuroda, K.; Bernik, S.; Mori, T. Improvement in the thermoelectric properties of porous networked Al-doped ZnO nanostructured materials synthesized via an alternative interfacial reaction and low-pressure SPS processing. *Inorg. Chem. Front.* **2020**, *7*, 4118–4132. [CrossRef]
55. Ohtaki, M.; Tsubota, T.; Eguchi, K.; Arai, H. High-temperature thermoelectric properties of $(Zn_{1-x}Al_x)O$. *J. Appl. Phys.* **1996**, *79*, 1816–1818. [CrossRef]
56. Prasad, K.S.; Rao, A.; Bhardwaj, R.; Johri, K.K.; Chang, C.; Kuo, Y. Spark plasma sintering technique: An alternative method to enhance ZT values of Sb doped Cu_2SnSe_3. *J. Mater. Sci. Mater. Electron.* **2018**, *29*, 13200–13208. [CrossRef]
57. Hedge, G.S.; Prabhu, A.N.; Huang, R.Y.; Kuo, Y.K. Reduction in thermal conductivity and electrical resistivity of indium and tellurium co-doped bismuth selenide thermoelectric system. *J. Mater. Sci. Mater. Electron.* **2020**, *31*, 19511–19525. [CrossRef]
58. Neeli, G.; Behara, D.K.; Kumar, M.K. State of the Art Review on Thermoelectric Materials. *Int. J. Sci. Res.* **2016**, *5*, 1833–1844. [CrossRef]
59. Gayner, C.; Kar, K.K. Recent advances in thermoelectric materials. *Prog. Mater. Sci.* **2016**, *83*, 330–382. [CrossRef]
60. Muthusamy, O.; Singh, S.; Hirata, K.; Kuga, K.; Harish, S.K.; Shimomura, M.; Adachi, M.; Yamamoto, Y.; Matsunami, M.; Takeuchi, T. Synergetic Enhancement of the Power Factor and Suppression of Lattice Thermal Conductivity via Electronic Structure Modification and Nanostructuring on a Ni- and B-Codoped p-Type Si–Ge Alloy for Thermoelectric Application. *ACS Appl. Electron. Mater.* **2021**, *3*, 5621–5631. [CrossRef]
61. Kavita; Gupta, V.; Ranjeet. Structural and morphological properties of nanostructured Bi2Te3 with Mn-doping for thermoelectric applications. *Mater. Today Proc.* **2021**, *54*, 820–826. [CrossRef]
62. Li, D.H.; He, S.F.; Chen, J.; Jiang, C.Y.; Yang, C. Solid-state Chemical Reaction Synthesis and Characterization of Lanthanum Tartrate Nanocrystallites under Ultrasonication Spectra. *IOP Conf. Ser. Mater. Sci. Eng.* **2017**, *242*, 012023. [CrossRef]
63. Yahya, N. *Carbon and Oxide Nanostructures: Synthesis, Characterisation and Applications*; Springer Link: Berlin/Heidelberg, Germany, 2011; ISBN 978-3-642-14673-2.
64. Jeevanandam, J.; Barhoum, A.; Chan, Y.S.; Dufresne, A.; Danquah, M.K. Review on nanoparticles and nanostructured materials: History, sources, toxicity and regulations. *J. Nanotechnol.* **2018**, *9*, 1050–1074. [CrossRef]
65. Zhang, D.-B.; Li, H.-Z.; Zhang, B.-P.; Liang, D.; Xia, M. Hybrid-structured ZnO thermoelectric materials with high carrier mobility and reduced thermal conductivity. *RSC Adv.* **2017**, *7*, 10855–10864. [CrossRef]
66. Tsubota, T.; Ohtaki, M.; Eguchi, K.; Arai, H. Thermoelectric properties of ZnO Doped with The Group 13 Elements. In Proceedings of the 16th International Conference on Thermoelectrics, Dresden, Germany, 26–29 August 1997; pp. 240–243. [CrossRef]
67. Cai, K.F.; Müller, E.; Drašar, C.; Mrotzek, A. Preparation and thermoelectric properties of Al-doped ZnO ceramics. *Mater. Sci. Eng. B Solid-State Mater. Adv. Technol.* **2003**, *104*, 45–48. [CrossRef]
68. Callaway, J. Model for Lattice Thermal Conductivity at Low Temperatures. *Phys. Rev.* **1959**, *113*, 1046–1051. [CrossRef]
69. Bui, C.T.; Sow, C.H.; Li, B.; Thong, J.T.L. Diameter-Dependent Thermal Transport in Individual ZnO Nanowires and its Correlation with Surface Coating and Defects. *Small* **2012**, *8*, 738–745. [CrossRef] [PubMed]
70. Kinemuchi, Y.; Nakano, H.; Mikami, M.; Kobayashi, K.; Watari, K.; Hotta, Y. Enhanced boundary-scattering of electrons and phonons in nanograined zinc oxide. *J. Appl. Phys.* **2010**, *108*, 053721. [CrossRef]

71. Luu, S.D.N.; Duong, T.A.; Phan, T.B. Effect of dopants and nanostructuring on the thermoelectric properties of ZnO materials. *Adv. Nat. Sci. Nanosci. Nanotechnol.* **2019**, *10*, 023001. [CrossRef]
72. Wu, X.; Lee, J.; Varshney, V.; Wohlwend, J.L.; Roy, A.K.; Luo, T. Thermal Conductivity of Wurtzite Zinc-Oxide from First-Principles Lattice Dynamics—A Comparative Study with Gallium Nitride. *Sci. Rep.* **2016**, *6*, 22504. [CrossRef] [PubMed]
73. Finn, P.A.; Asker, C.; Wan, K.; Bilotti, E.; Fenwick, O.; Nielsen, C.B. Thermoelectric Materials: Current Status and Future Challenges. *Front. Electron. Mater.* **2021**, *1*, 1–13. [CrossRef]
74. Prasad, R.; Bhame, S.D. Review on texturization effects in thermoelectric oxides. *Mater. Renew. Sustain. Energy.* **2020**, *9*, 3. [CrossRef]
75. Han, C.; Li, Z.; Dou, S. Recent progress in thermoelectric materials. *Chin. Sci. Bull.* **2014**, *59*, 2073–2091. [CrossRef]
76. Zak, A.K.; Majid, W.H.A.; Mahmoudian, M.R.; Darroudi, M.; Yousefi, R. Starch-stabilized synthesis of ZnO nanopowders at low temperature and optical properties study. *Adv. Powder Technol.* **2013**, *24*, 618–624. [CrossRef]
77. Takashiri, M.; Miyazaki, K.; Tanaka, S.; Kurosaki, J.; Nagai, D.; Tsukamoto, H. Effect of grain size on thermoelectric properties of n-type nanocrystalline bismuth-telluride based thin films. *J. Appl. Phys.* **2008**, *104*, 084302. [CrossRef]
78. Liang, X. Structure and Thermoelectric Properties of ZnO Based Materials Structure and Thermoelectric Properties of ZnO. Ph.D. Thesis, Havard University, Cambridge, MA, USA, 2013.
79. Zhang, X.; Zhao, L.-D. Thermoelectric materials: Energy conversion between heat and electricity. *J. Mater.* **2015**, *1*, 92–105. [CrossRef]
80. Guan, W.; Zhang, L.; Wang, C.; Wang, Y. Theoretical and experimental investigations of the thermoelectric properties of Al-, Bi- and Sn-doped ZnO. *Mater. Sci. Semicond. Proces.* **2017**, *66*, 247–252. [CrossRef]
81. Zhang, D.; Zhang, B.; Ye, D.; Liu, Y.; Li, S. Enhanced Al/Ni co-doping and power factor in textured ZnO thermoelectric ceramics prepared by hydrothermal synthesis and spark plasma sintering. *J. Alloys Compd.* **2016**, *656*, 784–792. [CrossRef]
82. Matiullah; Wang, C.L.; Su, W.B.; Zaman, A.; Ullah, I.; Zhai, J.Z.; Liu, D.K. Effects of sintering atmospheres on thermoelectric properties, phase, microstructure and lattice parameters c/a ratio of Al, Ga dual doped ZnO ceramics sintered at high temperature. *J. Mater. Sci. Mater. Electron.* **2018**, *29*, 9555–9563. [CrossRef]
83. Han, L.; Nong, N.V.; Hung, L.T.; Holgate, T.; Pryds, N.; Ohtaki, M.; Linderotha, S. The influence of alpha- and gamma-Al_2O_3 phases on the thermoelectric properties of Al-doped ZnO. *J. Alloys Compd.* **2013**, *555*, 291–296. [CrossRef]
84. Ohtaki, M.; Araki, K.; Yamamoto, K. High thermoelectric performance of dually doped ZnO ceramics. *J. Electron. Mater.* **2009**, *38*, 1234–1238. [CrossRef]
85. Wang, M.; Lee, K.E.; Hahn, S.H.; Kim, E.J.; Kim, S.; Chung, J.S.; Shin, E.W.; Park, C. Optical and photoluminescent properties of sol-gel Al-doped ZnO thin films. *Mater. Lett.* **2007**, *61*, 1118–1121. [CrossRef]
86. Maciá-Barber, E. *Thermoelectric Materials Advances and Applications*; Jenny Stanford Publishing: United Square, Singapore, 2015.
87. Jood, P.; Peleckis, G.; Wang, X.; Dou, S.X. Effect of gallium doping and ball milling process on the thermoelectric performance of n-type ZnO. *J. Mater. Res.* **2012**, *27*, 2278–2285. [CrossRef]
88. Lee, S.; Lee, J.; Choi, S.; Park, J. Studies of thermoelectric transport properties of atomic layer deposited gallium-doped ZnO. *Ceram. Int.* **2017**, *43*, 7784–7788. [CrossRef]

MDPI
St. Alban-Anlage 66
4052 Basel
Switzerland
Tel. +41 61 683 77 34
Fax +41 61 302 89 18
www.mdpi.com

Crystals Editorial Office
E-mail: crystals@mdpi.com
www.mdpi.com/journal/crystals

www.ingramcontent.com/pod-product-compliance
Lightning Source LLC
LaVergne TN
LVHW070726100526
838202LV00013B/1176